Lecture Notes in Computer Science 12283

Advanced Research in Computing and Software Science

Subline of Lecture Notes in Computer Science

More information about this series at http://www.springer.com/series/7409

Tobias Harks · Max Klimm (Eds.)

Algorithmic Game Theory

13th International Symposium, SAGT 2020 Augsburg, Germany, September 16–18, 2020 Proceedings

Editors
Tobias Harks
Institute of Mathematics
Augsburg University
Augsburg, Germany

Max Klimm
Institute of Mathematics
Technical University Berlin
Berlin, Germany

ISSN 0302-9743 ISSN 1611-3349 (electronic)
Lecture Notes in Computer Science
ISBN 978-3-030-57979-1 ISBN 978-3-030-57980-7 (eBook)
https://doi.org/10.1007/978-3-030-57980-7

LNCS Sublibrary: SL3 – Information Systems and Applications, incl. Internet/Web, and HCI

This Springer imprint is published by the registered company Springer Nature Switzerland AG
The registered company address is: Gewerbestrasse 11, 6330 Cham, Switzerland

Preface

This volume contains the papers and extended abstracts presented at the 13th International Symposium on Algorithmic Game Theory (SAGT 2020) held virtually during September 16–18, 2020.

The purpose of SAGT is to bring together researchers from Computer Science, Economics, Mathematics, Operations Research, Psychology, Physics, and Biology to present and discuss original research at the intersection of Algorithms and Game Theory.

The Program Committee (PC), consisting of 24 top researchers from the field, reviewed 53 submissions and decided to accept 24 papers. Each paper had three reviews, with additional reviews solicited as needed. We are very grateful to the PC for their insightful reviews and discussions. The review process was conducted entirely electronically via Easy Chair – we gratefully acknowledge this support.

The works accepted for publication in this volume cover most of the major aspects of Algorithmic Game Theory, including auction theory, mechanism design, two-sided markets, computational aspects of games, congestion games, dynamic equilibrium flows, resource allocation problems, and computational social choice.

To accommodate the publishing traditions of different fields, authors of accepted papers could ask that only a one-page abstract of the paper appeared in the proceedings. Among the 24 accepted papers, the authors of 3 papers selected this option.

Furthermore, due to the general support by Springer, we were able to provide a Best Paper Award. The PC decided to give the award to the paper "On the Approximability of the Stable Matching Problem with Ties of Constant Size up to the Integrality Gap" authored by Jochen Könemann, Kanstantsin Pashkovich, and Natig Tofigzade.

The program included three invited talks by leading researchers in the field: Dirk Bergemann (Yale University, USA), Paul Dütting (London School of Economics, UK), Ruta Mehta (University of Illinois at Urbana-Champaign, USA).

We would like to thank all the authors for their interest in submitting their work to SAGT 2020, as well as the PC members and the external reviewers for their great work in evaluating the submissions. We also want to thank EATCS, Springer, Facebook, and the COST Action GAMENET (CA16228) for their generous financial support. We are grateful to Monika Deininger at Augsburg University for her help with the conference website and organization.

Finally, we would also like to thank Anna Kramer at Springer for helping with the proceedings, and the EasyChair conference management system.

July 2020

Tobias Harks
Max Klimm

Organization

Program Committee

Umang Bhaskar	Tata Institute of Fundamental Research, India
Vittorio Bilò	University of Salento, Italy
Ozan Candogan	University of Chicago, USA
Jose Correa	Universidad de Chile, Chile
Ágnes Cseh	Hungarian Academy of Sciences, Hungary
Edith Elkind	University of Oxford, UK
John Fearnley	The University of Liverpool, UK
Aris Filos-Ratsikas	The University of Liverpool, UK
Dimitris Fotakis	National Technical University of Athens, Greece
Martin Gairing	The University of Liverpool, UK
Yiannis Giannakopoulos	Technical University of Munich, Germany
Tobias Harks	Augsburg University, Germany
Martin Hoefer	Goethe University Frankfurt, Germany
Max Klimm	Technical University Berlin, Germany
Maria Kyropoulou	University of Essex, UK
David Manlove	University of Glasgow, UK
Evangelos Markakis	Athens University of Economics and Business, Greece
Dario Paccagnan	University of California, Santa Barbara, USA
Georgios Piliouras	Singapore University of Technology and Design, Singapore
Guido Schaefer	CWI Amsterdam, The Netherlands
Orestis Telelis	University of Piraeus, Greece
Christos Tzamos	University of Wisconsin-Madison, USA
Marc Uetz	University of Twente, The Netherlands
Adrian Vetta	McGill University, Canada

Additional Reviewers

Adil, Deeksha
Bailey, James
Birmpas, Georgios
Boehmer, Niclas
Bullinger, Martin
Carvalho, Margarida
Cechlarova, Katarina
Chandan, Rahul
Cheung, Yun Kuen
Cristi, Andrés
de Haan, Ronald
de Keijzer, Bart
Deligkas, Argyrios
Ferguson, Bryce
Ganesh, Sai
Gergatsouli, Evangelia
Ghalme, Ganesh
Gourves, Laurent
Gupta, Sushmita
Harrenstein, Paul

Hoeksma, Ruben
Ismaili, Anisse
Kaiser, Marcus
Kanellopoulos, Panagiotis
Kenig, Batya
Kern, Walter
Kodric, Bojana
Kontonis, Vasilis
Kovacs, Annamaria
Lackner, Martin
Laraki, Rida
Leonardos, Stefanos
Lianeas, Thanasis
Misra, Neeldhara
Miyazaki, Shuichi
Molitor, Louise
Monaco, Gianpiero
Mouzakis, Nikos
Oosterwijk, Tim
Paarporn, Keith
Papasotiropoulos, Georgios
Patsilinakos, Panagiotis
Plaxton, Greg
Poças, Diogo
Psomas, Alexandros
Reiffenhäuser, Rebecca
Rubinstein, Aviad
Sakos, Iosif
Schmand, Daniel
Skopalik, Alexander
Skoulakis, Stratis
Terzopoulou, Zoi
Varloot, Estelle
Vera, Alberto
Vinci, Cosimo
Voudouris, Alexandros
Wilczynski, Anaëlle
Yokoi, Yu

Contents

Auctions and Mechanism Design

Two-Buyer Sequential Multiunit Auctions with No Overbidding

Mete Şeref Ahunbay[1(✉)], Brendan Lucier[2], and Adrian Vetta[3]

[1] Department of Mathematics and Statistics, McGill University, Montreal, Canada
mete.ahunbay@mail.mcgill.ca

[2] Microsoft Research New England, Cambridge, USA
brlucier@microsoft.com

[3] Department of Mathematics and Statistics, School of Computer Science, McGill University, Montreal, Canada
adrian.vetta@mcgill.ca

Abstract. We study equilibria in two-buyer sequential second-price (or first-price) auctions for identical goods. Buyers have weakly decreasing incremental values, and we make a behavioural no-overbidding assumption: the buyers do not bid above their incremental values. Structurally, we show equilibria are intrinsically linked to a greedy bidding strategy. We then prove three results. First, any equilibrium consists of three phases: a competitive phase, a competition reduction phase and a monopsony phase. In particular, there is a time after which one buyer exhibits monopsonistic behaviours. Second, the declining price anomaly holds: prices weakly decrease over time at any equilibrium in the no-overbidding game, a fact previously known for equilibria with overbidding. Third, the price of anarchy of the sequential auction is exactly $1 - 1/e$.

1 Introduction

In a two-buyer multiunit sequential auction a collection of T identical items are sold one after another. This is done using a single-item second-price (or first-price) auction in each time period. Due to their temporal nature, equilibria in sequential auctions are extremely complex and somewhat misunderstood objects [8,12,13]. This paper aims to provide a framework in which to understand two-buyer sequential auctions. Specifically, we study equilibria in the auction setting where both *duopsonists* have non-decreasing & concave valuation functions under the natural assumption of no-overbidding. Our main technical contribution is an in-depth analysis of the relationship between equilibrium bidding strategies and a greedy behavioural strategy. This similitude allows us to provide a characterization of equilibria with no-overbidding and to prove three results.

One, any equilibrium in a two-buyer sequential auction with no-overbidding induces three phases: a competitive phase, a competition reduction phase and a monopsony phase. In particular, there is a time after which one of the two duopsonists will behave as a *monoposonist*. Here monopsonistic behaviour refers

T. Harks and M. Klimm (Eds.): SAGT 2020, LNCS 12283, pp. 3–16, 2020.
https://doi.org/10.1007/978-3-030-57980-7_1

to the type of strategies expected from a buyer with the ability to clinch the entire market. Intriguingly, we show that this fact does not hold for equilibria where overbidding is permitted.

Two, the *declining price anomaly* holds for two-buyer sequential auctions with no-overbidding; the price weakly decreases over time for any equilibrium in the auction. This result shows that the seminal result of Gale and Stegeman [8], showing the declining price anomaly holds for equilibria in two-buyer sequential auctions with overbidding permitted, carries over to equilibria in auctions with no-overbidding. Notably, this declining price anomaly can fail to hold for three or more buyers, even with no-overbidding [13].

Three, the *price of anarchy* in two-buyer sequential auctions with no-overbidding is exactly $1 - \frac{1}{e} \simeq 0.632$. We remark that the same bound has been claimed in [3,4] for equilibria where overbidding is allowed but, unfortunately, there is a flaw in their arguments.

1.1 Related Work

The complete information model of two-buyer sequential auctions studied in this paper was introduced by Gale and Stegeman [8]. This was extended to multi-buyer sequential auctions by Paes Leme et al. [12] (see also [13]). Rodriguez [14] studied equilibria in the special case of identical items and identical buyers with endowments.

Ashenfelter [1] observed that the price of identical lots fell over time at a sequential auction for wine. This tendency for a decreasing price trajectory is known as the *declining price anomaly* [10]. Many attempts have been made to explain this anomaly and there is now also a plethora of empirical evidence showing its existence in practice; see [2,13,15] and the references within for more details. On the theoretical side, given complete information, Gale and Stegeman [8] proved that a weakly decreasing price trajectory is guaranteed in a two-buyer sequential auction for identical items. Prebet et al. [13] recently proved that declining prices are *not* assured in sequential multiunit auctions with three or more buyers, but gave experimental evidence to show that counter-examples to the anomaly appear extremely rare.

In the computer science community research has focussed on the welfare of equilibria in sequential auctions. Bae et al. [3,4] study the *price of anarchy* in two-buyer sequential auctions for identical items. There has also been a series of works bounding the price of anarchy in multi-buyer sequential auctions for non-identical goods; see, for example, [6,12,16].

Sequential auctions with incomplete information have also been studied extensively since the classical work of Milgrom and Weber [11,17]. We remark that to study the temporal aspects of the auction independent of informational aspects it is natural to consider the case of complete information. Indeed, our work is motivated by the fact that, even in the basic setting of complete information, the simplest case of two-buyers is not well understood.

1.2 Overview of the Paper

Section 2 presents the model of two-buyer sequential auctions with complete information. It also includes a collection of examples that illustrate some of the difficulties that arise in understanding sequential auctions and provide the reader with a light introduction to some of the technical concepts that will play a role in the subsequent analyses of equilibria. They will also motivate the importance and relevance of no-overbidding. This section concludes by incorporating tie-breaking rules in winner determination. Section 3 provides a measure for the *power of a duopsonist* and presents a natural greedy bidding strategy that a buyer with duopsony power may apply. Section 4 studies how prices and duopsony power evolve over time when the buyers apply the greedy bidding strategy.

The relevance of greedy bidding strategies is exhibited in Sect. 5 where we explain their close relationship with equilibria bidding strategies. This relationship allows us to provide a characterization of equilibria with no-overbidding. Key features of equilibria follow from these structural results. First, any equilibria induces three distinct phases with a time after which some buyer behaves as a monopsonist. Second, prices weakly decrease over time for any equilibrium. Finally, in Sect. 6 we prove the price of anarchy is exactly $1 - \frac{1}{e}$.

Due to space constraints, all proofs are deferred to the full version.

2 Two-Buyer Sequential Auctions

In this section we introduce two-buyer sequential auctions and illustrate their strategic aspects via a set of simple examples. There are T items to be sold, one per time period by a second-price auction.[1] Buyer $i \in \{1, 2\}$ has a *value* $V_i(k)$ for obtaining exactly k items and *incremental value* $v_i(k) = V_i(k) - V_i(k-1)$ for gaining a kth item. We assume $V_i(\cdot)$ is normalised at zero, non-decreasing and concave.

EXAMPLE 1: Consider a two-buyer auction with two items, with incremental valuations $(v_1(1), v_1(2)) = (10, 9)$ and $(v_2(1), v_2(2)) = (8, 5)$. The outcome that maximizes social welfare is for buyer 1 to receive both copies of the item. However, at equilibrium, buyer 2 wins the first item at a price of 6, and buyer 1 wins the second item at a price of 5. To see this, imagine that buyer 1 wins in the first period. Then in the second period she will have to pay 8 to beat buyer 2. Given this, buyer 2 will be willing to pay up to 8 to win in the first round. Thus, buyer 1 will win both permits for 8 each and obtain a *utility* (profit) of $(10 + 9) - 2 \cdot 8 = 3$. Suppose instead that buyer 2 wins in the first round. Now in the second period, buyer 1 will only need to pay 5 to beat buyer 2, yielding a profit of $10 - 5 = 5$. So, by bidding 6 in the first period, buyer 1 can guarantee herself a profit of 5. Given this bid, buyer 2 will maximize his own utility by winning the first permit for 6. Note that this outcome, the only rational solution, proffers suboptimal welfare.

[1] We present our results for second-price auctions. Given an appropriate formulation of the bidding space to ensure the existence of an equilibrium [12] these results also extend to the case of first-price auctions.

2.1 An Extensive-Form Game

We compactly model this sequential auction as an extensive-form game with complete information using a directed graph. The node set is given $\mathbb{H} = \{(x_1, x_2) \in \mathbb{Z}_+ | x_1 + x_2 \leq T\}$. Each node has a label $\mathbf{x} = (x_1, x_2)$ denoting how many items each buyer has currently won. There is a *source node*, $\mathbf{0} = (0, 0)$, corresponding to the initial round of the auction, and *terminal nodes* (x_1, x_2), where $x_1 + x_2 = T$. If $\mathbf{x}$ is a terminal node, we write $\mathbf{x} \in \mathbb{H}_0$, otherwise we say that $\mathbf{x}$ is a *decision node* and write $\mathbf{x} \in \mathbb{H}_+$. We also denote by $t(\mathbf{y}) = T - y_1 - y_2$ the number of items remaining to be sold at node $\mathbf{y}$; when the decision node is actually denoted $\mathbf{x}$, we simply write t for $t(\mathbf{x})$.

We also extend our notation for incremental valuations. Specifically, we denote the incremental value of buyer i of a kth additional item *given* endowment $\mathbf{x}$ (i.e. from decision node $\mathbf{x}$) as $v_i(k|\mathbf{x}) = V_i(\mathbf{x}_i + k) - V_i(\mathbf{x}_i + k - 1)$, for $k \in \mathbb{N}$. For valuations at the source node, we drop the explicit notation of the decision node: for example, $v_i(k) = v_i(k|\mathbf{0})$.

We find an equilibrium by calculating the *forward utility* of each buyer at every node. The forward utility is the profit a buyer will earn from that period in the auction onwards. There is no future profit at the end of the auction, so the forward utility of each buyer is zero at each terminal node. The forward utilities at decision nodes are then obtained by backwards induction: each decision node $\mathbf{x}$ has a left child $\mathbf{x} + \mathbf{e}_1$ and a right child $\mathbf{x} + \mathbf{e}_2$, respectively corresponding to buyer 1 and 2 winning an item. For the case of two-buyer second-price auctions, it is a weakly dominant strategy for each buyer to bid its *marginal value for winning*. This bid value is the incremental value plus the forward utility of winning minus the forward utility of losing. Thus, at the node $\mathbf{x}$, the bids of each buyer are

$$\begin{aligned} b_1(\mathbf{x}) &= v_1(1|\mathbf{x}) + u_1(\mathbf{x} + \mathbf{e}_1) - u_1(\mathbf{x} + \mathbf{e}_2), \\ b_2(\mathbf{x}) &= v_2(1|\mathbf{x}) + u_2(\mathbf{x} + \mathbf{e}_2) - u_2(\mathbf{x} + \mathbf{e}_1). \end{aligned}$$

If $b_1(\mathbf{x}) \geq b_2(\mathbf{x})$ then buyer 1 will win and the forward utilities at $\mathbf{x}$ are then

$$\begin{aligned} u_1(\mathbf{x}) &= v_1(1|\mathbf{x}) + u_1(\mathbf{x} + \mathbf{e}_1) - b_2(\mathbf{x} + \mathbf{e}_2) \\ &= (v_1(1|\mathbf{x}) - v_2(1|\mathbf{x})) + u_1(\mathbf{x} + \mathbf{e}_1) - u_2(\mathbf{x} + \mathbf{e}_2) + u_2(\mathbf{x} + \mathbf{e}_1), \\ u_2(\mathbf{x}) &= u_2(\mathbf{x} + \mathbf{e}_1). \end{aligned}$$

The forward utilities are defined symmetrically if $b_1(\mathbf{x}) \leq b_2(\mathbf{x})$ and buyer 2 wins. Given the forward utilities at every node, the iterative elimination of weakly dominated strategies then produces a unique equilibrium [3,8].

The auction of Example 1 is illustrated in Fig. 1. The first row in each node contains its label $\mathbf{x} = (x_1, x_2)$ and also the number of items, $t = T - x_1 - x_2$, remaining to be sold. The second row shows the forward utility of each buyer. Arcs are labelled by the bid value; here arcs for buyer 1 point left and arcs for buyer 2 point right. Solid arcs represent winning bids and dotted arcs represent losing bids. The equilibrium path, in bold, verifies our previous argument: buyer 2 wins the first item at price 6 and buyer 1 wins the second item at price 5.

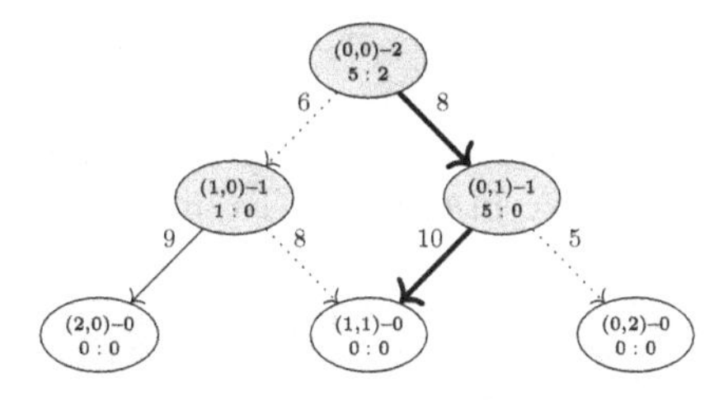

Fig. 1. The extensive form for Example 1. The set of histories has the structure of a rooted tree. Iteratively solving for an equilibrium gives the auction tree, with bidding strategies and forward utilities shown.

Consequently, in a two-buyer sequential auction, each individual auction corresponds to a standard *second-price auction*. We remark that for sequential auctions with three or more buyers each decision node in the extensive-form game corresponds to an *auction with interdependent valuations* (or an *auction with externalities*) [7,9]. As a result, equilibria in such multi-buyer sequential auctions are even more complex than for two buyers; see [12,13] for details.

2.2 No-Overbidding

Unfortunately, equilibria in sequential auctions can have undesirable and unrealistic properties. In particular they may exhibit severe *overbidding.*

EXAMPLE 2: Consider a sequential auction with T items for sale and incremental valuations shown in Fig. 2, where $0 < \epsilon \ll 1/T^2$. The key observation here is that if buyer 1 wins her first item before the final period she will then win in every subsequent period for a price $1 - \epsilon$. On the other hand, if buyer 1 wins her first item in the final period then the price will be 0. This is because buyer 2 must then have won the first $T - 1$ items and so has no value for winning in the final round.

Incremental Values	$v_i(1)$	$v_i(2)$	...	$v_i(T-1)$	$v_i(T)$
BUYER 1	1	1	...	1	1
BUYER 2	$1-\epsilon$	$1-\epsilon$	...	$1-\epsilon$	0

Fig. 2. Incremental values of each buyer which induce "severe" overbidding.

These observations imply that buyer 1 will bid $b_1(t) = (T - t) \cdot \epsilon$ in period t. At equilibrium, buyer 2 will beat these bids in the first $T - 1$ periods and make a profit of $(1 - \epsilon) \cdot (T - 1) - \frac{1}{2}(T - 1)T \cdot \epsilon = \Omega(T)$. But if buyer 2 loses the first item, he will win no items at all in the auction and thus his forward utility from losing is zero. Consequently, his marginal value for winning the first period is

$\Omega(T)$ and so he will bid $b_2(1) = \Omega(T) \gg 1-\epsilon$. Thus, at the equilibrium, buyer 2 will massively overbid in nearly every round.

Overbidding in a sequential auction is very risky and depends crucially on perfect information, so it is rare in practice. To understand some of these risks consider again Example 2. Equilibria are very sensitive to the valuation functions and any *uncertainty* concerning the payoff valuations could lead to major changes in the outcome. For instance, suppose buyer 1 is mistaken in her belief regarding the Tth incremental value of buyer 2. Then she will be unwilling to let buyer 2 win the earlier items at a low price. Consequently, if buyer 2 bids $b_2(1) = \Omega(T)$ then he will make a loss, and continuing to follow the equilibrium strategy will result in a huge loss. This is important even with complete information because, for computational or behavioural reasons, a buyer cannot necessarily assume with certainty that the other buyer will follow the equilibrium prescription; for example, the computation of equilibria in extensive-form games is hard. Likewise in competitive settings with externalities, where the a buyer may have an interest in limiting the profitability of its competitor, overbidding is an unappealing option. We address this wedge between theory and practice by imposing a non-overbidding assumption, and indeed such assumptions are common in the theoretical literature [5,16]. We leave the analysis of models that explicitly capture the risks described above as a direction for future research.

For our sequential auctions, given its valuation function, each buyer will naturally constrain its bid by its incremental value. So we will assume this *no-(incremental) overbidding* property:

$$b_i(\mathbf{x}) \leq v_i(1|\mathbf{x}) \tag{1}$$

In particular, at each stage a buyer will bid the *minimum* of its incremental value and its marginal value for winning.

We note that the no-overbidding property is especially well-suited to our setting of valuations that exhibit decreasing marginal values and free disposal. That is, valuations that are non-decreasing and weakly concave. Without these assumptions, sequential auctions can exhibit severe *exposure problems*[2] that introduce inefficiencies driven by the tension of overbidding. For this reason, sequential auctions are pathologically inappropriate mechanisms when valuations are not concave or monotone. Many practical sequential multiunit auctions therefore assume (or impose) that buyers declare concave non-decreasing valuations. For example, in cap-and-trade (sequential) auctions, such as the Western Climate Initiative (WCI) and the Regional Greenhouse Gas Initiative (RGGI), multiple items are sold in each time period but bids are constrained to be weakly decreasing. We follow the literature on multiunit sequential auctions and focus on concave and non-decreasing valuations, where a no-overbidding constraint is more natural.

[2] The exposure problem arises when a buyer has large value for a set S of items but much less value for strict subsets of S. Thus bidding for the items of S sold early in the auction exposes the buyer to a high risk if he fails to win the later items of S.

2.3 Tie-Breaking Rules

When overbidding is allowed the forward utilities at equilibria are unique (see also [8]), regardless of the tie-breaking rule. Surprisingly, this is **not** the case when overbidding is prohibited:

EXAMPLE 3: Take a four round auction, where $v_i(k) = 1$ for $k \leq 3$ and $v_i(k) = 0$ otherwise. Solving backwards, the forward utilities are the same for every *non-source node* whether or not overbidding is permitted. In particular, at the successor nodes of the source we have $u_i(\mathbf{e}_i) = 2$ and $u_i(\mathbf{e}_{-i}) = 1$. But now a difference occurs. Without the overbidding constraint, both buyers would bid 2 at the source node $\mathbf{0}$ and, regardless of the winner, each buyer has $u_i(\mathbf{0}) = 1$. But with the no-overbidding constraint both buyers will bid 1. Consequently, if this tie is broken in favour of buyer 1 with probability p, then buyer 1 has forward utility $u_1(\mathbf{0}) = 1 + p$ and buyer 2 has forward utility $u_2(\mathbf{0}) = 2 - p$, so buyers' payoffs depend on p.

Thus, under no-overbidding we need to account for the tie-breaking process. To do this, let $\mathbf{b} = (b_1, b_2)$ where $b_i : \mathbb{H}_+ \to \mathbb{R}$ is the bidding strategy of buyer i. Given the bids at the node $\mathbf{x}$, let $\pi_i(\mathbf{b}|\mathbf{x})$ denote the probability buyer i is awarded the item, where $\pi_i(\mathbf{b}|\mathbf{x}) = 1$ if $b_i(\mathbf{x}) > b_{-i}(\mathbf{x})$. This defines a tie-breaking rule at each node, and the *forward utility* of each buyer can again be calculated inductively. For any terminal node $\mathbf{x} \in \mathbb{H}_0$ the forward utility is zero: $u_i(\mathbf{b}|\mathbf{x}) = 0$. The forward utility of buyer i at decision node $\mathbf{x} \in \mathbb{H}_+$ is then:

$$u_i(\mathbf{b}|\mathbf{x}) = \pi_i(\mathbf{b}|\mathbf{x}) \cdot (v_i(1|\mathbf{x}) - b_{-i}(\mathbf{x}) + u_i(\mathbf{b}|\mathbf{x} + \mathbf{e}_i)) + (1 - \pi_i(\mathbf{b}|\mathbf{x})) \cdot u_i(\mathbf{b}|\mathbf{x} + \mathbf{e}_{-i})$$

With the tie-breaking rule defined, we may again compute an equilibrium that survives iterative elimination of weakly dominated strategies. By backwards induction, the (expected) forward utilities at equilibria are unique. Moreover, there is a unique bidding strategy $\mathbf{b}$ which survives the iterated elimination of weakly dominated strategies. Specifically, under the no-overbidding condition, at any node $\mathbf{x}$ each bidder should bid the minimum of its marginal value for winning and its incremental value:

$$b_i(\mathbf{x}) = \min\left[v_i(1|\mathbf{x}),\, v_i(1|\mathbf{x}) + u_i(\mathbf{b}|\mathbf{x} + \mathbf{e}_i) - u_i(\mathbf{b}|\mathbf{x} + \mathbf{e}_{-i})\right] \tag{2}$$

3 Greedy Bidding Strategies

To understand equilibria in two-buyer sequential auctions with no-overbidding, we need to consider greedy bidding strategies. At decision node $\mathbf{x}$, suppose buyer i attempts to win exactly k items by the following strategy: she waits (bids zero) for $t - k$ rounds and then outbids buyer $-i$ in the final k rounds. To implement this strategy, by the no-overbidding assumption, she must bid $\geq v_{-i}(1|(t-k) \cdot \mathbf{e}_{-i})$ in the final k rounds. For this strategy to be feasible, it must be that:

$$v_i(k|\mathbf{x}) \;\geq\; v_{-i}(t - k + 1|\mathbf{x})$$

This strategy would then give buyer i a utility of:

$$\bar{\mu}_i(k|\mathbf{x}) = \sum_{j=1}^{k} v_i(j|\mathbf{x}) - k \cdot v_{-i}(t-k+1|\mathbf{x}) \tag{3}$$

If buyer i attempts to apply this greedy strategy, it should select k to maximize its profit $\bar{\mu}_i(k|\mathbf{x})$. So in equilibrium, buyer i should earn at least the maximum of these utilities over all feasible k. Remarkably, this property need **not** be true for equilibria when overbidding is allowed; see Example 4 below.

Buyer i's *greedy utility* at decision node $\mathbf{x}$ is the resultant utility from applying its greedy strategy from $\mathbf{x}$:

$$\mu_i(\mathbf{x}) = \max_{k \in [t] \cup \{0\}} \bar{\mu}_i(k|\mathbf{x}) \tag{4}$$

In turn, buyer i's corresponding *greedy demand* at $\mathbf{x}$ is:

$$\kappa_i(\mathbf{x}) = \min \arg \max_{k \in [t] \cup \{0\}} \bar{\mu}_i(k|\mathbf{x}) \tag{5}$$

But when can buyer i profitably apply this greedy strategy? It can apply it whenever it has *duopsony power*. In a sequential auction this ability arises when $v_i(1|\mathbf{x}) > v_{-i}(t|\mathbf{x})$. Formally, let buyer i's *duopsony factor* at $\mathbf{x}$ be:

$$f_i(\mathbf{x}) = \max\{k \in [t] : v_i(k|\mathbf{x}) > v_{-i}(t-k+1|\mathbf{x})\} \cup \{0\} \tag{6}$$

Observe that if $f_i(\mathbf{x}) = 0$ then buyer i cannot apply the greedy strategy, and we then have $\mu_i(\mathbf{x}) = 0$ and $\kappa_i(\mathbf{x}) = 0$. On the other hand, if $f_i(\mathbf{x}) > 0$ then $\mu_i(\mathbf{x}) > 0$, and any maximizer of $\mu_i(\mathbf{x})$ is necessarily at most $f_i(\mathbf{x})$.

We say that a buyer is a *monopsonist* if the other buyer has no duopsony power, that is, if $f_{-i}(\mathbf{x}) = 0$. In turn, a buyer is a *strict monopsonist* if she has total duopsony power, i.e. $f_i(\mathbf{x}) = t$. So in a sequential auction with no-overbidding, a strict monopsonist can guarantee it gains at least its greedy utility. This is analogous to the corresponding *static market* setting. However, this simple fact can fail to hold when overbidding occurs:

EXAMPLE 4: Consider a three-item auction, where $v_1(k) = 1$ for any $k \in \{1,2,3\}$, and buyer 2 has incremental valuations $(v_2(1), v_2(2), v_2(3)) = (2/3 - \delta, 1/2 + \epsilon, 0)$, where we fix $\epsilon, \delta > 0$ to be small with $2\epsilon = 3\delta$. With overbidding permitted, in equilibrium with ties broken in favour of buyer 2, buyer 1 wins a single item.

Figure 3 illustrates this example. The key observation here is that $b_2(\mathbf{0}) = 2/3 - \delta + 2\epsilon > 2/3 - \delta = v_2(1)$, so buyer 2 overbids at the source node. Furthermore, buyer 1 obtains a profit of 1 in this equilibrium with overbidding, but $\bar{\mu}_1(3|\mathbf{0}) = 1 + 3\delta$. So in the equilibrium with overbidding, buyer 1 obtains less than her greedy utility. In contrast, under no-incremental overbidding, buyer 1 will win all three items and make exactly her greedy utility.

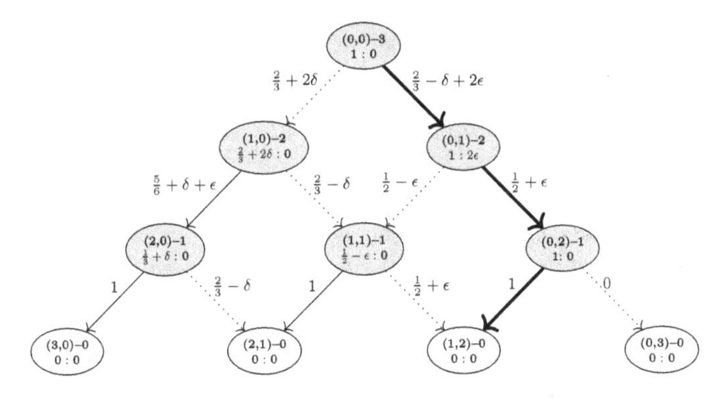

Fig. 3. A sequential auction with overbidding permitted where neither buyer exhibits monopolistic behaviours.

The greedy strategy induces two types of "price" that will be important. First, we say that the *baseline price* of buyer i at decision node $\mathbf{x}$ is:

$$\beta_i(\mathbf{x}, t) = \begin{cases} v_i(1|\mathbf{x}) & f_i(\mathbf{x}) = 0 \\ v_{-i}(t - \kappa_i(\mathbf{x}) + 1|\mathbf{x}) & f_i(\mathbf{x}) > 0 \end{cases} \tag{7}$$

Second, the *threshold price* of buyer i at decision node $\mathbf{x}$ is:

$$p_i(\mathbf{x}) = v_i(1|\mathbf{x}) + \mu_i(\mathbf{x} + \mathbf{e}_i) - \mu_i(\mathbf{x} + \mathbf{e}_{-i}) \tag{8}$$

The baseline price may be seen as the price a greedy buyer would post if it wanted to obtain its greedy utility. By posting a bid of $\beta_i(\mathbf{x}) + \epsilon$ on each node following $\mathbf{x}$, buyer i would be guaranteed, by the no-overbidding condition, to win at least $\kappa_i(\mathbf{x})$ items. The threshold price, in turn, arises from a behavioural rule: it is the bid a buyer would make on the assumption that it wins **exactly** its greedy utility through the rest of the auction.

4 Greedy Bidding Outcomes

Now imagine that buyers attempt to bid their threshold prices at each decision node, subject to the no-overbidding constraint. By definition (8) of threshold prices, this corresponds to the behavioural rule where buyers perceive their forward utilities to equal their greedy utilities, and bid accordingly. We will discover in Sect. 5 that such greedy bidding strategies are in some circumstances equivalent to equilibrium bidding strategies under the *no-overbidding assumption.*

Accordingly, to understand equilibria we must study the consequences of greedy bidding. So, in this section, we will inspect the properties of greedy outcomes. We begin by analysing the behaviour of greedy utilities:

Lemma 1. *The greedy utility of a buyer weakly decreases when the buyer loses. Specifically, for any decision node* $\mathbf{x}$ *and any buyer* i,

$$\begin{aligned} \mu_i(\mathbf{x}+\mathbf{e}_{-i}) &= \mu_i(\mathbf{x}) && \text{if } \kappa_i(\mathbf{x}) < t \\ \mu_i(\mathbf{x}+\mathbf{e}_{-i}) &< \mu_i(\mathbf{x}) && \text{if } \kappa_i(\mathbf{x}) = t \end{aligned}$$

In turn, if buyer i *has non-zero greedy demand at decision node* $\mathbf{x}$ *and wins an item, then its greedy utility decreases by at most the value it would have for purchasing an item at his baseline price. Formally,* $\forall \mathbf{x} \in \mathbb{H}_+, \forall i \in \{1,2\}$*:*

$$\kappa_i(\mathbf{x}) > 0 \Rightarrow \mu_i(\mathbf{x}+\mathbf{e}_i) \geq \mu_i(\mathbf{x}) - v_i(1|\mathbf{x}) + \beta_i(\mathbf{x})$$

Next, we turn attention to how the demand evolves. We show that, if the greedy demand of a buyer is less than the entire supply, then it remains constant upon losing the current item. Intuitively, we could assume that buyer i did not demand the item for sale at $\mathbf{x}$, so we could assume that the demand was a subset of the supply at $\mathbf{x}+\mathbf{e}_{-i}$. If instead buyer i wins an item, then its greedy demand can decrease by at most one. In particular, if buyer i demands the entire supply at decision node $\mathbf{x}$, upon winning an item, it will continue to demand the entire supply.

Lemma 2. *For any* $\mathbf{x} \in \mathbb{H}_+$ *and any* $i \in \{1,2\}$*:*

$$\begin{aligned} \kappa_i(\mathbf{x}) < t &\Rightarrow \kappa_i(\mathbf{x}+\mathbf{e}_{-i}) = \kappa_i(\mathbf{x}) \\ t > 1 &\Rightarrow \kappa_i(\mathbf{x}+\mathbf{e}_i) \geq \kappa_i(\mathbf{x}) - 1 \\ t > 1, \kappa_i(\mathbf{x}) = t &\Rightarrow \kappa_i(\mathbf{x}+\mathbf{e}_i) = t - 1 \end{aligned}$$

Finally, we inspect the evolution of baseline and threshold prices. The first lemma shows that the baseline price is a lower bound for the threshold price.

Lemma 3. *For any* $\mathbf{x} \in \mathbb{H}_+$ *and any* $i \in \{1,2\}$, $p_i(\mathbf{x}) \geq \beta_i(\mathbf{x})$.

Moreover, if buyer i's greedy demand corresponds to the entire supply, it should ensure that it wins every item while targeting his greedy utility. This implies the inequality of Lemma 3 should become strict.

Lemma 4. *If the greedy demand of a buyer is the entire supply then the threshold price is strictly greater than the baseline price. Specifically,*

$$\kappa_i(\mathbf{x}) = t \implies p_i(\mathbf{x}) > \beta_i(\mathbf{x})$$

Instead consider the case when some buyer i with duopsony power does not demand the entire supply. Suppose that, buyer i wins an item, and still has duopsony power after doing so. As its demand will not decrease significantly, its baseline price will be weakly higher. We would then presume that buyer i is in a situation that favours buying more items, hence it would be willing to pay higher prices.

Lemma 5. *Given* $\mathbf{x} \in \mathbb{H}_+$ *and* $i \in \{1,2\}$ *such that* $f_i(\mathbf{x}) > 1$ *and* $\kappa_i(\mathbf{x}) < t$. *Then* $\beta_i(\mathbf{x}+\mathbf{e}_i) \geq p_i(\mathbf{x})$, *where equality holds only if* $\bar{\mu}(\kappa_i(\mathbf{x}+\mathbf{e}_i)+1|\mathbf{x}) = \mu_i(\mathbf{x})$. *Moreover,* $p_i(\mathbf{x}+\mathbf{e}_i) \geq p_i(\mathbf{x})$.

If instead buyer $-i$ wins at $\mathbf{x}$, then buyer i loses the opportunity to apply its greedy strategy to win t items from $\mathbf{x}$. If buyer i still does not demand the entire supply at $\mathbf{x}+\mathbf{e}_{-i}$, then this loss of opportunity translates to a lesser incentive to purchase at a given price:

Lemma 6. *Given* $\mathbf{x} \in \mathbb{H}_+$ *and* $i \in \{1,2\}$ *such that* $f_i(\mathbf{x}) > 1$ *but* $\kappa_i(\mathbf{x}) < t-1$. *Then* $p_i(\mathbf{x}+\mathbf{e}_{-i}, t-1) \leq p_i(\mathbf{x}, t)$. *Moreover, the inequality is strict if and only if* $\kappa_i(\mathbf{x}+\mathbf{e}_i) = t-1$.

Finally, if buyer i with duopsony power targets his greedy utility and does not demand the entire supply, then incentives for buyer $-i$ are aligned such that it should want to purchase items without letting buyer i win. Buyer $-i$ would be able to do so if buyer i's bids do not exceed buyer $-i$'s incremental value. The following lemma shows that this is the case.

Lemma 7. *Given* $\mathbf{x} \in \mathbb{H}_+$ *and* $i \in \{1,2\}$. *If* $\kappa_i(\mathbf{x}) < t$ *and* $f_i(\mathbf{x}) > 0$ *then* $p_i(\mathbf{x}) \leq v_{-i}(t-\kappa_i(\mathbf{x}+\mathbf{e}_i)|\mathbf{x})$. *Moreover, the inequality is tight if only if* $\bar{\mu}_i(\kappa_i(\mathbf{x}+\mathbf{e}_i)+1|\mathbf{x}) = \mu_i(\mathbf{x})$.

Altogether, this implies the following for greedy outcomes, where realised quantities are those reached in the outcome with positive probability:

Theorem 1. *Suppose buyers implement their greedy bidding strategies. Then on any realised outcome path from some decision node* $\mathbf{x}$, *if there exists a monopsonist buyer* i *at* $\mathbf{x}$, *then her realised utility is* $\mu_i(\mathbf{x})$; *else some buyer* $i \in \operatorname{argmin}_{j\in\{1,2\}} p_i(\mathbf{x})$ *has realised utility equal to* $\mu_i(\mathbf{x})$. *Furthermore, buyer* i *purchases at least* $\kappa_i(\mathbf{x})$ *items. Finally, prices are equal to* p_i *alongside the realised outcome path until buyer* i *demands the entire supply, after which prices equal* β_i. *In particular, prices are declining along any realised outcome path.*

5 Characterisation of No-Overbidding Equilibria

In this section, we classify the equilibria of two-buyer sequential multiunit auctions under the no-overbidding condition. Structurally, we will see that any equilibrium is made up of three phases (a competitive phase, a competition reduction phase and a monopsony phase) characterized by very different strategic behaviours.

First, however, let's show that the *declining price anomaly* holds. Here, it is worth emphasizing declining prices do not follow as a direct consequence of the no-overbidding assumption. Indeed, for ≥ 3 buyers, the declining price anomaly can fail to hold given decreasing incremental valuations even with the no-overbidding assumption; see Prebet et al. [13].

Theorem 2. *In a two-buyer sequential multiunit auction with no-incremental overbidding, under equilibrium bidding strategies, prices are non-increasing along any realised equilibrium path.*

We now proceed to show when there necessarily is a direct equivalence between greedy bidding and equilibrium bidding strategies: it is exactly when there exists a monopsonist.

Theorem 3. *Suppose that at decision node* $\mathbf{x}$*, some buyer* i *is a monopsonist. Then for any decision node* $\mathbf{x}'$ *of the auction tree rooted at* $\mathbf{x}$*, prices and utilities are equal for equilibrium and greedy bidding strategies.*

Informally, suppose only buyer 1 has duopsony power; then buyer 1 is constrained by the equilibrium bidding strategies to make her greedy utility at every possible future node. Thus her bids equal to her threshold price at every round of the auction. Buyer 2 will then take advantage of buyer 1's bidding strategies by purchasing an item whenever possible.

But what happens in the more complex setting where both buyers have duopsony power? Call buyer i a *quasi-monopsonist* at decision node $\mathbf{x}$ if there exists a realised equilibrium path from $\mathbf{x}$ to a final round $\mathbf{y}$ such that $b_i(\mathbf{y}) \geq b_{-i}(\mathbf{y})$. Note that it is possible for both agents to be quasi-monopsonists at a node $\mathbf{x}$ if there is a randomized tie-breaking rule. By decreasing prices, a quasi-monopsonist may have a realised payoff weakly less than its greedy utility; moreover, a monopsonist is always a quasi-monopsonist. This definition, along with properties of greedy bidding, allows us to fully characterise equilibria.

Theorem 4. *For equilibrium bidding strategies, while no buyer demands the entire supply, prices at each node are no less than the minimum threshold price. Moreover, at every decision node* $\mathbf{x}$ *there exists a quasi-monopsonist buyer* i*. Finally, if buyer* i *is a quasi-monopsonist at decision node* $\mathbf{x}$ *and if* $\mathbf{x}+\mathbf{e}_{-i}-\mathbf{e}_i$ *is also a decision node, then* i *is again a quasi-monopsonist at decision node* $\mathbf{x}+\mathbf{e}_{-i}-\mathbf{e}_i$.

It may not be immediately apparent, but Theorem 4 gives us a very clear picture of what happens at an equilibrium. Specifically, each equilibrium consists of three phases. The first phase is the *competitive phase.* In this phase the identity of the "eventual monopsonist" may change depending on the winner of an item (and may be uncertain due to randomized tie-breaking). Consequently, the two buyers compete to buy items and drive prices above the threshold prices. The buyer who fails to win enough items in this phase retains sufficient duopsony power to become a monopsonist. The second phase, the *competition reduction* phase, begins once the identity of the monopsonist is established. The monopsonist then posts its threshold price in each round. The other buyer exploits the monopsonist's bidding strategy to purchase items. This phase ends when the competition from the other buyer has been weakened sufficiently enough for the monopsonist to desire winning all the remaining items. Thus we enter the third phase, the *monopsony phase*, where the monopsonist purchases all the remaining items at its current baseline price.

6 The Price of Anarchy

The *price of anarchy* of a sequential auction is the worst-case ratio between the social welfare attained at an equilibrium allocation and the welfare of the optimal allocation. In this section, we exploit our equilibrium characterization to prove that the price of anarchy is exactly $1-1/e$ in two-buyer sequential auctions with no-overbidding, assuming weakly decreasing incremental valuations.

To prove our efficiency result, we first show a result parallelling an argument in Theorem 2 of [3]: if efficiency along an equilibrium path is less than 1, then the efficiency is bounded below by that along a subpath, where a buyer (without loss of generality buyer 1) holds monopsony power. We then consider extending incremental valuations to the real line, where $\forall \tau \in [0,t]$:

$$\begin{aligned}\bar{v}_1(\tau|\mathbf{x}) &= v_1(\lceil \tau \rceil\,|\mathbf{x})\\ \bar{v}_2(\tau|\mathbf{x}) &= v_2(\lfloor \tau+1 \rfloor\,|\mathbf{x})\end{aligned} \tag{9}$$

By our equilibrium characterisation, the *social welfare* of the auction is at least

$$\int_0^{\kappa_1(\mathbf{0})} \bar{v}_1(\tau)d\tau + \int_0^{T-\kappa_1(\mathbf{0})} \bar{v}_2(\tau)d\tau.$$

However, as buyer 1 earns its greedy utility by winning $\kappa_1(\mathbf{0})$ items, it must be that for any $k \in (0,T)$

$$\int_0^{\kappa_1(\mathbf{0})} \bar{v}_1(\tau)d\tau \geq \int_0^{k} \bar{v}_1(\tau)d\tau - k\cdot \bar{v}_2(T-k).$$

This yields a lower bound for $\bar{v}_2(T-k)$, and combining the two expressions allows us to compute the following lower bound on the price of anarchy.

Theorem 5. *A two-buyer sequential multiunit auction with concave and non-decreasing valuations has price of anarchy at least* $(1 - 1/e)$*, given no-overbidding.*

To match this bound, for $T \in \mathbb{N}$, consider the equilibrium where all ties are broken in favour of buyer 2, and we let $v_1(i) = 1$ and

$$v_2(i) = \max\left\{\frac{\lfloor T(1-1/e)\rfloor - i + 1}{T-i+1}, 0\right\}.$$

Then the efficiency of the equilibrium in the limit $T \to \infty$ ends up being a Riemann integral which evaluates to $1 - 1/e$. This implies the following upper bound.

Theorem 6. *There exist two-buyer sequential multiunit auctions with concave and non-decreasing valuations and T items, whose efficiency tends to* $(1 - 1/e)$ *as T grows, given no-overbidding.*

Acknowledgements. We are very grateful to Rakesh Vohra for discussions on this topic. We thank the referees for their helpful comments and suggestions.

References

1. Ashenfelter, O.: How auctions work for wine and art. J. Econ. Perspect. **3**(3), 23–36 (1989)
2. Ashta, A.: Wine auctions: More explanations for the declining price anomaly. J. Wine Res. **17**(1), 53–62 (2006)
3. Bae, J., Beigman, E., Berry, R., Honig, M., Vohra, R.: Sequential bandwidth and power auctions for distributed spectrum sharing. J. Sel. Areas Commun. **26**(7), 1193–1203 (2008)
4. Bae, J., Beigman, E., Berry, R., Honig, M., Vohra, R.: On the efficiency of sequential auctions for power sharing. In: Proceedings of 2nd International Conference on Game Theory for Networks, pp 199–205 (2009)
5. Christodoulou, G., Kovacs, A., Schapira, M.: Bayesian combinatorial auctions. J. ACM **63**(2), 11 (2016)
6. Feldman, M., Lucier, B., Syrgkanis, V.: Limits of efficiency in sequential auctions. In: Proceedings of 9th Conference on Web and Internet Economics, pp 160–173 (2013)
7. Funk, P.: Auctions with interdependent valuations. Int. J. Game Theory **25**, 51–64 (1996)
8. Gale, I., Stegeman, M.: Sequential auctions of endogenously valued objects. Games Econ. Behav. **36**(1), 74–103 (2001)
9. Jehiel, P., Moldovanu, B.: Strategic nonparticipation. Rand J. Econ. **27**(1), 84–98 (1996)
10. McAfee, P., Vincent, D.: The declining price anomaly. J. Econ. Theory **60**, 191–212 (1993)
11. Milgrom, P., Weber, R.: A theory of auctions and competitive bidding. Econometrica **50**, 1089–1122 (1982)
12. Paes Leme, R., Syrgkanis, V., Tardos, E.: Sequential auctions and externalities. In: Proceedings of 23rd Symposium on Discrete Algorithms, pp 869–886 (2012)
13. Prebet, E., Narayan, V., Vetta, A.: The declining price anomaly is not universal in multibuyer sequential auctions (but almost is). In: Proceedings of 9th International Symposium on Algorithmic Game Theory, pp 109–122 (2019)
14. Rodriguez, G.: Sequential auctions with multi-unit demands. BE J. Theoret. Econ. **9**(1), 45 (2009)
15. Salladarre, F., Guilloteau, P., Loisel, P., Ollivier, P.: The declining price anomaly in sequential auctions of identical commodities with asymmetric bidders: empirical evidence from the Nephrops norvegicus market in France. Agric. Econ. **48**, 731–741 (2017)
16. Syrgkanis, V., Tardos, E.: Composable and efficient mechanisms. In: Proceedings of 45th Symposium on Theory of Computing, pp 211–220 (2013)
17. Weber, R.: Multiple object auctions. In: Engelbrecht-Wiggans, R., Shubik, M., Stark, R. (eds.) Auctions, Bidding and Contracting: Use and Theory, New York University Press, pp 165–191 (1983)

Asymptotically Optimal Communication in Simple Mechanisms

Ioannis Anagnostides(✉), Dimitris Fotakis, and Panagiotis Patsilinakos

School of Electrical and Computer Engineering,
National Technical University of Athens, 15780 Athens, Greece
{ioannis,patsilinak}@corelab.ntua.gr, fotakis@cs.ntua.gr

Abstract. In this work, we show how well-known mechanisms for extensively studied single-parameter environments can be implemented with asymptotically optimal communication complexity. Specifically, we first turn our attention to single-parameter domains in auctions, namely single item and multi-unit auctions. For the former case, we show that the Vickrey auction can be implemented with an expected communication complexity of at $1+\epsilon$ bits per bidder, for any $\epsilon > 0$, assuming that the valuations can be represented with a constant number of bits. As a corollary, we provide a compelling method to increment the price in English auctions. By employing an efficient encoding scheme, we show that the same bound can be obtained for multi-item auctions with additive bidders and a constant number of items, and for multi-unit auctions with unit demand bidders. Moreover, we apply our framework to games without monetary transfers and in particular, the canonical case of facility location games. We present an implementation of Moulin's generalized median mechanism that achieves an $1+\epsilon$ approximation of the optimal social welfare, for any $\epsilon > 0$, while extracting an arbitrarily small fraction of information. Our results follow from simple sampling schemes and do not require any prior knowledge on the agents' parameters.

1 Introduction and Motivation

Communication complexity has been a primary concern from the inception of Mechanism Design. The first consideration relates to the tractability of the communication exchange required to approximate an underlying objective function, such as the social welfare or the expected revenue; the domain of combinatorial auctions provides such an example where strong negative results have been established [16]. A second active area of research endeavors to design the interaction process so that efficient communication is an inherent feature of the mechanism. Following this line of work, we aim to establish a natural framework for

This work was supported by the Hellenic Foundation for Research and Innovation (H.F.R.I.) under the "First Call for H.F.R.I. Research Projects to support Faculty members and Researchers and the procurement of high-cost research equipment grant", project BALSAM, HFRI-FM17-1424.

T. Harks and M. Klimm (Eds.): SAGT 2020, LNCS 12283, pp. 17–31, 2020.
https://doi.org/10.1007/978-3-030-57980-7_2

developing asymptotically optimal mechanisms in well-studied single-parameter environments in Auction Theory and Social Choice.

This emphasis is strongly motivated for a number of reasons. First, there is a need to design mechanisms with strong performance guarantees in settings with communication restrictions and possibly truncated action spaces, due to technical, behavioral or regulatory purposes [1,7]. Moreover, extracting data from distributed parties can be burdensome, an impediment magnified in environments with vast participation. It has been also understood that the amount of communication captures the extent of information leakage from the participants. In this context, behavioral economists have recognized that soliciting information requires a high cognitive cost (e.g. [24,28]) and bidders may be even reluctant to completely reveal their private valuation. Finally, truncating the information disclosure would provide stronger information privacy guarantees [32] for the agents.

As a motivating example, we consider the single item auction and in particular, the shortcomings of the most well-established formats, namely the *sealed-bid* and the *English* auction. First, it is important to point out that although every mechanism can be simulated with direct revelation - as implied by the revelation principle, this equivalence has been criticized in the literature of Economics, not least due to the communication cost of revealing the entire valuation space. Indeed, our work will show that the communication complexity of Vickrey's sealed bid auction [34] is suboptimal. Moreover, despite the theoretical appeal of Vickrey's auction, the ascending or English auction exhibits superior performance in practice [2,3,21,22], for reasons that mostly relate to the simplicity, the transparency and the privacy guarantees of the latter format. However, a faithful implementation of Vickrey's rule through a standard English auction requires - in the worst case - exponential communication and indeed, time complexity since the auctioneer has to increment the price by a single bit. In principle, the lack of prior knowledge on the agents' valuations would dramatically impede its performance.

One of the issues we address is how to increment the price in an ascending auction, without any prior knowledge, so that the communication cost is minimized and the desirable properties of each format are retained. More broadly, we apply sampling techniques in order to establish mechanisms with asymptotically optimal communication complexity guarantees, without sacrificing the social welfare and the incentive properties of the interaction process. In particular, we employ random samples of agents and we either request the full information, or we query on whether their valuations exceed a particular threshold. In this way, our mechanism elicits - asymptotically - only the necessary information in order to implement the optimal allocation rule.

1.1 Previous Work

Communication efficiency has been a central desideratum in the literature of Algorithmic Mechanism Design. The first consideration relates to the interplay between communication constraints and incentive compatibility; in particular,

Van Zandt [33] articulated conditions under which they can be studied separately, while the authors in [17,30] investigated the communication overhead induced in truthful implementations, i.e. the communication cost of truthfulness. In a closely related direction, Blumrosen et al. [7] (see also [25]) considered the design of optimal single-item auctions under severely bounded communication: every bidder can only transmit a limited number of bits. One of their key results was a 0.648 social welfare approximation for 1-bit auctions and uniformly distributed valuations. In addition, the design of optimal - with respect to the obtained revenue - bid levels in English auctions was addressed in [15], where the authors had to posit on known distributions.

Turning to games without monetary transfers, the solution concept of efficient *preference elicitation* has also engendered significant amount of research; in particular, Segal [31] provided bounds on the communication required to realize a Social Choice rule through the notion of *budget sets*, with applications in resource allocation tasks and stable matching. Moreover, the boundaries of computational tractability and the strategic issues that arise were investigated by Conitzer and Sandholm in [10], while the same authors established in [11] the worst-case number of bits required to execute common voting rules. The trade-off between accuracy and information leakage in facility location games was tackled by Feldman et al. [18], where they investigated the behavior of truthful mechanisms with truncated input space - *ordinal* and voting information models - and constitutes the main focus of our work as well. Our approximation scheme is founded on Moulin's generalized median rule [26] (see also [5]).

1.2 Contributions

We develop a simple algorithmic framework for obtaining strong communication complexity guarantees in exemplar multi-agent interaction environments. More precisely, one of our main techniques consists of simulating a sub-auction - essentially as a black box - on a random sample of agents in order to determine the increment in an underlying ascending format. In addition, we develop an algorithm that yields tight upper and lower bounds on the market clearing price by querying random samples of agents in order to navigate on the search tree that represents the valuation space. From an algorithmic standpoint, our approach offers a communication efficient procedure to determine the jth highest number in an unordered list (see Subsect. 3.3).

These ideas are applied in Sect. 3 to implement the desirable allocation rule in several extensively studied single-parameter environments with an asymptotically optimal communication of $1 + \epsilon$ transmitted bits on average per bidder (Theorem 1), for any $\epsilon > 0$, assuming the valuations can be represented with a constant number of bits. Our work supplements the results of [17,30,33] by showing that for a series of fundamental domains, the incentive compatibility constraint does not increase the asymptotic communication requirements of the interaction process. We also corroborate on one of the main observations in the work of Blumrosen et al. [7]: *asymmetry helps* - in deriving tight communication bounds. More precisely, the winner in our English auction will have to transmit a logarithmic - with respect to the initial number of players - amount of bits, while most of the bidders will transmit a single bit. This asymmetry distinguishes from their model where a universal communication restriction was imposed on all of the agents. Moreover, inspired by techniques from Information Theory, we design efficient encoding schemes in simultaneous auctions (Theorem 2) in the domain of additive valuations.

In Sect. 4 we turn to games without monetary transfers and in particular, the canonical case of (single) facility location problems. In this context, our framework yields a $1 + \epsilon$ approximation of the social welfare achieved by Moulin's generalized median mechanisms, for every $\epsilon > 0$, and with an arbitrarily small information leakage - relatively to the full information mechanism. This result constitutes a natural continuation of research in preference elicitation with truncated input space by Feldman et al. [18]; however, while their approach reduces the input through an information-extraction model beyond direct revelation (e.g. an agent votes for her preferred location amongst a set of candidates), we differentiate on the use of a limited sample of agents, without sacrificing the obtained social welfare - up to some arbitrarily small error. In addition, our proof technique is fundamentally different from the existing ones in the literature and is based on the asymptotic characterization of a distribution derived from estimating the behavior of the underlying mechanism - in our case the generalized median - and could be of independent interest. We believe that our results can be applied in practical applications due to their simplicity and their communication efficiency.

1.3 Broader Context

More broadly, communication complexity has been a primary consideration in Game Theory. A series of works have established tractable communication procedures in order to reach an approximate Nash equilibrium in two-player games [4,14,19]. Moreover, important work by Nisan and Segal [27] has illustrated the limitations, and in particular the exponential communication requirements in the domain of submodular bidders, as well as in combinatorial allocation problems - even when 2 players compete for m indivisible items. There has been also extensive research devoted in designing incentive compatible and efficient preference elicitation mechanisms in combinatorial auctions [6,9,20]. For a comprehensive review on fundamental notions and problems in communication complexity we refer to [8,13,23].

2 Preliminaries

In our study we denote with n the number of participants in the game. In single parameter environments the *rank* of agent i corresponds to the index of her private valuation in ascending order (and indexed from 1 unless explicitly stated otherwise). In the case of identical valuation profiles we accept some arbitrary but fixed order among the agents - e.g. lexicographic order. In addition, throughout

Sect. 3 we assume that an agent remains active in the auction only when positive utility can be obtained; that is, if the announced price for the item is greater or equal to the valuation of some agent i, then i will withdraw from the forthcoming rounds of the auction. In Mechanism 2 we will assume that the agents' valuations are distinct.

A mechanism will be referred to as *strategyproof* or *incentive compatible* if truthful reporting is a *universally* dominant strategy - a best response under any possible action profile and randomized realization - for every agent. Moreover, we will require a weaker notion of incentive compatibility; in particular, a strategy profile $(s_1, \ldots, s_n)$ constitutes an *ex-post Nash equilibrium* if the action $s_i(v_i)$ is a best response to every action profile $\mathbf{s}_{-i}(\mathbf{v}_{-i})$ - for any agent i and valuation v_i. In this setting, a mechanism will be called *ex-post incentive compatible* if sincere bidding constitutes an ex-post Nash equilibrium. A strategy s_i is *obviously dominant* if, for any deviating strategy s_i', starting from any earliest information set where s_i' and s_i disagree, the best possible outcome from s_i' is no better than the worst possible outcome from s_i. A mechanism is *obviously strategyproof* (OSP) if it has an equilibrium in obviously dominant strategies.

We use the standard notation of $f(n) \sim g(n)$ if $\lim_{n \to +\infty} f(n)/g(n) = 1$ and $f(n) \lesssim g(n)$ if $\lim_{n \to +\infty} f(n)/g(n) \leq 1$, where n will be implied as the asymptotic parameter. Moreover, in order to analyze the bit complexity in Sect. 3 the valuation space will be assumed discretized and every valuation can be represented with k bits; we will mostly consider k to be a constant. For notational clarity we posit that $\binom{n}{m} = 0$, when $m > n$. Communication complexity is defined as the cumulative amount of bits elicited from the participants; our analysis will be worst-case with respect to the input - i.e. the agents' valuations - and average-case with respect to the introduced randomization in the procedure.

The Median Mechanism. Consider that we have to allocate a single facility on a metric space $(\mathbb{R}^d, ||\cdot||_1)$ and n agents, with $\mathbf{x}_i \in \mathbb{R}^d$ the preferred location of agent i. The social cost of an allocation $\mathbf{x}$ is defined as $\mathrm{SC} = \sum_{i=1}^{n} \mathrm{d}(\mathbf{x}, \mathbf{x}_i)$. In this context, the generalized median [26] is a strategyproof and optimal - with respect to the social cost in L^1 - mechanism that allocates the facility to the coordinate-wise median of the reported instance.

3 Auctions

We commence this section by presenting a sampling mechanism for the single item auction; then, analogous techniques will be employed in gradually more general environments. The more technical proofs of our claims can be found in the full version of our paper.

3.1 Single Item Auction

Our implementation is established based on a black-box algorithm; in particular, let $\mathcal{A}$ be an algorithm that interacts with a set of agents and faithfully simulates a second-price auction; that is, $\mathcal{A}$ returns the VCG outcome without actually allocating items and imposing payments. However, the agents that are excluded by $\mathcal{A}$ will be also automatically eliminated from the remainder of the auction. Our mechanism consists of the following steps.

Mechanism 1: Ascending Auction through Sampling

Result: Winner & VCG payment
Input: Set of agents N, size of sample c, algorithm $\mathcal{A}$
while $|N| > c$ **do**
  $S :=$ random sample of c agents from N
  w := winner in $\mathcal{A}(S)$
  Announce $p :=$ payment in $\mathcal{A}(S)$
  Update the active agents: $N := \{i \in N \setminus S \mid v_i > p\} \cup \{w\}$
end
if $|N| = 1$ **then**
  return w, p
else
  return $\mathcal{A}(N)$
end

Naturally, we assume that $c \geq 2$, so that the second-price rule is properly implemented. This mechanism induces a format that couples the auction that is simulated by $\mathcal{A}$ with an ascending auction. We shall establish the following properties.

Proposition 1. *Assuming truthful bidding, Mechanism 1 implements - with probability 1 - the VCG allocation rule.*

Proof. First, if after the termination of some round only a single agent i remains active, it follows that the announced price p - that coincides with the valuation of some player - exceeds the valuation of every player besides i; thus, by definition, the outcome implements the VCG allocation rule. Moreover, the claim when $2 \leq |N| \leq c$ follows given that $\mathcal{A}$ faithfully simulates a second-price auction. Otherwise, in a given round - with $|N| > c$ - only agents that are below or equal to the second-highest valuation will withdraw from the auction. Thus, the allocation rule over the active players remains invariant between rounds, concluding the proof. □

Proposition 2. *If $\mathcal{A}$ simulates a sealed-bid auction, Mechanism 1 is strategyproof.*

Proof. Consider some round of the auction and some agent i that has been selected in the sample S; if we fix the reports from the agents in the sample besides i we can identify the following two cases. First, if $v_i \geq x_j, \forall j \in S \setminus \{i\}$, with x_j representing the report of agent j, then sincere bidding is a best response for i given that her valuation exceeds the announced price. Indeed, note that since $\mathcal{A}$ simulates a second-price auction, the winner in the sample does not have any control over the announced price of the round. In the contrary case, agent i does not have an incentive to misreport and remain active in the auction given that the reserved price will be greater or equal to her valuation. Let p the market clearing price in $\mathcal{A}$ and i some agent that was not selected in the sample. It is clear that if $v_i \leq p$ then a best response for i is to withdraw from the auction, while if $v_i > p$ then i's best response is to remain active in the forthcoming round. □

Proposition 3. *If $\mathcal{A}$ simulates an English Auction, Mechanism 1 is OSP.*

Proof. The claim follows from the OSP property of the English auction. In particular, note that we simply perform an English auction without interacting with every active agent in each round, but instead with a small sample; when only a single player survives from the sample, we announce the price to the remainder of the agents. □

Before we establish the communication complexity of the induced auction, we should point out that a trivial lower bound to recover the optimal social welfare is n bits. Indeed, since the information is distributed to n parties and the goal is to allocate the item to the agent with the highest utility - with probability 1, every player has to commit at least 1 bit to the procedure. Through this prism, we will show that our mechanism reaches this lower bound with arbitrarily small error - assuming that k is a constant. We should also note that the information leakage in 1 is asymmetrical, in the sense that statistically, the agents that are closer to winning the item have to reveal relatively more bits from their private valuation. It is clear that in order to truncate the communication complexity of the mechanism, one has to guarantee small inclusion rate - in expectation - for each round of the auction; this property is implied by the following lemma.

Lemma 1. *Let X_a the proportion of the agents that remain active in a given round of the auction; then*

$$\mathbb{E}[X_a] \lesssim \frac{2}{c+1} \tag{1}$$

Let us assume that $Q(n;k)$ is the (deterministic) communication complexity of $\mathcal{A}$ with n players. In particular, when $\mathcal{A}$ is a sealed-bid auction it follows that $Q(n;k) = n \cdot k$. On the other hand, the worst-case communication cost of an English auction is $Q(n,k) = 2^k n$. Indeed, given that $\mathcal{A}$ faithfully simulates a second-price auction, the auctioneer has to cover every possible point on the valuation space. If $T(n;c,k)$ is the (randomized) communication complexity of the induced Mechanism 1, it follows that when $n > c$

$$\mathbb{E}[T(n;c,k)] = \mathbb{E}[T(nX_a;c,k)] + Q(c;k) + n - c \tag{2}$$

Solving recursions of such form are standard in the analysis of randomized algorithms (see [12]); in particular, we can establish the following theorem.

Theorem 1. *Let $t(n; c, k)$ the expected communication complexity of Mechanism 1 with k assumed constant; then, $\forall \epsilon > 0, \exists c_0 = c_0(\epsilon)$ such that $\forall c \geq c_0$*

$$t(n; c, k) \lesssim n(1 + \epsilon) \tag{3}$$

Note that our asymptotic guarantee is invariant on the communication complexity of the second-price algorithm $\mathcal{A}$, assuming that k is a constant. On the other hand, if we allow k to depend on n our guarantee crucially depends on $\mathcal{A}$ (see Theorem 3).

3.2 Multi-item Auctions with Additive Valuations

As a direct extension of the previous setting, let us assume that the auctioneer has to allocate m (indivisible) items and the valuation space is *additive*, that is for every agent i and for a bundle of items S, $v_i(S) = \sum_{j \in S} v_{ij}$. In this setting, we shall perform an auction for each item using Mechanism 1; it is clear that assuming truthful bidding, the induced auction will implement - with probability 1 - the VCG allocation rule, as implied by Proposition 1. Moreover, the following proposition holds.

Proposition 4. *The mechanism induced by employing m auctions as described in Mechanism 1 is ex-post incentive compatible.*

However, we will illustrate that a simultaneous implementation can significantly truncate the communication exchange - relatively to a sequential format - through an efficient encoding scheme. First, we assume that m is arbitrary and that we have to perform a separate and independent auction for each of the m items. Under this assertion, the optimality condition yields the lower bound of $n \cdot m$ bits, which can be again asymptotically reached with arbitrarily small error:

Proposition 5. *Let $t(n; m, c, k)$ the expected communication complexity of implementing m sequential auctions as described in Mechanism 1 with k assumed constant; then, $\forall \epsilon > 0, \exists c_0 = c_0(\epsilon)$ such that $\forall c \geq c_0$*

$$t(n; m, c, k) \lesssim nm(1 + \epsilon) \tag{4}$$

On the other hand, if we assume that m is constant and that we perform the auctions simultaneously, we will show that we can reach the bound of n bits with a very simple coding scheme. To be precise, recall that - asymptotically - the expected inclusion rate in Mechanism 1 is at most $2/(c+1)$ and thus, as the sample size increases the overwhelmingly most probable scenario is that some random agent will drop from the next round of the auction; we shall exploit this property by considering the following encoding. An agent i - that remains active in at least one auction - will transmit the bit 0 in the case of withdrawal from every auction; otherwise, i will transmit an m bit vector that will indicate the auctions that she wishes to remain active. Although the latter part of the encoding is clearly sub-optimal, we will show that in fact, we can asymptotically obtain an optimality guarantee. In particular, consider a round of the auction with n players that remain active in at least one auction and p the expected probability that a player will withdraw from every auction in the current round. Since every player is active in at most m auctions, it follows from the union bound that $1 - p \lesssim 2m/(c+1)$. As a result, if N_b denotes the total number of bits transmitted in the round, we have that

$$\mathbb{E}[N_b] = n\left(1 \cdot p + m \cdot (1-p)\right) \lesssim n\left(\left(1 - \frac{2m}{c+1}\right) + m\left(\frac{2m}{c+1}\right)\right) \quad (5)$$

As a result, since m is a constant it follows that $\mathbb{E}[N_b] \lesssim n(1+\delta)$, for any $\delta > 0$ and for a sufficiently large constant c. Moreover, the expected inclusion rate - the proportion of agents that remain active in at least one auction - is asymptotically at most $2m/(c+1)$ and thus, we can establish the following theorem.

Theorem 2. *Let $t(n; m, c, k)$ the expected communication complexity of implementing m simultaneous auctions as described in Mechanism 1 with the aforementioned encoding scheme and k and m assumed constant; then, $\forall \epsilon > 0, \exists c_0 = c_0(\epsilon)$ such that $\forall c \geq c_0$*

$$t(n; m, c, k) \lesssim n(1+\epsilon) \quad (6)$$

3.3 Multi-unit Auctions with Unit Demand

Consider that we have to allocate m identical items to n unit demand bidders. We are interested in the non-trivial case where $m \leq n$; in this setting, our approach will differ depending on the asymptotic value of m.

First, we consider the canonical case where m is constant. In this setting, we can extend Mechanism 1 as follows. In each round, we invoke an algorithm $\mathcal{A}$ that simulates the VCG outcome[1] for a random sample of active agents $c = \kappa m + 1$ for $\kappa \in \mathbb{N}$. Next, the market clearing price in the sample will be announced in order to 'prune' the active agents. Through parameter κ, we are able to restrain the inclusion rate in the following rounds. As a result, we can prove statements analogous to Propositions 1, 2, 3 and Theorem 1. The analysis is very similar to the single item mechanism and therefore it can be omitted.

Next, we study the case where $m = \gamma \cdot n$ for $\gamma \in (0,1)$; in this setting, we need to alter our approach. In particular, the main idea is to broadcast two separate prices; the players that exceed the high price transmit a single bit of 1 and are automatically declared winners (the price will be determined in the following rounds), whilst the players that are below the low price transmit a bit of 0 and are disqualified from the remainder of the auction. Therefore, the mechanism

[1] The VCG outcome for this auction consists of allocating a single unit to each of the m-highest bidders for a price coinciding with the $m+1$-highest bid.

will recurse on the agents that reside in the intermediate region - who have to channel 2 arbitrary bits so that the encoding is non-singular. On a high level, our mechanism consists of the following steps.

Mechanism 2: $\mathcal{M}(N, m)$: Multi-Unit Auction through Sampling

---

Result: Winners & VCG payment
Input: Set of agents N, number of items $m := \gamma n$
Initialize the winners $W := \emptyset$ and the losers $L := \emptyset$
$p_h :=$ estimated upper bound on the price
$p_\ell :=$ estimated lower bound on the price
Announce p_ℓ and p_h
Update the winners: $W := W \cup \{i \in N \mid v_i > p_h\}$
Update the losers: $L := L \cup \{i \in N \mid v_i < p_\ell\}$
if $p_h = p_\ell$ **then**
  return W, p_h
else
  $N := N \setminus (W \cup L)$
  Update the number of items m
  return $\mathcal{M}(N, m)$
end

It is clear that assuming that the estimators are valid, Mechanism 2 implements the VCG allocation rule. The crux of this algorithm is to efficiently estimate the bounds p_h and p_ℓ, so that the in-between players are very limited. We introduce the following algorithm. We commence from the root of the tree that represents the valuation space and we make decisions in each branch by querying a random sample of agents. More precisely, the query informs us on whether an agent's valuation exceeds a particular threshold. Our goal is to reach a node on the tree - a particular price - such that the agents who exceed the price are - approximately - as many as the available items. As a randomized process, there is a non-zero probability that the estimates are inconsistent - e.g. the winners outnumber the available items; in this case, we simply repeat the sampling process.

Proposition 6. *Mechanism 2 is ex-post incentive compatible.*

However, we remark that Mechanism 2 is not strategyproof and in particular, answering sincerely to the queries is not necessarily a dominant strategy for the agents in the sample due to potential retaliation strategies. Finally, we establish the main theorem of this subsection.

Theorem 3. *Let $t(n; c, k)$ the communication complexity of Mechanism 2 with $k \in \mathcal{O}(n^{1-\ell})$ for some $\ell > 0$; then, $\forall \epsilon > 0, \exists c_0 = c_0(\epsilon, k)$ such that $\forall c \geq c_0$*

$$t(n; c, k) \lesssim n(1 + \epsilon) \tag{7}$$

Note that for this theorem we allowed k to depend on the (initial) number of agents.

4 Facility Location Games

Having established strong communication complexity guarantees in a series of environments in Auction Theory, we apply our framework to games without monetary transfers, and in particular the canonical case of facility location games with a single resource. Our main result in this section is a $1+\epsilon$ approximation scheme of the generalized median mechanism with very limited information from the participants (see Corollary 1). We remark that when k is assumed constant, our result can be obtained with an iterative process - analogously to Subsect. 3.3 - and Chernoff bounds in order to correlate the accuracy of the approximation with the size of the sample. Nonetheless, our approach is more robust since we do not even need the discretized valuation space hypothesis. More precisely, our mechanism will simply employ the generalized median scheme $\mathcal{M}$ for a random sample of c agents.

Proposition 7. *The approximate median Mechanism 3 is strategyproof.*

Proof. The claim follows from the incentive compatibility of the median mechanism. □

Our analysis commences with the one-dimensional case - i.e. allocating a single facility on the line; the extension to any metric space $(\mathbb{R}^d, ||\cdot||_1)$ will then follow easily. We conclude this section by illustrating why our sampling approach cannot be efficiently applied for allocating multiple facilities. In order to make the analysis more concise - and without any loss of generality - we assume that $n = 2\kappa + 1$ and $c = 2\rho + 1$ for some $\kappa, \rho \in \mathbb{N}$. Let X_r be the rank - among all of the agents - of the sample's median; in this section we shall assume that X_r is normalized in the domain $[-1, 1]$. Thus, when $X_r = 0$ the median of the sample coincides with the median among the entire instance. Through this prism, we can determine the probability mass function with simple combinatorial arguments as follows:

$$\Pr\left(X_r = \frac{i}{\kappa}\right) = \frac{\binom{\kappa - i}{\rho}\binom{\kappa + i}{\rho}}{\binom{2\kappa + 1}{2\rho + 1}} \tag{8}$$

Mechanism 3: Approximate Median through Sampling

Result: Facility's Location $\in \mathbb{R}^d$
Input: Set of agents N, size of sample c
$S :=$ random sample of c agents from N
return $\mathcal{M}(S)$

As a result, we note that the normalization constraint of the probability mass function (8) yields a variation of the Chu-Vandermonde identity:

$$\sum_{i=-\kappa}^{\kappa} \binom{\kappa-i}{\rho}\binom{\kappa+i}{\rho} = \sum_{i=0}^{2\kappa} \binom{i}{\rho}\binom{2\kappa-i}{\rho} = \binom{2\kappa+1}{2\rho+1} \tag{9}$$

For this reason, the distribution defined in Eq. 8 shall be referred to as Chu-Vandermonde distribution. One of the key aspects of our analysis is that we are oblivious to the agents' individual valuations and instead, we rely solely on their relative rank. This approach is justified by the following lemma.

Lemma 2. *Let $x_{opt} \in \mathbb{R}$ be the optimal location - i.e. the median of the instance - and $x \in \mathbb{R}$ some location, such that only at most $\epsilon \cdot n$ agents reside in the interval from x to x_{opt}. Then, if D_{opt} is the minimum social cost, allocating a facility on x yields a social cost D such that*

$$D \leq D_{opt}\left(1 + \frac{4\epsilon}{1-2\epsilon}\right) \tag{10}$$

Proof. Let $d = \text{dis}(x, x_{opt}) = |x - x_{opt}|$; shifting the facility from x to x_{opt} can only reduce the social cost by at most $2\epsilon n d$, that is $D \leq D_{opt} + 2\epsilon n d$. Moreover, it is clear that

$$D_{opt} \geq \left(\frac{n}{2} - \epsilon n\right) d \iff d \leq D_{opt}\frac{2}{n(1-2\epsilon)} \tag{11}$$

As a result, if we combine the previous bounds the lemma will follow. We should mention that the analysis and subsequently the obtained bound is tight for certain instances. □

As a corollary, obtaining a strong approximation ratio is tantamount to accumulating the probability mass close to the median. The main challenge is to quantify this concentration as a function of the sample's size. To this end, we prove that for $\kappa \to +\infty$ the Chu-Vandermonde distribution converges to a continuous function, allowing for a concise characterization of the concentration.

Theorem 4. *If we let $\kappa \to \infty$, the Chu-Vandermonde distribution converges to a transformed beta distribution with the following probability density function:*

$$f(t) = \frac{(2\rho+1)!}{(\rho!)^2 2^{2\rho+1}}(1-t^2)^{\rho} \tag{12}$$

We let X represent a random variable that follows distribution (12). Next, we correlate the concentration of the distribution with parameter ρ.

Theorem 5. *For any $\epsilon > 0$ and for any $\delta > 0$, there exists some constant $\rho_0 = \rho_0(\epsilon, \delta)$ such that $\forall \rho \geq \rho_0$*

$$\Pr(|X| \geq \epsilon) \leq \delta \tag{13}$$

Having established the concentration of the distribution, we apply Lemma 2 to prove the following theorem.

Theorem 6. *The approximate one-dimensional median Mechanism 3 obtains in expectation a* $1+\epsilon$ *approximation of the optimal social welfare, for any* $\epsilon > 0$ *and with constant input* $c = c(\epsilon)$*, while* $n \to \infty$.

This result can be easily extended for the generalized median scheme that applies to any metric space $(\mathbb{R}^d, ||\cdot||_1)$; to be precise, let us consider some basis for the metric space. Then, we can invoke the one-dimensional sampling approximation for each of the principal axes individually. As a result, we can prove the following proposition.

Corollary 1. *The approximate generalized median Mechanism 3 obtains in expectation a* $1+\epsilon$ *approximation of the optimal social welfare, for any* $\epsilon > 0$ *and with constant input* $c = c(\epsilon)$*, while* $n \to \infty$.

Finally, we illustrate why a sampling approach - with a constant sample size - cannot provide meaningful guarantees when allocating multiple facilities. In particular, we consider the family of the *percentile* mechanisms, namely strategyproof allocation rules on the line that partition the agents' reports into particular percentiles; the median can be clearly classified in this family. We will also assume that at least 2 facilities are to be allocated and that the leftmost percentile contains at most $(1-\alpha)\cdot n$ of the agents, for some $\alpha > 0$. Let us imagine a dynamic instance where the agents from the entire leftmost percentile - including the pivotal agent - have gradually smaller valuations $x \to -\infty$, while the other agents remain fixed at a finite distance; then, any sampling approximation has in expectation an unbounded competitive ratio with respect to the full information mechanism. Indeed, there will always be a positive probability, albeit exponentially small, that we fail to sample a single agent from $-\infty$, whilst the full information percentile mechanism will allocate a facility to accommodate the divergent agents. Thus, a sampling approach cannot provide a meaningful approximation of the percentile mechanisms - at least with respect to the expected social cost. An interesting open question is whether this limitation can be overcome if we impose additional restrictions on the instance, such as stability conditions or bounded valuation space.

References

1. Agogino, A.K., Tumer, K.: Handling communication restrictions and team formation in congestion games. Auton. Agents Multi Agent Syst. **13**(1), 97–115 (2006)
2. Ausubel, L.M.: An efficient ascending-bid auction for multiple objects. Am. Econ. Rev. **94**(5), 1452–1475 (2004)
3. Ausubel, L.M., Milgrom, P.: The lovely but lonely vickrey auction. In: Combinatorial Auctions, Chapter 1. MIT Press (2006)

4. Babichenko, Y., Rubinstein, A.: Communication complexity of approximate nash equilibria. In: Proceedings of the 49th Annual ACM SIGACT Symposium on Theory of Computing, pp. 878–889. STOC 2017, Association for Computing Machinery, New York, NY, USA (2017)
5. Black, D.: On the rationale of group decision-making. J. Polit. Econ. **56**(1), 23–34 (1948)
6. Blum, A., Jackson, J., Sandholm, T., Zinkevich, M.: Preference elicitation and query learning. J. Mach. Learn. Res. **5**, 649–667 (2004)
7. Blumrosen, L., Nisan, N., Segal, I.: Auctions with severely bounded communication. J. Artif. Intell. Res. **28**, 233–266 (2007)
8. Chattopadhyay, A., Pitassi, T.: The story of set disjointness. SIGACT News **41**(3), 59–85 (2010)
9. Conen, W., Sandholm, T.: Preference elicitation in combinatorial auctions. In: Proceedings of the 3rd ACM Conference on Electronic Commerce, pp. 256–259. EC 2001, Association for Computing Machinery, New York, NY, USA (2001)
10. Conitzer, V., Sandholm, T.: Vote elicitation: Complexity and strategy-proofness. In: Proceedings of the Eighteenth National Conference on Artificial Intelligence and Fourteenth Conference on Innovative Applications of Artificial Intelligence, pp. 392–397. AAAI Press/The MIT Press (2002)
11. Conitzer, V., Sandholm, T.: Communication complexity of common voting rules. In: Proceedings of the 6th ACM Conference on Electronic Commerce, pp. 78–87. EC 2005, Association for Computing Machinery, New York, NY, USA (2005)
12. Cormen, T.H., Leiserson, C.E., Rivest, R.L., Stein, C.: Introduction to Algorithms, 3rd edn. The MIT Press, Cambridge (2009)
13. Cover, T.M., Thomas, J.A.: Elements of Information Theory (Wiley Series in Telecommunications and Signal Processing). Wiley, USA (2006)
14. Czumaj, A., Deligkas, A., Fasoulakis, M., Fearnley, J., Jurdziński, M., Savani, R.: Distributed methods for computing approximate equilibria. Algorithmica **81**(3), 1205–1231 (2018)
15. David, E., Rogers, A., Schiff, J., Kraus, S., Jennings, N.R.: Optimal design of English auctions with discrete bid levels. In: ACM Conference on Electronic Commerce (EC 2005), pp. 98–107 (2005)
16. Dobzinski, S., Vondrák, J.: Communication complexity of combinatorial auctions with submodular valuations. In: Khanna, S. (ed.) Proceedings of the Twenty-Fourth Annual ACM-SIAM Symposium on Discrete Algorithms, SODA 2013, pp. 1205–1215. SIAM (2013)
17. Fadel, R., Segal, I.: The communication cost of selfishness. J. Econ. Theory **144**(5), 1895–1920 (2009)
18. Feldman, M., Fiat, A., Golomb, I.: On voting and facility location. In: Conitzer, V., Bergemann, D., Chen, Y. (eds.) Proceedings of the 2016 ACM Conference on Economics and Computation, EC 2016, pp. 269–286. ACM (2016)
19. Goldberg, P.W., Pastink, A.: On the communication complexity of approximate Nash equilibria. In: Serna, M. (ed.) SAGT 2012. LNCS, pp. 192–203. Springer, Heidelberg (2012). https://doi.org/10.1007/978-3-642-33996-7_17
20. Hudson, B., Sandholm, T.: Effectiveness of query types and policies for preference elicitation in combinatorial auctions. In: Proceedings of the Third International Joint Conference on Autonomous Agents and Multiagent Systems, vol. 1, p. 386–393. AAMAS 2004, IEEE Computer Society, USA (2004)
21. Kagel, J.H., Harstad, R.M., Levin, D.: Information impact and allocation rules in auctions with affiliated private values: A laboratory study. Econometrica **55**(6), 1275–1304 (1987)

22. Kagel, J.H., Levin, D.: Independent Private Value Auctions: Bidder Behaviour in First-, Second- and Third-Price Auctions with Varying Numbers of Bidders. Econ. J. **103**(419), 868–879 (1993)
23. Kushilevitz, E., Nisan, N.: Communication Complexity. Cambridge University Press, Cambridge (1996)
24. Li, S.: Obviously strategy-proof mechanisms. Am. Econ. Rev. **107**(11), 3257–3287 (2017)
25. Mookherjee, D., Tsumagari, M.: Mechanism design with communication constraints. J. Polit. Econ. **122**(5), 1094–1129 (2014)
26. Moulin, H.: On strategy-proofness and single peakedness. Public Choice **35**(4), 437–455 (1980)
27. Nisan, N., Segal, I.: The communication requirements of efficient allocations and supporting prices. J. Econ. Theory **129**(1), 192–224 (2006)
28. Parkes, D.C., Ungar, L.H., Foster, D.P.: Accounting for cognitive costs in on-line auction design. In: Noriega, P., Sierra, C. (eds.) AMET 1998. LNCS (LNAI), vol. 1571, pp. 25–40. Springer, Heidelberg (1999). https://doi.org/10.1007/3-540-48835-9_2
29. Procaccia, A.D., Tennenholtz, M.: Approximate mechanism design without money. ACM Trans. Econ. Comput. **1**(4), 1–26 (2013)
30. Reichelstein, S.: Incentive compatibility and informational requirements. J. Econ. Theory **34**(1), 32–51 (1984)
31. Segal, I.: The communication requirements of social choice rules and supporting budget sets. J. Econ. Theory **136**(1), 341–378 (2007)
32. Smith, H., Dinev, T., Xu, H.: Information privacy research: an interdisciplinary review. MIS Q. **35**, 989–1015 (2011)
33. Van Zandt, T.: Communication complexity and mechanism design. J. Eur. Econ. Assoc. **5**(2/3), 543–553 (2007)
34. Vickrey, W.: Counter speculation, auctions, and competitive sealed tenders. J. Finan. **16**(1), 8–37 (1961)

Finding Fair and Efficient Allocations When Valuations Don't Add Up

Nawal Benabbou[1], Mithun Chakraborty[2](✉), Ayumi Igarashi[3], and Yair Zick[4]

[1] Sorbonne Université, CNRS, Laboratoire d'Informatique de Paris 6, LIP6, 75005 Paris, France
nawal.benabbou@lip6.fr

[2] University of Michigan, Ann Arbor, USA
dcsmc@umich.edu

[3] National Institute of Informatics, Tokyo, Japan
ayumi_igarashi@nii.ac.jp

[4] University of Massachusetts, Amherst, USA
yzick@umass.edu

Abstract. In this paper, we present new results on the fair and efficient allocation of indivisible goods to agents whose preferences correspond to *matroid rank functions*. This is a versatile valuation class, with several desirable properties (monotonicity, submodularity) which naturally models several real-world domains. We use these properties to our advantage; first, we show that when agent valuations are matroid rank functions, a socially optimal (i.e. utilitarian social welfare-maximizing) allocation that achieves envy-freeness up to one item (EF1) exists and is computationally tractable. We also prove that the Nash welfare-maximizing and the leximin allocations both exhibit this fairness/efficiency combination, by showing that they can be achieved by minimizing any symmetric strictly convex function over utilitarian optimal outcomes. Moreover, for a subclass of these valuation functions based on maximum (unweighted) bipartite matching, we show that a leximin allocation can be computed in polynomial time.

Keywords: Fair division · Envy-freeness · Submodularity · Dichotomous preferences · Matroid rank functions · Optimal welfare

1 Introduction

Suppose that we are interested in allocating seats in courses to prospective students. How should this be done? On the one hand, courses offer limited seats and have scheduling conflicts; on the other, students have preferences over the

This research was funded by an MOE Grant (no. R-252-000-625-133) and a Singapore NRF Research Fellowship (no. R-252- 000-750-733). Most of this work was done when Chakraborty and Zick were employed at National University of Singapore (NUS), and Igarashi was a research visitor at NUS.

T. Harks and M. Klimm (Eds.): SAGT 2020, LNCS 12283, pp. 32–46, 2020.
https://doi.org/10.1007/978-3-030-57980-7_3

classes that they take, which must be accounted for. Course allocation can be thought of as a problem of allocating a set of *indivisible goods* (course slots) to *agents* (students). How should we divide goods among agents with subjective valuations? Can we find a "good" allocation in polynomial time?

These questions have been the focus of intense study in the CS/Econ community in recent years; several justice criteria as well as methods for computing allocations that satisfy them have been investigated. Generally speaking, there are two types of justice criteria: *efficiency* and *fairness*. *Efficiency* criteria are chiefly concerned with maximizing some welfare criterion, e.g. *Pareto optimality* (PO). *Fairness* criteria require that agents do not perceive the resulting allocation as mistreating them; for example, one might want to ensure that no agent wants another agent's assigned bundle [18]. This criterion is known as *envy-freeness* (EF); however, envy-freeness is not always achievable with indivisibilities: consider, for example, two students competing for a single course slot. Any student receiving this slot would envy the other (in our stylized example, there is just the one course with the one seat).

A simple solution ensuring envy-freeness would be to withhold the seat altogether, not assigning it to either student. This solution, however, violates most efficiency criteria. Indeed, as observed by Budish [12], envy-freeness is not always achievable, even with the weakest efficiency criterion of completeness requiring that each item is allocated to some agent. However, a less stringent fairness notion—*envy-freeness up to one good* (EF1)—can be attained. An allocation is EF1 if for any two agents i and j, there is some item in j's bundle whose removal results in i not envying j. EF1 complete allocations always exist, and in fact, can be found in polynomial time [26].

While trying to efficiently achieve individual criteria is challenging in itself, things get really interesting when trying to simultaneously achieve multiple justice criteria.Caragiannis et al. [13] show that when agent valuations are *additive*—i.e. every agent i values its allocated bundle as the sum of values of individual items—there exist allocations that are both PO and EF1. Specifically, these are allocations that maximize the product of agents' utilities—also known as the *max Nash welfare* (MNW). Further work [6] shows that such allocations can be found in pseudo-polynomial time. While encouraging, these results are limited to agents with additive valuations. In particular, they do not apply to settings such as the course allocation problem described above (e.g. being assigned two courses with conflicting schedules will not result in additive gain), or other settings we describe later on. In fact, Caragiannis et al. [13] left it open whether their result extends to other natural classes of valuation functions, such as the class of submodular valutions.[1] At present, little is known about other classes valuation functions; this is where our work comes in.

[1] There is an instance of two agents with monotone supermodular/subadditive valuations where no allocation is PO and EF1 [13].

1.1 Our Contributions

We focus on monotone submodular valuations with binary (or dichotomous) marginal gains, which we refer to as *matroid rank valuations*. In this setting, the added benefit of receiving another item is binary and obeys the law of diminishing marginal returns. This is equivalent to the class of valuations that can be captured by matroid constraints; namely, each agent has a different matroid constraint over the items, and the value of a bundle is determined by the size of a maximum independent set included in the bundle.

Matroids offer a highly versatile framework for describing a variety of domains [29]. This class of valuations naturally arises in many practical applications, beyond the course allocation problem described above (where students are limited to either approving/disapproving a class). For example, suppose that a government body wishes to fairly allocate public goods to individuals of different minority groups (say, in accordance with a diversity-promoting policy). This could apply to the assignment of kindergarten slots to children from different neighborhoods/socioeconomic classes[2] or of flats in public housing estates to applicants of different ethnicities [8,9]. A possible way of achieving group fairness in this setting is to model each minority group as an agent consisting of many individuals: each agent's valuation function is based on *optimally matching* items to its constituent individuals; envy naturally captures the notion that no group should believe that other groups were offered better bundles (this is the fairness notion studied by Benabbou et al. [8]). Such assignment/matching-based valuations (known as OXS valuations [25]) are non-additive in general, and constitute an important subclass of submodular valuations. Matroid rank functions correspond to submodular valuations with binary (i.e. $\{0, 1\}$) marginal gains. The binary marginal gains assumption is best understood in context of matching-based valuations—in this scenario, it simply means that individuals either approve or disapprove of items, and do not distinguish between items they approve (we call OXS functions with binary individual preferences $(0, 1)$-OXS valuations). This is a reasonable assumption in kindergarten slot allocation (all approved/available slots are identical), and is implicitly made in some public housing mechanisms (e.g. Singapore housing applicants are required to effectively approve a subset of flats by selecting a block, and are precluded from expressing a more refined preference model).

In addition, imposing certain constraints on the underlying matching problem retains the submodularity of the agents' induced valuation functions: if there is a hard limit due to a *budget* or an exogenous *quota* (e.g. ethnicity-based quotas in Singapore public housing; socioeconomic status-based quotas in certain U.S. public school admission systems) on the number of items each group is able or allowed to receive, then agents' valuations are *truncated* matching-based valuations. Such valuation functions are not OXS, but are still matroid rank functions. Since agents still have binary/dichotomous preferences over items even with the quotas in place, our results apply to this broader class as well.

[2] see, e.g. https://www.ed.gov/diversity-opportunity.

Using the matroid framework, we obtain a variety of positive existential and algorithmic results on the compatibility of (approximate) envy-freeness with welfare-based allocation concepts. The following is a summary of our main results (see also Table 1):

(a) For matroid rank valuations, we show that an EF1 allocation that also maximizes the utilitarian social welfare or USW (hence is Pareto optimal) always exists and can be computed in polynomial time.
(b) For matroid rank valuations, we show that leximin[3] and MNW allocations both possess the EF1 property.
(c) For matroid rank valuations, we provide a characterization of the leximin allocations; we show that they are identical to the minimizers of *any* symmetric strictly convex function over utilitarian optimal allocations. We obtain the same characterization for MNW allocations.
(d) For $(0, 1)$-OXS valuations, we show that both leximin and MNW allocations can be computed efficiently.

Table 1. Summary of our computational complexity results.

	MNW	Leximin	max-USW+EF1
$(0, 1)$-OXS	poly-time (Th. 5)	poly-time (Th. 5)	poly-time (Th. 1)
Matroid rank	?	?	poly-time (Th. 1)

All proofs omitted from the body of the paper due to space constraints as well as clarifying examples remarks, extensions, and additional references are available in the online full version with appendices at https://git.io/JJYdW.

Result (a) is remarkably positive: the EF1 and USW objectives are incompatible in general, even for additive valuations. Result (b) is reminiscent of Thm. 3.2 by Caragiannis et al. [13], showing that any MNW allocation is PO and EF1 under *additive* valuations. The PO+EF1 existence question beyond additive valuations, which they left open, has seen little progress. To our knowledge, the class of matroid rank valuations is the first valuation class not subsumed by additive valuations for which the EF1 property of the MNW allocation have been established.

1.2 Related Work

Our paper is related to the vast literature on the fairness and efficiency issue in resource allocation. Early work on divisible resource allocation provides an elegant answer: an allocation that satisfies envy-freeness and Pareto optimality

[3] Roughly speaking, a leximin allocation is one that maximizes the realized valuation of the worst-off agent and, subject to that, maximizes that of the second worst-off agent, and so on.

always exists under mild assumptions on valuations [34], and can be computed via convex programming of Eisenberg and Gale [17] for additive valuations. Four decades later, Caragiannis et al. [13] prove the discrete analogue of Eisenberg and Gale [17]: MNW allocation satisfies EF1 and Pareto optimality for additive valuations. Subsequently, Barman et al. [6] provide a pseudo-polynomial-time algorithm for computing allocations satisfying EF1 and PO.

While computing leximin/MNW allocations of indivisible items is hard in general, several positive results are known when agents have binary additive valuations. Darmann and Schauer [14] and Barman et al. [7] show that the maximum Nash welfare can be computed efficiently for binary additive valuations. Further, the equivalence between leximin and MNW for binary additive valuations has been obtained in several recent papers. Aziz and Rey [3] show that the algorithm proposed by Darmann and Schauer outputs a leximin optimal allocation; in particular this implies that the leximin and MNW solutions coincide for binary additive valuations. This is implied by our results. Similar results are shown by Halpern et al. [21], who also show that the leximin/MNW optimal allocation is group-strategyproof for agents with binary additive valuations. In the context of *divisible* goods, Aziz and Ye [4] show the leximin and MNW solutions also coincide for dichotomous preferences.

From a technical perspective, our work makes extensive use of matroid theory; while some papers have explored the application of matroid theory to the fair division problem [10,20], we believe that ours is the first to demonstrate its strong connection with fairness and efficiency guarantees.

One motivation for our paper is recent work by Benabbou et al. [8] on promoting diversity in assignment problems through efficient, EF1 allocations of items to groups in a population. Similar works study quota-based fairness/diversity [2,9,33, and references therein], or by the optimization of carefully constructed functions [1,15,23, and references therein] in allocation/subset selection.

Finally, Babaioff et al. [5] present a set of results similar to our own; they further explore strategyproof mechanisms for matroid rank valuations, showing that such mechanisms exist. Our work was developed independently, and is very different from a technical perspective.

2 Model and Definitions

Throughout the paper, given a positive integer r, let $[r]$ denote the set $\{1, 2, \ldots, r\}$. We are given a set $N = [n]$ of *agents*, and a set $O = \{o_1, \ldots, o_m\}$ of *items* or *goods*. Subsets of O are referred to as *bundles*, and each agent $i \in N$ has a *valuation function* $v_i : 2^O \to \mathbb{R}_+$ over bundles where $v_i(\emptyset) = 0$, i.e all valuations are *normalized*. We further assume polynomial-time oracle access to the valuation v_i of all agents. Given a valuation function $v_i : 2^O \to \mathbb{R}$, we define the *marginal gain* of an item $o \in O$ w.r.t. a bundle $S \subseteq O$, as $\Delta_i(S; o) \triangleq v_i(S \cup \{o\}) - v_i(S)$. A valuation function v_i is *monotone* if $v_i(S) \leq v_i(T)$ whenever $S \subseteq T$.

An *allocation* A of items to agents is a collection of n disjoint bundles $A_1, \ldots, A_n$, such that $\bigcup_{i \in N} A_i \subseteq O$; the bundle A_i is allocated to agent i.

Given an allocation A, we denote by A_0 the set of unallocated items, also referred to as *withheld items*. We may refer to agent i's valuation of its bundle $v_i(A_i)$ under the allocation A as its *realized valuation* under A. An allocation is *complete* if every item is allocated to *some* agent, i.e. $A_0 = \emptyset$. We admit incomplete, but *clean* allocations: a bundle $S \subseteq O$ is *clean* for $i \in N$ if it contains no item $o \in S$ for which agent i has zero marginal gain (i.e., $\Delta_i(S \setminus \{o\}; o) = 0$, or equivalently $v_i(S \setminus \{o\}) = v_i(S)$); an allocation A is *clean* if each allocated bundle A_i is clean for the agent i that receives it. It is easy to 'clean' any allocation without changing any realized valuation by iteratively revoking items of zero marginal gain from respective agents and placing them in A_0 (see Example 1 in Appendix A).

2.1 Fairness and Efficiency Criteria

Our fairness criteria are based on the concept of *envy*. Agent i *envies* agent j under an allocation A if $v_i(A_i) < v_i(A_j)$. An allocation A is *envy-free* (EF) if no agent envies another. We will use the following relaxation of the EF property due to Budish [12]: we say that A is *envy-free up to one good* (EF1) if, for every $i, j \in N$, i does not envy j or there exists o in A_j such that $v_i(A_i) \geq v_i(A_j \setminus \{o\})$.

The efficiency concept that we are primarily interested in is *Pareto optimality*. An allocation A' is said to *Pareto dominate* the allocation A if $v_i(A'_i) \geq v_i(A_i)$ for all agents $i \in N$ and $v_j(A'_j) > v_j(A_j)$ for some agent $j \in N$. An allocation is *Pareto optimal* (PO) if it is not Pareto dominated by any other allocation.

There are several ways of measuring the welfare of an allocation [31]. Specifically, given an allocation A, (i) its *utilitarian social welfare* is $\texttt{USW}(A) \triangleq \sum_{i=1}^{n} v_i(A_i)$; (ii) its *egalitarian social welfare* is $\texttt{ESW}(A) \triangleq \min_{i \in N} v_i(A_i)$; and (iii) its *Nash welfare* is $\texttt{NW}(A) \triangleq \prod_{i \in N} v_i(A_i)$. An allocation A is said to be *utilitarian optimal* (respectively, *egalitarian optimal*) if it maximizes $\texttt{USW}(A)$ (respectively, $\texttt{ESW}(A)$) among all allocations. Since it is possible that the maximum attainable Nash welfare is 0 (say, if there are less items than agents then one agent must have an empty bundle), we use the following refinement of the maximum Nash social welfare (MNW) used in [13]: we find a maximal subset of agents, say $N_{\max} \subseteq N$, to which we can allocate bundles of positive values, and compute an allocation to agents in $N_{\max}$ that maximizes the product of their realized valuations. If $N_{\max}$ is not unique, we choose the one that results in the highest product of realized valuations.

The *leximin* welfare is a lexicographic refinement of egalitarian optimality. Formally, for real n-dimensional vectors x and y, x is *lexicographically greater than or equal to* y (denoted by $x \geq_L y$) if and only if $x = y$, or $x \neq y$ and for the minimum index j such that $x_j \neq y_j$ we have $x_j > y_j$. For each allocation A, we denote by $\theta(A)$ the vector of the components $v_i(A_i)$ ($i \in N$) arranged in non-decreasing order. A *leximin* allocation A is one that maximizes the egalitarian welfare in a lexicographic sense, i.e., $\theta(A) \geq_L \theta(A')$ for any other allocation A'.

2.2 Submodular Valuations

The main focus of this paper is on fair allocation when agent valuations are *submodular.* A valuation function v_i is *submodular* if single items contribute more to smaller sets than to larger ones, namely: for all $S \subseteq T \subseteq O$ and all $o \in O \setminus T$, $\Delta_i(S; o) \geq \Delta_i(T; o)$.

One important subclass of submodular valuations is *assignment valuations,* introduced by Shapley [32] and also called OXS valuations [24]. Fair allocation in this setting was explored by Benabbou et al. [8]. Here, each agent $h \in N$ represents a group of individuals N_h (such as ethnic groups and genders); each individual $i \in N_h$ (also called a *member*) has a fixed non-negative weight $u_{i,o}$ for each item o. An agent h values a bundle S via a *matching* of the items to its individuals (i.e. each item is assigned to at most one member and vice versa) that maximizes the sum of weights [27]; namely, $v_h(S) = \max\{\sum_{i \in N_h} u_{i,\pi(i)} \mid \pi \in \Pi(N_h, S)\}$, where $\Pi(N_h, S)$ is the set of matchings $\pi : N_h \to S$ in the complete bipartite graph with bipartition (N_h, S).

Our particular focus is on submodular functions with *binary marginal gains.* We say that v_i has *binary marginal gains* if $\Delta_i(S; o) \in \{0, 1\}$ for all $S \subseteq O$ and $o \in O \setminus S$. The class of submodular valuations with binary marginal gains includes the classes of binary additive valuations [7] and of assignment valuations where the weight is binary [8]. We say that v_i is a *matroid rank* valuation if it is a submodular function with binary marginal gains (these are equivalent definitions [29]), and (0, 1)-OXS if it is an assignment valuation with binary marginal gains.

3 Matroid Rank Valuations

The main theme of all results in this section is that, when all agents have matroid rank valuations, fairness and efficiency properties are compatible with one another, and there exist allocations that satisfy all three welfare criteria we consider. We start by introducing some notions from matroid theory. Formally, a *matroid* is an ordered pair $(E, \mathcal{I})$, where E is some finite set and $\mathcal{I}$ is a family of its subsets (referred to as the *independent sets* of the matroid), which satisfies the following three axioms:

(I1) $\emptyset \in \mathcal{I}$,
(I2) if $Y \in \mathcal{I}$ and $X \subseteq Y$, then $X \in \mathcal{I}$, and
(I3) if $X, Y \in \mathcal{I}$ and $|X| > |Y|$, then there exists $x \in X \setminus Y$ such that $Y \cup \{x\} \in \mathcal{I}$.

The rank function $r : 2^E \to \mathbb{Z}$ of a matroid returns the *rank* of each set X, i.e. the maximum size of an independent subset of X. Another equivalent way to define a matroid is to use the axiom systems for a rank function. We require that (R1) $r(X) \leq |X|$, (R2) r is monotone, and (R3) r is submodular. Then, the pair $(E, \mathcal{I})$ where $\mathcal{I} = \{X \subseteq E \mid r(X) = |X|\}$ is a matroid [29]. In other words, if r satisfies properties (R1)–(R3) then it induces a matroid. In the fair allocation terminology, if an agent has a matroid rank valuation, then the set of *clean* bundles forms the set of independent sets of a matroid. Before proceeding further, we state some useful properties of the matroid rank valuation class.

Proposition 1. *A valuation function v_i with binary marginal gains is monotone and takes values in $[|S|]$ for any bundle S (hence $v_i(S) \leq |S|$).*

Proposition 2. *For matroid rank valuations, A is a clean allocation if and only if $v_i(A_i) = |A_i|$ for each $i \in N$.*

Even for binary additive valuations, EF and PO allocations may not exist (as a simple example of two agents and a single good valued at 1 by each of them demonstrates); thus, we turn our attention to EF1 and PO allocations.

3.1 Utilitarian Optimal and EF1 Allocation

For non-negative additive valuations, Caragiannis et al. [13] prove that every MNW allocation is Pareto optimal and EF1. However, the existence question of an allocation satisfying both the PO and EF1 properties remains open for submodular valuations. We show that the existence of a PO+EF1 allocation [13] extends to the class of matroid rank valuations. In fact, we provide a surprisingly strong relation between efficiency and fairness: utilitarian optimality (stronger than Pareto optimality) and EF1 turn out to be compatible under matroid rank valuations. Moreover, such an allocation can be computed in polynomial time!

Theorem 1. *For matroid rank valuations, a utilitarian optimal allocation that is also EF1 exists and can be computed in polynomial time.*

Our result is constructive: we provide a way of computing the above allocation in Algorithm 1. The proof of Theorem 1 and those of the latter theorems utilize Lemmas 1 and 2 which shed light on the interesting interaction between envy and matroid rank valuations.

Lemma 1 (Transferability property). *For monotone submodular valuation functions, if agent i envies agent j under an allocation A, then there is an item $o \in A_j$ for which i has a positive marginal gain with respect to A_i.*

Lemma 1 holds for submodular functions with arbitrary real-valued marginal gains, and is trivially true for (non-negative) additive valuations. However, there exist non-submodular valuation functions that violate the transferability property, even when they have binary marginal gains (see Example 2 in Appendix A).

Below, we show that if i's envy towards j cannot be eliminated by removing one item, then the sizes of their *clean* bundles differ by at least two. Formally, we say that agent i envies j up to more than 1 item if $A_j \neq \emptyset$ and $v_i(A_i) < v_i(A_j \setminus \{o\})$ for every $o \in A_j$.

Lemma 2. *For matroid rank valuations, if agent i envies agent j up to more than 1 item under an allocation A and j's bundle A_j is clean, then $v_j(A_j) \geq v_i(A_i) + 2$.*

We are now ready to show that under matroid rank valuations, utilitarian social welfare maximization is polynomial-time solvable (2).

Theorem 2. *For matroid rank valuations, one can compute a clean utilitarian optimal allocation in polynomial time.*

Proof. We prove the claim by a reduction to the matroid intersection problem. Let E be the set of pairs of items and agents, i.e., $E = \{\, \{o, i\} \mid o \in O \wedge i \in N \,\}$. For each $i \in N$ and $X \subseteq E$, we define X_i to be the set of edges incident to i, i.e., $X_i = \{\, \{o, i\} \in X \mid o \in O \,\}$. Note that taking $E = X$, E_i is the set of all edges in E incident to $i \in N$. For each $i \in N$ and for each $X \subseteq E$, we define $r_i(X)$ to be the valuation of i, under function $v_i(\cdot)$, for the items $o \in O$ such that $\{o, i\} \in X_i$; namely,

$$r_i(X) = v_i(\{\, o \in O \mid \{o, i\} \in X_i \,\}).$$

Clearly, r_i is also a submodular function with binary marginal gains; combining this with Proposition 1 and the fact that $r_i(\emptyset) = 0$, it is easy to see that each r_i is a rank function of a matroid. Thus, the set of clean bundles for i, i.e $\mathcal{I}_i = \{\, X \subseteq E \mid r_i(X) = |X| \,\}$, is the set of independent sets of a matroid. Taking the union $\mathcal{I} = \mathcal{I}_1 \cup \cdots \cup \mathcal{I}_n$, the pair $(E, \mathcal{I})$ is known to form a matroid [22], often referred to as a *union matroid*. By definition, $\mathcal{I} = \{\, \bigcup_{i \in N} X_i \mid X_i \in \mathcal{I}_i \wedge i \in N \,\}$, so any independent set in $\mathcal{I}$ corresponds to a union of clean bundles for each $i \in N$ and vice versa. To ensure that each item is assigned at most once (i.e. bundles are disjoint), we will define another matroid $(E, \mathcal{O})$ where the set of independent sets is given by

$$\mathcal{O} = \{\, X \subseteq E \mid |X \cap E_o| \leq 1, \forall o \in O \,\}.$$

Here, $E_o = \{e = \{o, i\} \mid i \in N\}$ for $o \in O$. The pair $(E, \mathcal{O})$ is known as a *partition matroid* [22].

Now, observe that a common independent set of the two matroids $X \in \mathcal{O} \cap \mathcal{I}$ corresponds to a clean allocation A of our original instance where each agent i receives the items o with $\{o, i\} \in X$; indeed, each item o is allocated at most once because $|E_o \cap X| \leq 1$, and each A_i is clean because the realized valuation of agent i under A is exactly the size of the allocated bundle. Conversely, any clean allocation A of our instance corresponds to an independent set $X = \bigcup_{i \in N} X_i \in \mathcal{I} \cap \mathcal{O}$, where $X_i = \{\, \{o, i\} \mid o \in A_i \,\}$: for each $i \in N$, $r_i(X_i) = |X_i|$ by Proposition 2, and hence $X_i \in \mathcal{I}_i$, which implies that $X \in \mathcal{I}$; also, $|X \cap E_o| \leq 1$ as A is an allocation, and hence $X \in \mathcal{O}$.

Thus, the maximum utilitarian social welfare is the same as the size of a maximum common independent set in $\mathcal{I} \cap \mathcal{O}$. It is well known that one can find a largest common independent set in two matroids in time $O(|E|^3 \theta)$ where θ is the maximum complexity of the two independence oracles [16]. Since the maximum complexity of checking independence in two matroids $(E, \mathcal{O})$ and $(E, \mathcal{I})$ is bounded by $O(mnF)$ where F is the maximum complexity of the value query oracle, we can find a set $X \in \mathcal{I} \cap \mathcal{O}$ with maximum $|X|$ in time $O(|E|^3 mnF)$. □

We are now ready to prove Theorem 1.

Proof (Proof of Theorem 1). Algorithm 1 maintains optimal `USW` as an invariant and terminates on an EF1 allocation. Specifically, we first compute a clean allocation that maximizes the utilitarian social welfare. The `EIT` subroutine in the algorithm iteratively diminishes envy by transferring an item from the envied bundle to the envious agent; Lemma 1 ensures that there is always an item in the envied bundle for which the envious agent has a positive marginal gain.

Algorithm 1: Algorithm for finding utilitarian optimal EF1 allocation

1 Compute a clean, utilitarian optimal allocation A.
2 **/*Envy-Induced Transfers (`EIT`)*/**
3 **while** *there are two agents* i, j *such that* i *envies* j *more than* 1 *item.* **do**
4 | Find item $o \in A_j$ with $\Delta_i(A_i; o) = 1$.
5 | $A_j \leftarrow A_j \setminus \{o\}$; $A_i \leftarrow A_i \cup \{o\}$.
6 **end**

Correctness: Each `EIT` step maintains the optimal utilitarian social welfare as well as cleanness: an envied agent's valuation diminishes exactly by 1 while that of the envious agent increases by exactly 1. Thus, if it terminates, the `EIT` subroutine retains the initial (optimal) `USW` and, by the stopping criterion, induces the EF1 property. To show that the algorithm terminates in polynomial time, we define the potential function $\phi(A) \triangleq \sum_{i \in N} v_i(A_i)^2$. At each step of the algorithm, $\phi(A)$ strictly decreases by 2 or a larger integer. To see this, let A' denote the resulting allocation after reallocation of item o from agent j to i. Since A is clean, we have $v_i(A'_i) = v_i(A_i) + 1$ and $v_j(A'_j) = v_j(A_j) - 1$; since all other bundles are untouched, $v_k(A'_k) = v_k(A_k)$ for every $k \in N \setminus \{i, j\}$. Also, since i envies j up to more than one item under allocation A, $v_i(A_i) + 2 \leq v_j(A_j)$ by Lemma 2. Combining these, simple algebra gives us $\phi(A') - \phi(A) \leq -2$.

Complexity: By Theorem 2, computing a clean utilitarian optimal allocation can be done in polynomial time. The value of the non-negative potential function has a polynomial upper bound: $\sum_{i \in N} v_i(A_i)^2 \leq (\sum_{i \in N} v_i(A_i))^2 \leq m^2$. Thus, Algorithm 1 terminates in polynomial time. □

An interesting implication of the above analysis is that a utilitarian optimal allocation that minimizes $\sum_{i \in N} v_i(A_i)^2$ is always EF1.

Corollary 1. *For matroid rank valuations, any clean, utilitarian optimal allocation A that minimizes $\phi(A) \triangleq \sum_{i \in N} v_i(A_i)^2$ among all utilitarian optimal allocations is EF1.*

Despite its simplicity, Algorithm 1 significantly generalizes that of Benabbou et al. [8]'s Theorem 4 (which ensures the existence of a *non-wasteful* EF1 allocation for (0, 1)-OXS valuations) to matroid rank valuations. We note, however, that the resulting allocation may be neither MNW nor leximin even when agents have $(0,1)$-OXS valuations: Example 3 in Appendix A illustrates this and also shows that the converse of Corollary 1 does not hold.

3.2 MNW and Leximin Allocations for Matroid Rank Functions

We characterize the set of leximin and MNW allocations under matroid rank valuations. We start by showing that Pareto optimal allocations coincide with utilitarian optimal allocations when agents have matroid rank valuations. Intuitively, if an allocation is not utilitarian optimal, one can find an 'augmenting' path that makes at least one agent happier but no other agent worse off. The full proof, which is more involved and relies on the concept of *circuits* of matrices, is available online in Appendix A.

Theorem 3. *For matroid rank valuations, PO allocations are utilitarian optimal.*

Since leximin and MNW allocations are Pareto optimal [11,13], Theorem 3 implies that such allocations are utilitarian optimal as well. Next, we show that for the class of matroid rank valuations, leximin and MNW allocations are identical to each other; further, they can be characterized as the minimizers of any symmetric strictly convex function among all utilitarian optimal allocations.

A function $\Phi : \mathbb{Z}^n \to \mathbb{R}$ is *symmetric* if for any permutation $\pi : [n] \to [n]$,

$$\Phi(z_1, z_2, \dots, z_n) = \Phi(z_{\pi(1)}, z_{\pi(2)}, \dots, z_{\pi(n)}),$$

and is *strictly convex* if for any $x, y \in \mathbb{Z}^n$ with $x \neq y$ and $\lambda \in (0,1)$ where $\lambda x + (1-\lambda)y$ is an integral vector, $\lambda\Phi(x) + (1-\lambda)\Phi(y) > \Phi(\lambda x + (1-\lambda)y)$. Examples of symmetric, strictly convex functions include: $\Phi(z_1, z_2, \dots, z_n) \triangleq \sum_{i=1}^n z_i^2$ for $z_i \in \mathbb{Z} \quad \forall i$; $\Phi(z_1, z_2, \dots, z_n) \triangleq \sum_{i=1}^n z_i \ln z_i$ for $z_i \in \mathbb{Z}_{\geq 0}$ $\forall i$. For an allocation A, we define $\phi(A) \triangleq \phi(v_1(A_1), v_2(A_2), \dots, v_n(A_n))$.

Theorem 4. *Let $\Phi : \mathbb{Z}^n \to \mathbb{R}$ be a symmetric strictly convex function; let A be some allocation. For matroid rank valuations, the following are equivalent:*

1. *A is a minimizer of Φ over all the utilitarian optimal allocations; and*
2. *A is a leximin allocation; and*
3. *A maximizes Nash welfare.*

The proof is highly technical and is hence relegated to Appendix A online. To summarize, we first establish the equivalence of statements 1 and 2 by showing: (i) Lemma 4: given a non-leximin utilitarian optimal allocation A, there exists an "adjacent" utilitarian optimal allocation A' which is the result of transferring one item from a 'happy' agent j to a less 'happy' agent i (the underlying submodularity guarantees the existence of such an allocation); (ii) Lemma 5: such an adjacent allocation A' has a strictly higher value of any symmetric strictly convex function than A. We complete the three-way equivalence by noting that maximizing Nash welfare is identical to minimizing the symmetric, strictly convex function $\phi(x) = -\sum_{i=1}^{n} \log x_i$ (carefully accounting for the possibility that some agents may realize zero valuations).

Theorem 4 does not generalize to the non-binary case: Example 5 in Appendix A presents an instance where the leximin and MNW allocation are not USW optimal.

Combining the above characterization with the results of Sect. 3.1, we get the following fairness-efficiency guarantee for matroid rank valuations.

Corollary 2. *For matroid rank valuations, any clean leximin or MNW allocation is EF1.*

4 Assignment Valuations with Binary Gains

We now consider the practically important special case where valuations come from maximum matchings. For this valuation class, we show that invoking Theorem 3, one can find a leximin or MNW allocation in polynomial time, by a reduction to the network flow problem. We note that the complexity of the problem remains open for general matroid rank valuations.

Theorem 5. *For assignment valuations with binary marginal gains, one can find a leximin or MNW allocation in polynomial time.*

The proof, available in Appendix A, is based on the following key idea: given any instance with (0, 1)-OXS valuations, we construct a flow network such that the problem of finding a leximin allocation in the original instance reduces to that of finding a *increasingly-maximal integer-valued flow* on the induced network for which Frank and Murota [19] recently gave a polynomial-time algorithm.

In contrast with (0, 1)-OXS valuations, computing a leximin or MNW allocation becomes NP-hard for weighted assignment valuations, even for two agents.

Theorem 6. *Computing a leximin/MNW allocation for two agents with general assignment valuations is NP-hard.*

The proof is available in Appendix A. We give a Turing reduction from Partition. The reduction is similar to the hardness reduction for two agents with identical additive valuations [28,30].

5 Discussion

We study allocations of indivisible goods under matroid rank valuations in terms of the interplay among envy, efficiency, and various welfare concepts. Since the class of matroid rank functions is rather broad, our results can be immediately applied to settings where agents' valuations are induced by a matroid structure. Beyond the domains described in this work, these include several others. For example, *partition matroids* model instances where agents' have access to different item types, but can only hold a limited number of each type (their utility is the total number of items they hold); a variety of other domains, such as spanning trees, independent sets of vectors, coverage problems and more admit a matroid structure (see Oxley [29] for an overview). Indeed, a well-known result in combinatorial optimization states that *any* agent valuation structure where the greedy algorithm can be used to find the (weighted) optimal bundle, is induced by some matroid [29, Theorem 1.8.5].

There are several known extensions to matroid structures, with deep connections to submodular optimization [29, Chapter 11]. Matroid rank functions are submodular functions with binary marginal gains; however, general submodular functions admit some matroid structure which may potentially be used to extend our results to more general settings. Finally, it would be interesting to explore other fairness criteria such as proportionality, the maximin share guarantee, equitability. etc. (see, e.g. [11] and references therein) for matroid rank valuations. We present some of our attempts along these lines in Appendices B through D.

References

1. Ahmed, F., Dickerson, J.P., Fuge, M.: Diverse weighted bipartite B-matching. In: Proceedings of the 26th International Joint Conference on Artificial Intelligence (IJCAI), pp. 35–41 (2017)
2. Aziz, H., Gaspers, S., Sun, Z., Walsh, T.: From matching with diversity constraints to matching with regional quotas. In: Proceedings of the 18th International Conference on Autonomous Agents and Multi-Agent Systems (AAMAS), pp. 377–385 (2019)
3. Aziz, H., Rey, S.: Almost group envy-free allocation of indivisible goods and chores. CoRR abs/1907.09279 (2019)
4. Aziz, H., Ye, C.: Cake cutting algorithms for piecewise constant and piecewise uniform valuations. In: Proceedings of the 10th Conference on Web and Internet Economics (WINE), pp. 1–14 (2014)
5. Babaioff, M., Ezra, T., Feige, U.: Fair and truthful mechanisms for dichotomous valuations. CoRR abs/2002.10704 (2020)
6. Barman, S., Krishnamurthy, S.K., Vaish, R.: Finding fair and efficient allocations. In: Proceedings of the 19th ACM Conference on Economics and Computation (EC), pp. 557–574. ACM (2018)
7. Barman, S., Krishnamurthy, S.K., Vaish, R.: Greedy algorithms for maximizing Nash social welfare. In: Proceedings of the 17th International Conference on Autonomous Agents and Multi-Agent Systems (AAMAS), pp. 7–13 (2018)
8. Benabbou, N., Chakraborty, M., Elkind, E., Zick, Y.: Fairness towards groups of agents in the allocation of indivisible items. In: Proceedings of the 28th International Joint Conference on Artificial Intelligence (IJCAI), pp. 95–101 (2019)
9. Benabbou, N., Chakraborty, M., Xuan, V.H., Sliwinski, J., Zick, Y.: Diversity constraints in public housing allocation. In: Proceedings of the 17th International Conference on Autonomous Agents and Multi-Agent Systems (AAMAS), pp. 973–981 (2018)
10. Biswas, A., Barman, S.: Fair division under cardinality constraints. In: Proceedings of the 27th International Joint Conference on Artificial Intelligence (IJCAI), pp. 91–97 (2018)

11. Bouveret, S., Chevaleyre, Y., Maudet, N.: Fair allocation of indivisible goods. In: Felix, B., Vincent, C., Ulle, E., Jérôme, L., Procaccia, A.D. (ed.) Handbook of Computational Social Choice, Chap. 12, pp. 284–310. Cambridge University Press (2016)
12. Budish, E.: The combinatorial assignment problem: approximate competitive equilibrium from equal incomes. J. Polit. Econ. **119**(6), 1061–1103 (2011)
13. Caragiannis, I., Kurokawa, D., Moulin, H., Procaccia, A.D., Shah, N., Wang, J.: The unreasonable fairness of maximum Nash welfare. In: Proceedings of the 17th ACM Conference on Economics and Computation (EC), pp. 305–322. ACM (2016)
14. Darmann, A., Schauer, J.: Maximizing Nash product social welfare in allocating indivisible goods. Eur. J. Oper. Res. **247**(2), 548–559 (2015)
15. Dickerson, J.P., Sankararaman, K.A., Srinivasan, A., Xu, P.: Balancing relevance and diversity in online bipartite matching via submodularity. In: Proceedings of the 33rd AAAI Conference on Artificial Intelligence (AAAI), pp. 1877–1884 (2019)
16. Edmonds, J.: Matroid intersection. Ann. Discrete Math. **4**, 39–49 (1979). Discrete Optimization I
17. Eisenberg, E., Gale, D.: Consensus of subjective probabilities: the pari-mutuel method. Ann. Math. Statist. **30**(1), 165–168 (1959)
18. Foley, D.: Resource allocation and the public sector. Yale Econ. Essays **7**, 45–98 (1967)
19. Frank, A., Murota, K.: Discrete decreasing minimization, Part III: Network Flows. arXiv e-prints arXiv:1907.02673v2, September 2019
20. Gourvès, L., Monnot, J.: Approximate maximin share allocations in Matroids. In: Fotakis, D., Pagourtzis, A., Paschos, V.T. (eds.) CIAC 2017. LNCS, vol. 10236, pp. 310–321. Springer, Cham (2017). https://doi.org/10.1007/978-3-319-57586-5_26
21. Halpern, D., Procaccia, A.D., Psomas, A., Shah, N.: Fair division with binary valuations: One rule to rule them all (2020), unpublished Manuscript
22. Korte, B., Vygen, J.: Combinatorial Optimization: Polyhedra and Efficiency. Algorithms and Combinatorics. Springer, Heidelberg (2006)
23. Lang, J., Skowron, P.K.: Multi-attribute proportional representation. In: Proceedings of the 30th AAAI Conference on Artificial Intelligence (AAAI), pp. 530–536 (2016)
24. Lehmann, B., Lehmann, D., Nisan, N.: Combinatorial auctions with decreasing marginal utilities. Games Econ. Behav. **55**(2), 270–296 (2006)
25. Leme, R.P.: Gross substitutability: an algorithmic survey. Games Econ. Behav. **106**, 294–316 (2017)
26. Lipton, R.J., Markakis, E., Mossel, E., Saberi, A.: On approximately fair allocations of indivisible goods. In: Proceedings of the 5th ACM Conference on Electronic Commerce (EC), pp. 125–131. ACM (2004)
27. Munkres, J.: Algorithms for the assignment and transportation problems. J. Soc. Ind. Appl. Math. **5**(1), 32–38 (1957)
28. Nguyen, T.T., Roos, M., Rothe, J.: A survey of approximability and inapproximability results for social welfare optimization in multiagent resource allocation. Ann. Math. Artif. Intell. **68**(1), 65–90 (2013). https://doi.org/10.1007/s10472-012-9328-4
29. Oxley, J.: Matroid Theory. Oxford Graduate Texts in Mathematics, 2nd edn. Oxford Univerity Press, Oxford (2011)
30. Ramezani, S., Endriss, U.: Nash social welfare in multiagent resource allocation. In: Proceedings of the 11th International Workshop on Agent-Mediated Electronic Commerce and Trading Agents Design and Analysis (AMEC/TADA), pp. 117–131 (2010)

31. Sen, A.: Collective Choice and Social Welfare. Holden Day, San Francisco (1970)
32. Shapley, L.S.: Complements and substitutes in the optimal assignment problem. ASTIA Document (1958)
33. Suzuki, T., Tamura, A., Yokoo, M.: Efficient allocation mechanism with endowments and distributional constraints. In: Proceedings of the 17th International Conference on Autonomous Agents and Multi-Agent Systems (AAMAS), pp. 50–58 (2018)
34. Varian, H.R.: Equity, envy, and efficiency. J. Econ. Theory **9**(1), 63–91 (1974)

Mechanism Design for Perturbation Stable Combinatorial Auctions

Giannis Fikioris and Dimitris Fotakis(✉)

School of Electrical and Computer Engineering,
National Technical University of Athens, 15780 Athens, Greece
fikioris@corelab.ntua.gr, fotakis@cs.ntua.gr

Abstract. Motivated by recent research on combinatorial markets with endowed valuations by (Babaioff et al., EC 2018) and (Ezra et al., EC 2020), we introduce a notion of perturbation stability in Combinatorial Auctions (CAs) and study the extend to which stability helps in social welfare maximization and mechanism design. A CA is γ-*stable* if the optimal solution is resilient to inflation, by a factor of $\gamma \geq 1$, of any bidder's valuation for any single item. On the positive side, we show how to compute efficiently an optimal allocation for 2-stable subadditive valuations and that a Walrasian equilibrium exists for 2-stable submodular valuations. Moreover, we show that a Parallel 2nd Price Auction (P2A) followed by a demand query for each bidder is truthful for general subadditive valuations and results in the optimal allocation for 2-stable submodular valuations. To highlight the challenges behind optimization and mechanism design for stable CAs, we show that a Walrasian equilibrium may not exist for 2-stable XOS valuations, that a polynomial-time approximation scheme does not exist for $(2-\varepsilon)$-stable submodular valuations, and that any DSIC mechanism that computes the optimal allocation for stable CAs and does not use demand queries must use exponentially many value queries. We conclude with analyzing the Price of Anarchy of P2A and Parallel 1st Price Auctions (P1A) for CAs with stable submodular and XOS valuations. Our results indicate that the quality of equilibria of simple non-truthful auctions improves only for γ-stable instances with $\gamma \geq 3$.

1 Introduction

Combinatorial auctions appear in many different contexts (e.g., spectrum auctions [26], network routing auctions [22], airport time-slot auctions [29], etc.) and have been studied extensively (and virtually from every possible aspect) for a few decades (see e.g., [28] and the references therein).

This work was supported by the Hellenic Foundation for Research and Innovation (H.F.R.I.) under the "First Call for H.F.R.I. Research Projects to support Faculty members and Researchers and the procurement of high-cost research equipment grant", project BALSAM, HFRI-FM17-1424.

T. Harks and M. Klimm (Eds.): SAGT 2020, LNCS 12283, pp. 47–63, 2020.
https://doi.org/10.1007/978-3-030-57980-7_4

In a combinatorial auction, a set M of m items (or goods) is to be allocated to n bidders. Each bidder i has a *valuation* function $v_i : 2^M \to \mathbb{R}_{\geq 0}$ that assigns a non-negative value $v_i(S)$ to any $S \subseteq M$ and quantifies i's preferences over item subsets. Valuation functions are assumed to be non-decreasing (free disposal), i.e., $v_i(S) \geq v_i(S')$, for all $S' \subseteq S$, and normalized, i.e., $v(\emptyset) = 0$. The goal is to compute a partitioning (a.k.a. *allocation*) $\mathcal{S} = (S_1, \ldots, S_n)$ of M that maximizes the *social welfare* $\text{sw}(\mathcal{S}) = \sum_{i=1}^{n} v_i(S_i)$.

Most of the previous work has focused on CAs with either submodular (and XOS) or complement-free valuations. A set function $v : 2^M \to \mathbb{R}_{\geq 0}$ is *submodular* if for all $S, T \subseteq M$, $v(S) + v(T) \geq v(S \cap T) + v(S \cup T)$, and *subadditive* (a.k.a. *complement-free*) if $v(S) + v(T) \geq v(S \cup T)$. A set function v is *XOS* (a.k.a. *fractionally subadditive*, see [17]) if there are additive functions $w_k : 2^M \to \mathbb{R}_{\geq 0}$ such that for every $S \subseteq M$, $v(S) = \max_k\{w_k(S)\}$. The class of submodular functions is a proper subset of the class of XOS functions, which in turn is a proper subset of subadditive functions.

Since bidder valuations have exponential size in n and m, algorithmic efficiency requires that the bidders communicate their preferences through either value or demand queries. A *value query* specifies a bidder i and a set (or bundle) $S \subseteq M$ and receives its value $v_i(S)$. A *demand query* specifies a bidder i, a set T of available items and a price p_j for each available item $j \in T$, and receives a bundle $S \subseteq T$ that maximizes i's *utility* $v_i(S) - \sum_{j \in S} p_j$ from the set of available items at these prices. Demand queries are strictly more powerful than value queries, in the sense that value queries can be simulated by polynomially many demand queries, and in terms of communication cost, demand queries are exponentially stronger than value queries [7].

The approximability of social welfare maximization by polynomial-time algorithms and truthful mechanisms for CAs with submodular and subadditive bidders has been extensively studied by the communities of Approximation Algorithms and Algorithmic Mechanism Design in the last two decades and are practically well understood (see e.g., Sect. 1.3 for a selective list of references most relevant to our work).

1.1 Perturbation Stability in Combinatorial Auctions

Motivated by recent work on beyond worst-case analysis of algorithms [31] and on endowed valuations for combinatorial markets [4,16], in this work, we investigate whether strong performance guarantees for social welfare maximization (by polynomial time algorithms and truthful mechanisms, or even at the equilibrium of simple auctions) can be achieved for a very restricted (though still natural) class of CAs with *perturbation stable* valuations, where the optimal solution is resilient to a small increase of any bidder's valuation for any single item.

From a bird's-eye view, we follow the approach of *beyond worst-case analysis* (see e.g., [31]), where we seek a theoretical understanding of the superior practical performance of certain algorithms by formally analyzing them on practically relevant instances. Hence, researchers restrict their attention to instances that satisfy certain application-area-specific assumptions, which are likely to be satisfied in practice. Such assumptions may be of stochastic (e.g., smoothed analysis of Simplex and local search [15,34,35]) or deterministic nature (e.g., perturbation stability in clustering [1,3,5,6]).

The beyond worst-case approach is not anything new for (Algorithmic) Mechanism Design. *Bayesian* analysis, where the bidder valuations are drawn as independent samples from an arbitrary distribution known to the mechanism, is standard in revenue maximization [33] and has led to many strong and elegant results for social welfare maximization by truthful posted price mechanisms (see e.g., [14,19]). However, in this work, we significantly deviate from Bayesian analysis, where the mechanism has a relatively accurate knowledge of the distribution of bidder valuations. Instead, we suggest a deterministic restriction on the class of instances (namely, perturbation stability) and investigate if there is a natural class of mechanisms (e.g., Parallel 2nd Price Auctions (P2A)) that are incentive-compatible and achieve optimality for CAs with stable submodular valuations.

Our focus on perturbation stable valuations was actually motivated by the recent work of Babaioff et al. [4] and Ezra et al. [16] on combinatorial markets where the valuations exhibit the endowment effect. The *endowment effect* was proposed by the Nobel Laureate Richard Thaler [37] to explain situations where owning a bundle of items causes its value to increase. Babaioff et al. [4] defined that if an allocation $\mathcal{S} = (S_1, \ldots, S_n)$ is α-endowed, for some $\alpha > 1$, in a CA with valuations $(v_1, \ldots, v_n)$, then the valuation function of each bidder i becomes

$$v'_i(T) = v_i(T) + (\alpha - 1)v_i(S_i \cap T)\,, \qquad (1)$$

for all item sets $T \subseteq M$. Roughly speaking, the value of S_i (and its subsets) is inflated by a factor of α due to the endowment effect. The main result of [4] is that for any combinatorial market with submodular valuations $(v_1, \ldots, v_n)$, any locally optimal allocation $\mathcal{S}$ and any $\alpha \geq 2$, the market with α-endowed valuations $(v'_1, \ldots, v'_n)$ for $\mathcal{S}$ admits a Walrasian equilibrium (see Sect. 2 for the definition) where each bidder i receives S_i. In simple words, social welfare maximization in combinatorial markets with endowed valuations $(v'_1, \ldots, v'_n)$ is polynomially solvable and the optimal allocation is supported by item prices. Subsequently, Ezra et al. [16] presented a general framework for endowed valuations and extended the above result to XOS valuations and general valuations, for a sufficiently large endowment (see also previous work on bundling equilibrium and conditional equilibrium [12,21]).

Inflated valuations due to the endowment effect naturally occur in auctions that take place regularly over time. Imagine auctions for e.g., season tickets of an athletic club, spots in a parking lot, reserving timeslots for airport gates, vacation packages at resorts, etc., where regular participants tend to value more the bundles allocated to them in the past, due to the endowment effect (see also [37] for more examples). Given the strong positive results of [4,16], a natural question is whether CAs with valuations inflated due to the endowment effect allow for stronger approximation guarantees in social welfare maximization and mechanism design.

Stable Combinatorial Auctions. To investigate the question above, we adopt a slightly stronger condition on valuation profiles, namely perturbation stability, which is inspired by (and bears a resemblance to) the definition of perturbation stable clustering instances (see e.g., [1,3,5,6]).

Definition 1. *For a constant $\gamma \geq 1$, a γ-perturbation of a valuations profile $\boldsymbol{v} = (v_1, \ldots, v_n)$ on a bidder i and an item j is a new valuations profile $\boldsymbol{v}' = (v'_1, \ldots, v'_n)$, where $v'_k = v_k$ for all bidders $k \neq i$, and for all $S \subseteq M$,*

$$v'_i(S) = v_i(S) + (\gamma - 1)v_i(S \cap \{j\}) \tag{2}$$

A CA with valuations $\boldsymbol{v} = (v_1, \ldots, v_n)$ is γ-perturbation stable *(or* γ-stable*) if the optimal allocation for $\boldsymbol{v}$ is unique and remains unique for all γ-perturbations $\boldsymbol{v}'$ of $\boldsymbol{v}$.*

Example 1. Let Alice and Bob compete for 2 items, a and b, and have valuations $v_A(\{a\}) = v_A(\{a,b\}) = 2$ and $v_A(\{b\}) = 1$, and $v_B(\{b\}) = v_B(\{a,b\}) = 2$ and $v_B(\{a\}) = 1$. The (unique) optimal allocation is to give a to Alice and b to Bob, with social welfare 4. A perturbation with most potential to change the optimal solution is to inflate Alice's value of b by $\gamma \geq 1$. Then, we get $v'_A(\{a\}) = 2$, $v'_A(\{b\}) = \gamma$ and $v'_A(\{a,b\}) = 1 + \gamma$. The optimal solution remains unique for any $\gamma < 3$. Hence the above CA is $(3 - \varepsilon)$-stable, for any $\varepsilon > 0$. □

At the conceptual level, we feel that the condition of γ-stability is easier to grasp and to think about in the context of mechanism design for CAs (compared against considering valuation profiles $\boldsymbol{v}$ resulting from the α-endowment of an optimal solution to an initial valuations profile $\boldsymbol{x}$)[1]. From an algorithmic and mechanism design viewpoint, we remark that for any $\gamma \geq 2$, CAs with γ-stable submodular valuations can be treated (to a certain extent) as multi-item auctions with additive bidders. In fact, this is the technical intuition behind several of our positive results.

[1] For a better understanding of the two conditions at a technical level, we note that a (technically very useful) necessary condition for a valuations profile $\boldsymbol{v}$ to be γ-stable is that for the optimal allocation $(O_1, \ldots, O_n)$, any bidders $i \neq k$ and any item $j \in O_i$,

$$v_i(O_i) - v_i(O_i \setminus \{j\}) > v_k(O_k \cup \{j\}) - v_k(O_k) + (\gamma - 1)v_k(\{j\}) \geq (\gamma - 1)v_k(\{j\}).$$

For this condition, we use (local) optimality of $(O_1, \ldots, O_n)$ for both $\boldsymbol{v}$ and its γ-perturbation on bidder k and item j (see also Lemma 1).

A similar (technically useful) condition satisfied by any valuations profile $\boldsymbol{v}$ that has resulted from the α-endowment of an optimal (or locally optimal) solution $(O_1, \ldots, O_n)$ to an initial valuations profile $\boldsymbol{x}$ is that for any bidders $i \neq k$ and any item $j \in O_i$,

$$v_i(O_i) - v_i(O_i \setminus \{j\}) \geq \alpha\big(v_k(O_k \cup \{j\}) - v_k(O_k)\big).$$

For this condition, we use local optimality of $(O_1, \ldots, O_n)$ for $\boldsymbol{x}$, multiply the resulting inequality by α, and observe that $v_i(O_i) - v_i(O_i \setminus \{j\}) = \alpha\big(x_i(O_i) - x_i(O_i \setminus \{j\})\big)$ and that $v_k(O_k \cup \{j\}) - v_k(O_k) = x_k(O_k \cup \{j\}) - x_k(O_k)$.

1.2 Contributions

We focus on deterministic algorithms and mechanisms. We first show that a simple greedy algorithm (Algorithm 1) that allocates each item j to the bidder i with maximum $v_i(\{j\})$ finds the optimal allocation for CAs with 2-stable subadditive valuations (Theorem 1). Moreover, similarly to [4], we show that for 2-stable submodular valuations, combining the optimal allocation with a second price approach, where each item j gets a price of $p_j = \max_{k \neq i} v_k(\{j\})$, results in a Walrasian equilibrium (Theorem 3).

On the negative side, we prove that our positive results above cannot be significantly strengthened. We first show that there is a simple $(2-\varepsilon)$-stable CA with submodular bidders where approximating the social welfare within any factor larger than $1 - \frac{1}{2k}$ requires at least $\binom{m}{k}$ value queries, for any integer $k \geq 1$ (Theorem 2). Thus, a polynomial-time approximation scheme does not exist for $(2-\varepsilon)$-submodular valuations. Moreover, we show that for any $\gamma \geq 1$, there is a γ-stable CA with a XOS bidder and a unit demand bidder that does not admit a Walrasian equilibrium (Lemma 2).

On the mechanism design part, in a nutshell, we show that (possibly appropriately modified) Parallel 2nd Price Auctions (P2A) behave very well for stable CAs. We should highlight that despite the fact that maximizing the social welfare for 2-stable subadditive CAs is easy, VCG is not an option for the design of computationally efficient incentive compatible mechanisms. The reason is that removing a single bidder from a 2-stable CA may result in an NP-hard (and hard to approximate) (sub)instance.

In Sect. 5, we show that a P2A followed by a demand query for each bidder is dominant strategy incentive compatible (DSIC) for all CAs with subadditive bidders and maximizes the social welfare if the valuations profile is submodular and 2-stable (Theorem 4). If demand queries are not available, the mechanism boils down to a simple P2A. We show that P2A is ex-post incentive compatible (EPIC) for 2-stable submodular valuations and that truthful bidding leads to the optimal allocation.

On the negative side and rather surprisingly, we show that demand queries are indeed necessary for computationally efficient mechanisms that are DSIC for all submodular valuations and maximize the social welfare if the instance is γ-stable (even if γ is arbitrarily large, Theorem 6). Our construction is an insightful adaptation of the elegant lower bound in [10, Theorem 1] to the case of stable submodular valuations. We show that any DSIC mechanism that computes the optimal allocation for stable CAs and does not use demand queries must use exponentially many value queries. The crux of the proof is that in certain instances, the bidders may find profitable to misreport and switch from a non-stable instance to a stable one.

In Sect. 6, we analyze the Price of Anarchy (PoA) of P2A and Parallel 1st Price Auctions (P1A). Our results demonstrate that the quality of equilibria of simple non-truthful auctions improves only for γ-stable valuations, with $\gamma \geq 3$. We show that the PoA of P2A for CAs with 3-stable submodular valuations is 1 (Theorem 7), while there are $(3-\varepsilon)$-stable CAs with PoA equal to 1/2

(Lemma 3), which matches the PoA for CAs with general submodular valuations (see e.g., [32]). Moreover, we show that the PoA of both P2A and P1A for CAs with γ-stable XOS valuations is at least $\frac{\gamma-2}{\gamma-1}$, for any $\gamma \geq 2$ (Theorem 8 and Theorem 9).

The technical details and the proofs omitted from this extended abstract, due to lack of space, can be found at the full version of our work [20].

1.3 Previous Work

Social welfare maximization with submodular and subadditive valuations has been studied extensively. Submodular Welfare Maximization (SMOD-WM) is known to be $(1-1/e)$-approximable with polynomially many value queries [38] and $(1-1/e+\varepsilon)$-approximable, for a fixed constant $\varepsilon > 0$, with polynomially many demand queries [18]. Moreover, a simple and natural greedy algorithm achieves an approximation ratio of $1/2$ using only value queries [25]. The results about polynomial-time approximability with value queries are best possible, in the sense that approximating SMOD-WM within a factor of $1-1/e+\varepsilon$, for any constant $\varepsilon > 0$, is NP-hard [23] and requires exponentially many value queries [27]. Furthermore, there is a constant $\varepsilon > 0$, such that approximating SMOD-WM within a factor of $1-\varepsilon$ with demand queries is NP-hard [18]. Subadditive Welfare Maximization (SADD-WM) is $m^{-1/2}$-approximable with polynomially many value queries (and this is best possible [27]) and $1/2$-approximable with polynomially many demand queries [17].

Truthful maximization of social welfare in CAs with submodular (or XOS) bidders has been a central problem in Algorithmic Mechanism Design. In the *worst-case* setting, where we do not make any further assumptions on bidder valuations, Dobzinski et al. [13] presented the first truthful mechanism that uses polynomially many demand queries and achieves a non-trivial approximation guarantee of $O((\log m)^{-2})$. Dobzinski [9] improved the approximation ratio to $O(\frac{1}{\log m \log\log m})$ for the more general class of subadditive valuations. Subsequently, Krysta and Vöcking [24] provided an elegant randomized online mechanism with an approximation ratio of $O(\frac{1}{\log m})$ for XOS valuations. Dobzinski [11] broke the logarithmic barrier for XOS valuations, by showing an approximation guarantee of $O((\log m)^{-1/2})$, which was recently improved to $O((\log\log m)^{-3})$ by Assadi and Singla [2]. Accessing valuations through demand queries is essential for these strong positive results. Dobzinski [10] proved that any truthful mechanism for CAs with submodular bidders with approximation ratio better than $m^{-\frac{1}{2}+\varepsilon}$ must use exponentially many value queries. Truthful $\Theta(m^{-1/2})$-approximate mechanisms that use polynomially many value queries are known even for the more general class of subadditive valuations (see e.g., [13]).

In the Bayesian setting, Feldman et al. [19] showed how to obtain item prices that provide a constant approximation ratio for XOS valuations. These results were significantly extended and strengthened by Düetting et al. [14].

Previous work has also shown strong Price of Anarchy (PoA) guarantees for CAs with submodular, XOS and subadditive bidders that can be achieved by simple (non-truthful) auctions, such as P2A and P1A (see e.g., [8,30,32,36]).

Our notion of perturbation stability for CAs is inspired by conceptually similar notions of perturbation stability in clustering [3,6]. Angelidakis et al. [1] presented a polynomial-time algorithm for 2-stable clustering instances with center-based objectives (e.g., k-median, k-means, k-center), while Balcan et al. [5] proved that there is no polynomial-time algorithm for $(2-\varepsilon)$-stable instances of k-center, unless NP = RP. To the best of our knowledge, this is the first time that the notion of perturbation stability has been applied to social welfare maximization and to algorithmic mechanism design for Combinatorial Auctions.

2 Notation and Preliminaries

The key notion of γ-perturbation stability (Definition 1) and a significant part of the terminology and the notation are introduced in Sect. 1. In this section, we introduce some additional terminology, notation and conventions used in the technical part.

We always let $\mathcal{O} = (O_1, \ldots, O_n)$ denote the optimal allocation for the instance at hand, and let O_i be the bundle of bidder i in $\mathcal{O}$. For convenience, we usually let an index j also denote the singleton set $\{j\}$ (we write $v_i(j)$, $v_i(S \cup j)$, $v_i(S \setminus j)$, etc., instead of $v_i(\{j\})$, $v_i(S \cup \{j\})$, $v_i(S \setminus \{j\})$). We use both $S_1 \setminus S_2$ and $S_1 - S_2$ for the set difference. We denote the marginal contribution of a bundle S wrt. T as $v(S|T) = v(S \cup T) - v(T)$.

In addition to submodular, XOS and subadditive valuations, we consider *additive* and *unit-demand* valuations $v : 2^M \to \mathbb{R}_{\geq 0}$, where there exist $b_1, \ldots, b_m \in \mathbb{R}_{\geq 0}$, such that for any $S \subseteq M$, $v(S) = \sum_{j \in S} b_j$ and $v(S) = \max_{j \in S} b_j$, respectively. A useful property of an XOS valuation v is that for any $S \subseteq M$, there is an additive valuation q that *supports* S, in the sense that $v(S) = q(S)$ and for any $T \subseteq M$, $v(T) \geq q(T)$.

We focus on deterministic algorithms and mechanisms and consider bidders with *quasi-linear utilities*, where the utility of bidder i with valuation v_i for a bundle S at price $p(S)$ is $u_i(S) = v_i(S) - p(S)$. For a price vector $(p_1, \ldots, p_m)$, we often let $p(S) = \sum_{j \in S} p_j$ denote the price of a bundle $S \subseteq M$.

An allocation $\mathcal{S} = (S_1, \ldots, S_n)$ and a price vector $(p_1, \ldots, p_m)$ form a *Walrasian Equilibrium* if all items j with $p_j > 0$ are allocated and each bidder i gets a utility maximizing bundle (or, his *demand*) in $\mathcal{S}$, i.e., $\forall S \subseteq M$, $v_i(S_i) - p(S_i) \geq v_i(S) - p(S)$.

A mechanism is *dominant-strategy incentive compatible* (DSIC) (or *truthful*) if for any valuations profile $\boldsymbol{v}$, answering (value or demand) queries truthfully is a dominant strategy and guarantees non-negative utility for all bidders. A mechanism is called *ex-post incentive compatible* (EPIC) if truthful bidding is an ex-post Nash equilibrium and guarantees non-negative utility for all bidders.

Let $\boldsymbol{D} = (D_1, ..., D_n)$ be a profile of distributions over valuation functions (i.e., over possible bids). In a mechanism with allocation rule $\boldsymbol{S}(\cdot) = (S_1(\cdot), \ldots, S_n(\cdot))$ and item pricing rule $\boldsymbol{p} = (p_1(\cdot), \ldots, p_m(\cdot))$, $\boldsymbol{D}$ forms a *Mixed Nash Equilibrium* (MNE), if no bidder has an incentive to unilaterally deviate from $\boldsymbol{D}$, i.e., for any bidder i and any distribution D'_i over valuation functions,

Algorithm 1: Algorithm for 2-Stable Subadditive Valuations

Input: Value query access to subadditive valuations $v_1(\cdot), ..., v_n(\cdot)$
Set $O_1 = O_2 = ... = O_n = \emptyset$
for $j \in M$ **do**
  Let i be the bidder that maximizes $v_i(j)$, and set $O_i \leftarrow O_i \cup \{j\}$.
return *Allocation* $(O_1, ...O_n)$

$$\mathop{\mathbb{E}}_{b \sim D}\left[v_i(S_i(b)) - \sum_{j \in S_i(b)} p_j(b)\right] \geq \mathop{\mathbb{E}}_{b \sim (D'_i, D_{-i})}\left[v_i(S_i(b)) - \sum_{j \in S_i(b)} p_j(b)\right]$$

If instead of distributions over valuation functions, we restrict each D_i and D'_i to valuation functions (i.e., to pure strategies over possible bids), we get the definition of a *Pure Nash Equilibrium* (PNE).

In Sect. 6, we consider the *Price of Anarchy* (PoA) of Parallel 2nd Price Auctions (P2A) and Parallel 1st Price Auctions (P1A). In a Combinatorial Auction, the PoA of a mechanism is the ratio of (resp. expected) social welfare at the worst Pure (resp. Mixed) Nash Equilibrium to the social welfare of the optimal allocation. Formally, focusing on the more general case of Mixed Nash Equilibria:

$$\text{PoA} = \min_{D \text{ is a MNE}} \frac{\mathbb{E}_{b \sim D}\left[\sum_i v_i(S_i(b))\right]}{\sum_i v_i(O_i)}$$

Properties of Stable Valuations. The following shows a technically useful property of γ-stable CAs (see also Footnote 1).

Lemma 1 (Valuation Stability). *Let v be γ-stable and subadditive valuations. Then, for all bidders $i \neq k$ and all items $j \in O_i$, $v_i(j) \geq v_i(O_i) - v_i(O_i \setminus j) > (\gamma - 1)v_k(j)$.*

3 Social Welfare Maximization for Stable Valuations

We next consider the problem of social welfare maximization for 2-stable CAs. We first show that for 2-stable subadditive valuations, we can compute the optimal solution with value queries in polynomial time.

Theorem 1. *Let v be a 2-stable subadditive valuations profile. Then Algorithm 1 outputs the optimal allocation $(O_1, ..., O_n)$ using nm value queries.*

Proof. The number of value queries follows directly from the description of the algorithm. As for optimality, we fix an item j and let i be the bidder that gets j in the optimal solution. Because of Lemma 1 and the fact that v is 2-stable, we know that $v_i(j) > v_k(j)$, for any other bidder $k \neq i$. Because Algorithm 1 allocates j to the bidder with the highest singleton value, i gets item j in the allocation of Algorithm 1. □

On the negative side, we next show that a polynomial-time approximation scheme does not exist even for $(2-\varepsilon)$-stable submodular valuations.

Theorem 2. *For any $\varepsilon > 0$, there exists a submodular $(2-\varepsilon)$-stable valuations profile $\boldsymbol{v}$ such that for any integer $k \geq 1$, approximating the optimal allocation in $\boldsymbol{v}$ within any factor larger than $1 - \frac{1}{2k}$ requires at least $\binom{m}{k}$ value queries.*

Proof (Sketch). Inspired by [10, Lemma 3.10], we consider the following class of submodular valuations:

$$v^O(S) = \begin{cases} |S|, & \textit{if } |S| \leq |O| - 1 \\ |O| - 1/2, & \textit{if } |S| = |O| \textit{ and } S \neq O \\ |O|, & \textit{otherwise } (S = O \textit{ or } S \geq |O| + 1) \end{cases}$$

We consider 2 and any allocation (O_1, O_2), with $|O_1| = |O_2|$. We can show that the valuations v^{O_1} and v^{O_2} are submodular and $(2-\varepsilon)$-stable. Moreover, finding the optimal allocation requires $\binom{m}{|O_1|}$ queries. By generalizing this argument to n bidders, we get the inapproximability ratio. □

4 Existence of Walrasian Equilibrium

Similarly to [4, Theorem 4.2], we next show that combinatorial markets with 2-stable submodular valuations admit a Walrasian Equilibrium.

Theorem 3. *Let $\boldsymbol{v}$ be 2-stable submodular valuations. For every bidder i every item $j \in O_i$, let $\max_{k \neq i} v_k(j) \leq p_j \leq v_i(j|O_i - j)$. Then, the prices $p_1, \ldots, p_m$ form a Walrasian Equilibrium.*

Proof. Fix a bidder i and his optimal bundle O_i. We note that the price p_j of each item is well defined. Because of Lemma 1, 2-stability and $j \in O_i$, $v_i(j|O_i - j) > v_k(j)$.

We next show that for any $j \notin O_i$, bidder i is not interested in getting j. Because of subadditivity, i's additional utility due to item j is at most $v_i(j) - p_j$. Because $j \notin O_i$, $p_j \geq v_i(j)$, making i's utility from j non-positive. Hence, i's demand is a subset of O_i.

To conclude the proof, we show that for any item $j \in O_i$, i gets non-negative utility due to j. Fix a bundle S in the demand of bidder i with $j \notin S$. Note that we have already proven that $S \subseteq O_i$. The utility gained by taking j is $v_i(j|S) - p_j$, which is non-negative; submodularity makes $v_i(j|S) \geq v_i(j|O_i - j) \geq p_j$. Hence, O_i is the demand of bidder i. □

We next show that Theorem 3 cannot be extended to stable XOS valuations. In the proof of Theorem 3, 2-stability and submodularity ensure that the prices cannot exceed the marginal increase $v_i(j|S)$ of adding an item j to some $S \subset O_i$. For XOS valuations, however, the marginals $v_i(j|S)$ may not be decreasing with S. As a result, the utility of a bidder i may be maximized by a strict subset of his optimal bundle O_i.

Lemma 2. *For every $\gamma \geq 1$, there exists a γ-stable valuations profile with a XOS bidder and a unit demand bidder which does not admit a Walrasian Equilibrium.*

Mechanism 2: Extended Parallel 2nd Price Auction (EP2A)

Input: Value and demand query access to valuations $\boldsymbol{v} = (v_1, \ldots, v_n)$.
For all bidders i and items j, query $v_i(j)$ and let b_{ij} denote the response.
Set the price p_j of each item j to its second highest bid.
For each bidder i, let S_i be the set of items for which i has the highest bid.
Bidder i receives his demand from S_i, where each item has price p_j.
Bidder i pays the total price for his demand from S_i.

5 Mechanism Design for Stable Combinatorial Auctions

In this section, we investigate truthful mechanism design for CAs with stable submodular valuations. We should emphasize that despite Theorem 1, VCG cannot be used as a computationally efficient DSIC mechanism for stable subadditive CAs, because the subinstances $\boldsymbol{v}_{-i}$, whose optimal solutions determine the payments, may not be stable and may be NP-hard to solve optimally (e.g., adding a bidder with additive valuation that has a huge value for each singleton to any CA results in a stable instance).

We first present a truthful extension of Algorithm 1, which is implemented as a Parallel 2nd Price Auction (P2A) and also uses a demand query for each bidder.

Theorem 4. *Mechanism 2 uses nm value queries and n demand queries, and is DSIC for any CA with subadditive valuations. Moreover, if the valuations profile $\boldsymbol{v}$ is 2-stable submodular, Mechanism 2 returns the optimal allocation.*

Proof. First, we show that Mechanism 2 is DSIC for subadditive bidders. We focus on the bidding step, because assuming that each set S_i is determined in a truthful way, it is always in each bidder's best interest to respond to his demand query truthfully.

We observe that no bidder has incentive to bid lower than his singleton value for an item. Bidding lower could only lead to the bidder losing some items, thus restricting the set of items available tom him through his demand query. Moreover, no bidder has incentive to bid higher than his singleton value for an item. This would only entail having access to an item that has price at least his actual singleton value. However, because of subadditivity, the bidder does not include such an item in his demand set.

The fact that Mechanism 2 computes an optimal allocation for 2-stable submodular valuations is an immediate consequence of Theorem 1 and Theorem 3. □

Next, we show that a P2A (Mechanism 3), that uses only value queries, is ex-post incentive compatible when restricted to 2-stable submodular valuations profiles. The proof of the following is an immediate consequence of Theorem 3 and Theorem 1.

Mechanism 3: Parallel 2nd Price Auction (P2A)

Input: Value query access to valuations $\boldsymbol{v} = (v_1, \ldots, v_n)$.

For all bidders i and items j, query $v_i(j)$ and let b_{ij} denote the response.

Set the price p_j of each item j to its second highest bid.

For each bidder i, let S_i be the set of items for which i has the highest bid.

Bidder i receives S_i and pays $\sum_{j \in S_i} p_j$.

Theorem 5. *Mechanism 3 uses nm value queries and is EPIC for any CA with 2-stable submodular valuations. Moreover, under truthful bidding, Mechanism 3 computes the optimal allocation.*

Interestingly, Mechanism 3 is not DSIC even when restricted to submodular CAs. The reason is that the bidder valuations profile may be 2-stable, but their bids might be not. Hence, it may happen that bidder k bids higher than his real singleton value on some item j, but j is allocated to different bidder i. This may increase p_j to a level that is no longer profitable for bidder i to get item j (which is exactly the reason that we employ the demand queries in Mechanism 2).

The remark above naturally motivates the question about existence of a computationally efficient DISC mechanism that computes the optimal allocation for 2-stable submodular CAs using only value queries. Rather surprisingly, the following answers this question in the negative.

Theorem 6. *Let $\mathcal{A}$ be any mechanism that is DSIC, uses only value queries and finds the optimal solution for γ-stable submodular valuations, for some $\gamma \geq 1$. Then $\mathcal{A}$ makes exponentially many value queries.*

Proof. The proof is an interesting adaptation of the proof of [10, Theorem 3.1]. For the proof, we use instances with just 2 bidders. Fixing one to be additive, the other may bid "stably" and get any bundle. However, due to the structure of his (hidden) valuation, finding his demand may be intractable, which makes misreporting a profitable strategy.

To reach a contradiction, we assume that $\mathcal{A}$ is DSIC, makes polynomially many value queries and always finds the optimal solution for γ-stable submodular valuations, for some fixed $\gamma \geq 1$. First we establish the following, which helps determining whether a set of additive valuations is stable.

Proposition 1. *Let $\boldsymbol{v}$ be a profile with additive valuations. Then, for any $\gamma \geq 1$, $\boldsymbol{v}$ is γ-stable if for any item $j \in M$, the largest value for j in $\boldsymbol{v}$ differs from the second largest value for j in $\boldsymbol{v}$ by a factor larger than γ. Namely, if $i = \arg\max_{k \in [n]}\{v_k(j)\}$, then $v_i(j) > \gamma v_k(j)$, for all bidders $k \neq i$.*

Proof (of Proposition 1). The proposition follows directly from the fact that endowing an additive bidder k for an item j keeps him additive and inflates his singleton value by a factor of γ. □

For the rest of the proof, we consider 2 bidders and fix the valuation according to which bidder 1 makes his bids: $v_1(S) = |S|/m$. We fix his (hidden) valuation to be also additive, with the value of each item large enough. This valuation, together with any other bounded and submodular valuation, results in a valuations profile that is submodular and stable (for a large enough stability factor).

We next prove that bidder 1 can get any bundle.

Proposition 2 *For any bundle O, bidder 2 will be allocated O, if he bids according to*

$$v_2(S) = |S \cap O| + \frac{|S - O|}{m^2} \tag{3}$$

Proof (of Proposition 2*).* First we fix any bundle $O \subseteq M$. By Proposition 1, valuations (v_1, v_2) are $(m - \varepsilon)$-stable. Taking m large enough makes the valuations γ-stable. Given that they are also additive, and thus submodular, we get that the mechanism $\mathcal{A}$ (which, by hypothesis, computes the optimal allocation for γ-stable instances) allocates O to bidder 2. □

Next, we show that the prices set by $\mathcal{A}$ for bidder 2 are bounded and increasing.

Proposition 3. *For any bundle T and any $S \subseteq T$, it holds that*

$$\frac{|T| - |S|}{m^2} \leq p_T - p_S \leq |T| - |S| \tag{4}$$

where p_S and p_T are the prices of bundles S and T assigned from $\mathcal{A}$ for bidder 2.

Proof (of Proposition 3*).* We examine what happens when bidder 2 bids according to (3). First we set $O = T$, which means that bidder 2 will receive T. Since $\mathcal{A}$ is DSIC, bidder 2 should not prefer S over T, i.e., $v_2(T) - p_T \geq v_2(S) - p_S$. This implies the rhs of (4), because $v_2(T) = |T|$ and $v_2(S) = |S|$ (since $S \subseteq T$).

The argument for the lhs of (4) is symmetric. We let bidder 2 bid according to (3), where $O = S$. Then, bidder 2 should not prefer T over S, i.e., $v_2(S) - p_S \geq v_2(T) - p_T$. Since $v(S) = |S|$ and $v(T) = m \cdot |S| + |T - S|/m^2$, we get the lhs of (4). □

We also note that setting $S = \emptyset$ in (4), we get that $|T|/m^2 \leq p_T \leq |T|$.

We now create an exponentially large structured submenu, as in [10, Definition 3.2]. This concludes the proof, since the existence of such a submenu entails that $\mathcal{A}$ requires exponentially many value queries to find the demand of bidder 2, as shown in [10, Lemma 3.10]. For completeness, we recall that a collection of bundles $\mathcal{S}$ comprises a *structured submenu* for bidder 2 if:

1. For all $S \in \mathcal{S}$, bidder 2 can be allocated S.
2. For each $S, T \in \mathcal{S}$: $|S| = |T|$ and $|p_S - p_T| \leq \frac{1}{m^5}$.
3. For all $S, T \subseteq M$ such that $S \in \mathcal{S}$ and $S \subset T$: $p_T - p_S \geq \frac{1}{m^3}$.
4. For all $S \in \mathcal{S}$: $p_S \leq m$.

Since bidder 2 can get any bundle, the first property is satisfied. Also by Proposition 3, the third property is satisfied, because for any $S \subset T$: $p_T - p_S > 1/m^3$.

To create the structured submenu, we fix $k = m/2$ and consider all the $\binom{m}{k}$ different bundles of size k. Our submenu is a subset of those bundles, which immediately satisfies the first part of the second property. Also since $|T|/m^2 \leq p_T \leq |T|$, the price of each bundle is at most k, which implies the last property.

We need to show that last part of the second property. To this end, following the construction of [10, Section 3.1], we split the interval $[0, m]$ into m^5 bins. For each bundle S of size k, we put S in the i-th bin if $p_S \in \left[i/m^5, (i+1)/m^5\right)$. Since there are m^5 bins and $\binom{m}{k}$ bundles, one bin must have exponentially many bundles. Let $\mathcal{S}$ be the set of bundles in such a bin. Notice that the bundles of the same bin have prices which differ less $1/m^5$, thus satisfying the last part of the second property.

Proof. This completes the proof that there is an exponentially large collection $\mathcal{S}$ of bundles that comprises a structured submenu. The last step is to apply [10, Lemma 3.10]. □

A natural question is whether one could also follow the first part of the proof of [10, Theorem 3.1], in order to get a much stronger inapproximability bound of $m^{-1/2+\varepsilon}$, for any $\varepsilon > 0$. Unfortunately the answer is negative, because the polar additive valuation profiles in [10, Section 3.1] are far from stable. This explains the necessity of our careful construction of stable valuation profiles, in the first part of the proof of Theorem 6.

6 Price of Anarchy in Stable Combinatorial Auctions

For XOS valuations, the PoA of P2A is at least 1/2 [8] . We next show that even for $(3 - \varepsilon)$-stable valuations, the PoA of P2A does not improve.

Lemma 3. *There exists a* $(3 - \varepsilon)$*-stable profile with unit-demand valuations for which the PoA of P2A is* 1/2.

Proof. The instance in Example 1 is $(3 - \varepsilon)$-stable and has been used to show that the PoA of P2A is at most 1/2. More precisely, we observe that there is an equilibrium with social welfare 2: Alice bids 0 for a and 1 for b and Bob bids 1 for a and 0 for b. □

Interestingly, the previous result is tight. With 3-stable submodular valuations, every equilibrium is optimal. To prove this, we introduce a no-overbidding assumption, which is weaker than the usual Strong No Overbidding assumption (where each bidder's value for any bundle S is at most the sum of his bids for S). We call our assumption Singleton No Overbidding (SiNO), as it restricts the bids to be below each corresponding singleton value. We also note that the bidding profile in Lemma 3 SiNO.

Definition 2 (Singleton No Overbidding). *A bidding profile* $(\boldsymbol{b}_1, ..., \boldsymbol{b}_n)$ *satisfies Singleton No Overbidding (SiNO) if for any bidder* i *and item* j: $v_i(j) \geq b_{ij}$.

The Price of Anarchy in Parallel 2nd Price Auctions. Now we are ready to prove that with SiNO, PoA is always 1 for CAs with 3-stable submodular valuations.

Theorem 7. *Let* $\boldsymbol{v}$ *be a 3-stable submodular valuations profile, and let* $\boldsymbol{b}$ *be a bidding profile that forms a Pure Nash Equilibrium for P2A and satisfies SiNO. Then the allocation at the equilibrium coincides with the optimal allocation.*

Intuitively, because the gap between the highest and the second highest singleton values has gotten large enough, the bidder who is optimally allocated an item has much more incentive to outbid the other bidders for that item. The intuitive that the valuations must be at least 3 stable is that bidders should value their optimal items at least twice than the other bidders value them. Otherwise, if the other bidders bid their maximum value, the bidder might prefer to get other items at low prices, which is what happens in Lemma 3.

We proceed to study the PoA of P2A for the more general class of XOS valuations. For general XOS valuations, the PoA is at least 1/2, which cannot be improved for $(3 - \varepsilon)$-stable XOS valuations, due to Lemma 3. We can show that as valuation stability increases, the PoA improves.

Theorem 8. *For any* $\gamma \geq 2$, *let* $\boldsymbol{v}$ *be a* γ*-stable profile with XOS valuations. Let* $\boldsymbol{b}$ *be a bidding profile that forms a Pure Nash Equilibrium for P2A and satisfies SiNO. Then the PoA is larger than* $\frac{\gamma-2}{\gamma-1}$.

The intuition is similar to that in the proof of Theorem 7. Even if the prices are as high as possible, as γ gets larger, each bidder has more incentive to prefer the items in his optimal bundle than any other items.

The Price of Anarchy in Parallel 1st Price Auctions. We conclude with a lower bound on the PoA of Parallel 1st Price Auctions (P1A) for CAs with stable valuations. If bidders are restricted to a mixed Nash equilibrium, the PoA of P1A for bidders with XOS valuations is at least $1 - \frac{1}{e}$. Similarly to Theorem 8, we show that the PoA of P1A increases, as the stability of a XOS valuations profile increases.

Theorem 9. *For any* $\gamma \geq 2$, *let* $\boldsymbol{v}$ *be a* γ*-stable profile with XOS valuations. Let* $\boldsymbol{b}$ *be a bidding profile that forms a Mixed Nash Equilibrium for P1A. Then the PoA is larger than* $\frac{\gamma-2}{\gamma-1}$.

For the proof, we observe that as the stability factor γ increases, the valuation of a bidder i for each item j in his optimal bundle becomes considerably larger than the second highest singleton valuation for item j. Hence, if bidder i bids the second highest singleton valuation for each item in his optimal bundle, i's utility should be large enough to establish that the allocation of the equilibrium achieves a large enough social welfare.

Acknowledgements. We wish to thank Kyriakos Lotidis and Grigoris Velegkas for many helpful discussions on combinatorial markets with endowed valuations and on the possibility of exploiting endowed valuations in mechanism design.

References

1. Angelidakis, H., Makarychev, K., Makarychev, Y.: Algorithms for stable and perturbation-resilient problems. In: Proceedings of the 49th ACM Symposium on Theory of Computing (STOC 2017), pp. 438–451 (2017)
2. Assadi, S., Singla, S.: Improved Truthful Mechanisms for Combinatorial Auctions with Submodular Bidders. In: Proceedings of the 60th IEEE Symposium on Foundations of Computer Science (FOCS 2019), pp. 233–248 (2019)
3. Awasthi, P., Blum, A., Sheffet, O.: Center-based clustering under perturbation stability. Inf. Process. Lett. **112**(1–2), 49–54 (2012)
4. Babaioff, M., Dobzinski, S., Oren, S.: Combinatorial auctions with endowment effect. In: Proceedings of the 2018 ACM Conference on Economics and Computation (EC 2018), pp. 73–90 (2018)
5. Balcan, M., Haghtalab, N., White, C.: k-center clustering under perturbation resilience. In: Proc. of the 43rd International Colloquium on Automata, Languages and Programming (ICALP 2016). LIPIcs, vol. 55, pp. 68:1–68:14 (2016)
6. Bilu, Y., Linial, N.: Are stable instances easy? In: Proc. of the 1st Symposium on Innovations in Computer Science (ICS 2010), pp. 332–341. Tsinghua University Press (2010)
7. Blumrosen, L., Nisan, N.: On the computational power of demand queries. SIAM J. Comput. **39**(4), 1372–1391 (2009)
8. Christodoulou, G., Kovács, A., Schapira, M.: Bayesian combinatorial auctions. In: Aceto, L., Damgård, I., Goldberg, L.A., Halldórsson, M.M., Ingólfsdóttir, A., Walukiewicz, I. (eds.) ICALP 2008. LNCS, vol. 5125, pp. 820–832. Springer, Heidelberg (2008). https://doi.org/10.1007/978-3-540-70575-8_67
9. Dobzinski, S.: Two Randomized Mechanisms for Combinatorial Auctions. In: Charikar, M., Jansen, K., Reingold, O., Rolim, J.D.P. (eds.) APPROX/RANDOM -2007. LNCS, vol. 4627, pp. 89–103. Springer, Heidelberg (2007). https://doi.org/10.1007/978-3-540-74208-1_7
10. Dobzinski, S.: An impossibility result for truthful combinatorial auctions with submodular valuations. In: Proceedings of the 43rd ACM Symposium on Theory of Computing (STOC 2011), pp. 139–148 (2011)
11. Dobzinski, S.: Breaking the logarithmic barrier for truthful combinatorial auctions with submodular bidders. In: Proceedings of the 48th ACM Symposium on Theory of Computing (STOC 2016), pp. 940–948 (2016)
12. Dobzinski, S., Feldman, M., Talgam-Cohen, I., Weinstein, O.: Welfare and revenue guarantees for competitive bundling equilibrium. In: Markakis, E., Schäfer, G. (eds.) WINE 2015. LNCS, vol. 9470, pp. 300–313. Springer, Heidelberg (2015). https://doi.org/10.1007/978-3-662-48995-6_22
13. Dobzinski, S., Nisan, N., Schapira, M.: Truthful randomized mechanisms for combinatorial auctions. J. Comput. Syst. Sci. **78**(1), 15–25 (2012)
14. Düetting, P., Feldman, M., Kesselheim, T., Lucier, B.: Prophet inequalities made easy: stochastic optimization by pricing non-stochastic inputs. In: Proceedings of the 58th Symposium on Foundations of Computer Science (FOCS 2017), pp. 540–551 (2017)

15. Englert, M., Röglin, H., Vöcking, B.: Smoothed analysis of the 2-Opt algorithm for the general TSP. ACM Trans. Algorithms **13**(1), 10:1–10:15 (2016)
16. Ezra, T., Feldman, M., Friedler, O.: A general framework for endowment effects in combinatorial markets. In: Proceedings of the 2020 ACM Conference on Economics and Computation (EC 2020) (2020)
17. Feige, U.: On maximizing welfare when utility functions are subadditive. SIAM J. Comput. **39**(1), 122–142 (2009)
18. Feige, U., Vondrák, J.: The submodular welfare problem with demand queries. Theor. Comput. **6**(1), 247–290 (2010)
19. Feldman, M., Gravin, N., Lucier, B.: Combinatorial auctions via posted prices. In: Proceedings of the 26th ACM-SIAM Symposium on Discrete Algorithms, pp. 123–135 (2014)
20. Fikioris, G., Fotakis, D.: Mechanism design for perturbation stable combinatorial auctions. CoRR abs/2006.09889 (2020). https://arxiv.org/abs/2006.09889
21. Fu, H., Kleinberg, R., Lavi, R.: Conditional equilibrium outcomes via ascending price processes with applications to combinatorial auctions with item bidding. In: Proceedings of the 13th ACM Conference on Electronic Commerce (EC 2012), p. 586 (2012)
22. Hershberger, J., Suri, S.: Vickrey prices and shortest paths: what is an edge worth? In: Proceedings of the 42nd Symposium on Foundations of Computer Science (FOCS 2001), pp. 252–259 (2001)
23. Khot, S., Lipton, R.J., Markakis, E., Mehta, A.: Inapproximability results for combinatorial auctions with submodular utility functions. In: Deng, X., Ye, Y. (eds.) WINE 2005. LNCS, vol. 3828, pp. 92–101. Springer, Heidelberg (2005). https://doi.org/10.1007/11600930_10
24. Krysta, P., Vöcking, B.: Online mechanism design (randomized rounding on the fly). In: Czumaj, A., Mehlhorn, K., Pitts, A., Wattenhofer, R. (eds.) ICALP 2012. LNCS, vol. 7392, pp. 636–647. Springer, Heidelberg (2012). https://doi.org/10.1007/978-3-642-31585-5_56
25. Lehmann, B., Lehmann, D., Nisan, N.: Combinatorial auctions with decreasing marginal utilities. Games Econ. Behav. **55**(2), 270–296 (2006)
26. Milgrom, P.: Putting Auction Theory to Work. Churchill Lectures in Economics. Cambridge University Press, Cambridge (2004)
27. Mirrokni, V., Schapira, M., Vondrák, J.: Tight information-theoretic lower bounds for welfare maximization in combinatorial auctions. In: Proceedings 9th ACM Conference on Electronic Commerce (EC 2008), pp. 70–77 (2008)
28. Cramton, P., Shoham, Y., Steinberg, R.: Combinatorial Auctions. MIT Press, Cambridge (2006)
29. Rassenti, S., Smith, V., Bulfin, R.: A combinatorial auction mechanism for airport time slot allocation. Bell J. Econ. **13**(2), 402–417 (1982)
30. Roughgarden, T.: Barriers to near-optimal equilibria. In: Proceedings of the 55th IEEE Symposium on Foundations of Computer Science (FOCS 2014), pp. 71–80 (2014)
31. Roughgarden, T.: Beyond worst-case analysis. Commun. ACM **62**(3), 88–96 (2019)
32. Roughgarden, T., Syrgkanis, V., Tardos, É.: The price of anarchy in auctions. J. Artif. Intell. Res. **59**, 59–101 (2017)
33. Roughgarden, T., Talgam-Cohen, I.: Approximately optimal mechanism design. CoRR abs/1812.11896 (2018). http://arxiv.org/abs/1812.11896
34. Spielman, D., Teng, S.: Smoothed analysis of algorithms: why the simplex algorithm usually takes polynomial time. J. ACM **51**(3), 385–463 (2004)

35. Spielman, D., Teng, S.: Smoothed analysis: an attempt to explain the behavior of algorithms in practice. Commun. ACM **52**(10), 76–84 (2009)
36. Syrgkanis, V., Tardos, É.: Composable and efficient mechanisms. In: Proceedings of the 45th Symposium on Theory of Computing (STOC 2013), pp. 211–220 (2013)
37. Thaler, R.: Toward a positive theory of consumer choice. J. Econ. Behav. Organ. **1**(1), 39–60 (1980)
38. Vondrák, J.: Optimal approximation for the submodular welfare problem in the value oracle model. In: Proceedings of the 40th ACM Symposium on Theory of Computing (STOC 2008), pp. 67–74 (2008)

Congestion Games and Flows over Time

Congestion Games with Priority-Based Scheduling

Vittorio Bilò[1(✉)] and Cosimo Vinci[2]

[1] Department of Mathematics and Physics, University of Salento, Lecce, Italy
vittorio.bilo@unisalento.it
[2] Department of Computer Science, Gran Sasso Science Institute, L'Aquila, Italy
cosimo.vinci@gssi.it

Abstract. We reconsider atomic and non-atomic affine congestion games under the assumption that players are partitioned into p priority classes and resources schedule their users according to a priority-based policy, breaking ties uniformly at random. We derive tight bounds on both the price of anarchy and the price of stability as a function of p, revealing an interesting separation between the general case of $p \geq 2$ and the priority-free scenario of $p = 1$. In fact, while non-atomic games are more efficient than atomic ones in absence of priorities, they share the same price of anarchy when $p \geq 2$. Moreover, while the price of stability is lower than the price of anarchy in atomic games with no priorities, the two metrics become equal when $p \geq 2$. Our results hold even under singleton strategies. Besides being of independent interest, priority-based scheduling shares tight connections with online load balancing and finds a natural application within the theory of coordination mechanisms and cost-sharing policies for congestion games. Under this perspective, a number of possible research directions also arises.

1 Introduction

In *priority-based scheduling*, requests issued by some users are favoured over others, differently than what happens in fair policies, such as round-robin or first-in first-out, where all users are treated equally. Priority-based scheduling is effectively used in a variety of domains, ranging from manufacturing processes to socio-economic activities. Simple examples are the largest-job-first algorithm and the boarding strategies of airline companies.

To the best of our knowledge, despite this widespread diffusion, the impact of priority-based scheduling has been considered only marginally in state-of-the-art game-theoretical models. In this work, we try to fill this gap by investigating *congestion games with priority-based scheduling*.

This work was partially supported by the Italian MIUR PRIN 2017 Project ALGADIMAR "Algorithms, Games, and Digital Markets".

T. Harks and M. Klimm (Eds.): SAGT 2020, LNCS 12283, pp. 67–82, 2020.
https://doi.org/10.1007/978-3-030-57980-7_5

In a congestion game [40], there is a finite set of players competing for the usage of a finite set of resources and all players require the same effort on every resource. We assume that players are partitioned into p priority classes and that, on every resource, all players of priority class c are scheduled before any player of class $c' > c$ (the lower the class, the higher the priority), while players of the same class are scheduled in a random order. Hence, the cost that player i experiences on resource r is a function of two parameters: *(i)* the position that i occupies in the schedule of r, and *(ii)* the *latency function* of r, i.e., how fast r processes its requests. Parameter *(i)*, which depends on i's priority class, is a random variable. Thus, the cost of i becomes the sum of the expected costs she experiences on every selected resource. There are two fundamental models of congestion games, namely *atomic* and *non-atomic games*, which differ on the way in which players and requests are interpreted. In atomic games, the effort that every player requires on a resource is non-negligible and normalized to one, whereas, in the non-atomic variant, each player asks for an infinitesimally small effort. Roughly speaking, a non-atomic game can be thought as an atomic game in which players are allowed to arbitrarily split their requests along different sets of resources.

The majority of the literature devoted to congestion games (e.g. [3,18,36,40, 43] and subsequent work) assumes that all users experience the same cost on a same resource (the *makespan model*). This can be interpreted as the outcome of the round-robin scheduling under the assumption that requests are processed according to a time-sharing policy organized in such a way that all requests are completed (almost) simultaneously. Note that this assumption requires preemption of the requests. Other approaches, for which preemption is not necessary, consider Smith's Rule [22], the first-in first-out policy [28] and the random policy [11,22,36,39]. The first and the last model, in particular, can be seen as the specialization of priority-based scheduling obtained when $p = n$ and $p = 1$, respectively. With this respect, our model extends and generalizes previously considered non-preemptive scheduling policies by simultaneously incorporating the presence of priorities and a source of uncertainty due to randomness.

We are interested in characterizing the performance of priority-based congestion games by studying the *price of anarchy* (PoA) [36] and *price of stability* (PoS) [2] of *pure Nash equilibria* [38] with respect to the utilitarian social welfare. Pure Nash equilibria constitute an ideal solution concept in congestion games as they are guaranteed to exist in almost all of their variants. These metrics have been precisely characterized in several subclasses of congestion games, the majority of them focusing on the makespan model. When the resources have affine latency functions, the PoA is $5/2$ [18] and the PoS is $1 + 1/\sqrt{3} \approx 1.577$ [13,17] for atomic games; for non-atomic ones both metrics are equal to $4/3$ [43]. Scheduling policies departing from the makespan model generate a richer cost model yielding a wide range of possibilities with significantly different outcomes. For instance, the PoA under non-preemptive policies based on a total ordering of the players gets equal to 4 and $17/3$, respectively, in non-atomic and atomic games [28]. However, much better bounds are possible when either preemption

p	PoA	p	PoA	p	PoA	p	PoA
1	$\{4/3, 5/3\}$	6	2.9683	11	3.4576	20	3.7625
2	2	7	3.1063	12	3.5137	30	3.8756
3	2.3248	8	3.2196	13	3.5617	40	3.9239
4	2.5875	9	3.3133	14	3.603	50	3.9487
5	2.7984	10	3.3916	15	3.6389	∞	4

Fig. 1. The PoA of priority-based affine congestion games for some values of p. For $p = 1$, the bound 4/3 [43] holds for non-atomic games, while the bound 5/3 [39] holds for atomic ones. For $p \geq 2$, the bounds hold also for the PoS under singleton strategies.

or randomization is allowed. In fact, for atomic games, a preemptive scheduling policy yielding a PoA of 5/2 is derived in [22], while, under the random policy, the PoA drops to 5/3 [39] and the PoS to $1 + 1/\sqrt{5} \approx 1.447$ [11]. Thus, as in this setting the efficiency of pure Nash equilibria is tremendously influenced by the chosen strategy, the study of the PoA/PoS induced by different scheduling policies, despite being interesting per se, plays a fundamental role also in the theory of coordination mechanisms and cost-sharing policies for congestion games. A *coordination mechanism* [19] is a local policy rule that each resource applies to schedule its assigned requests, while a cost-sharing policy [32,33,46] is a rule determining how the cost of a resource has to be shared among its users. Both machineries are usually used with the aim of mitigating the inefficiencies caused by selfish behavior.

Our Results. First of all, we prove that both atomic and non-atomic priority-based affine congestion games admit pure Nash equilibria. As for non-atomic games it also turns out that all pure Nash equilibria share the same social welfare, it follows that the PoA and the PoS coincide within this class.

Having shown the existence of pure Nash equilibria in both models, our main result is the derivation of tight bounds for both the PoA and the PoS of priority-based affine congestion games as a function of p. These bounds, which are reported in Fig. 1, are tight even under singleton strategies and reveal an interesting separation between the general case of $p \geq 2$ and the priority-free scenario of $p = 1$. In fact, while non-atomic games are more efficient than atomic ones in absence of priorities, they share the same PoA when $p \geq 2$. Moreover, while the PoS is lower than the PoS in atomic games with no priorities, the two metrics become equal when $p \geq 2$. To the best of our knowledge, this is the first example of such a unified behavior. The technique we exploit to derive the upper bounds is the primal-dual method introduced in [7] and based on pairs of primal/dual formulations.

An interesting application of our results falls within the problem of online scheduling with related machines and identical jobs. Assume we have an online scheduling problem P, where we are given a set of m related machines, with machine i having speed s_i, and an input sequence of n unit-length jobs (with n not known in advance), each coming with an associated set of machines where it can be processed. Suppose also that the input sequence is divided into p subsequences of jobs (p does not need to be known) and that, for each $1 \leq c \leq p$, when the c-th subsequence arrives, the sets of allowable machines of all jobs in the subsequence are immediately revealed (thus, the traditional setting in which jobs arrive one at time coincides with the case of $p = n$). Now interpret P as an atomic affine congestion game with p priority classes, where each subsequence is seen as a priority class and each machine of speed s is a resource having latency function equal to $\ell(x) = x/s$. As the cost of a job belonging to class c is not influenced by jobs of higher classes, it is easy to see that one can inductively construct a pure Nash equilibrium for the congestion game yielded by the jobs of class c upon a pure Nash equilibrium for the game induced by all jobs belonging to classes smaller than c, so as to obtain a pure Nash equilibrium for the whole game P. As a pure Nash equilibrium for singleton congestion games can be computed in polynomial time [34], our results provide an efficient algorithm for online scheduling with related machines and identical jobs arranged into p subsequences and characterize its competitive ratio as a function of p.

Related Work. The study of the efficiency of (pure) Nash equilibria in congestion games initiated with the seminal papers [2,3,18,36,43]. Since then, many results have been obtained in the literature under different generalizations or specializations [1,5–7,9,10,13,15–17,20,23–25,27,29–31,37,41,42,44]. All of these contributions, however, focus on the makespan model.

To the best of our knowledge, priority-based congestion games have been previously addressed in [28] only. They consider the case in which each resource applies an independent non-preemptive scheduling policy which, however, always assumes a total ordering of the players (i.e., as in the case of $p = n$ in our model). For affine latencies, they show a PoA of 4 for non-atomic games and a PoA of 17/3 for atomic ones. Observe that, while the PoA of non-atomic games coincides with the one we derive in this case when $p = n$, this is not the case for atomic games, where the PoA in our model is much lower. This is due to the fact that, in [28], a different cost function is considered. In fact, while we assume that a player scheduled at position k on a resource with latency function $\ell(x)$ pays a cost of $\ell(k)$, they assume a cost of $\int_{k-1}^{k} \ell(x)dx$. While this diversity is inconsequential in non-atomic games, for atomic ones a different cost model, with different efficiency bounds, arises. Finally, [28] also considers generalizations to polynomial latency functions and to weighted players.

Tight connections between (singleton) affine congestion games and greedy algorithms for (online) scheduling problems have been noted and investigated in several papers [8,9,12–14,21,22,35,44,45].

Paper Organization. Next section introduces the model and definitions. Section 3 contains the existential results of pure Nash equilibria, while Sect. 4 presents the characterization of their efficiency. Finally, we conclude in Sect. 5 by discussing possible future research directions. Due to space constraints, some proofs have been omitted.

2 Model

For an integer $k \geq 1$, denote by $[k] := \{1, \ldots, k\}$ the set of the first k positive integers. Moreover, set $[0] := \emptyset$.

Atomic Games. For any integer $p \geq 1$, a *priority-based affine atomic congestion game with p priority classes* $\Gamma_p^a = ([n], R, (\mathcal{S}_i)_{i\in[n]}, (\alpha_r, \beta_r)_{r\in R}, (P_c)_{c\in[p]})$ is defined by a finite set $[n]$ of $n \geq 2$ players, a finite set R of resources, a strategy set $\mathcal{S}_i \subseteq 2^R \setminus \emptyset$ for each player $i \in [n]$, two coefficients $\alpha_r \geq 0$ and $\beta_r \geq 0$ for each resource $r \in R$ and a priority class $P_c \subseteq [n]$ for each $c \in [p]$ such that $\cup_{c\in[p]} P_c = [n]$ and $P_c \cap P_{c'} = \emptyset$ for each $c, c' \in [p]$ with $c \neq c'$, i.e., the sets $P_1, \ldots, P_p$ realize a partition of $[n]$. We use $c(i)$ to refer to the priority class of player i, i.e., $c(i) = j$ if and only if $i \in P_j$.

Denote by $\boldsymbol{\sigma} = (\sigma_1, \ldots, \sigma_n)$ the strategy profile in which each player $i \in [n]$ chooses strategy $\sigma_i \in \mathcal{S}_i$. For a strategy profile $\boldsymbol{\sigma}$, a priority class $c \in [p]$ and a resource $r \in R$, let $n_r^c(\boldsymbol{\sigma}) = |\{i \in P_c : r \in \sigma_i\}|$ be the number of players belonging to class c selecting resource r in $\boldsymbol{\sigma}$, $n_r^{<c}(\boldsymbol{\sigma}) = \sum_{c'\in[c-1]} n_r^{c'}(\boldsymbol{\sigma})$ be the number of players belonging to any class $c' < c$ selecting resource r in $\boldsymbol{\sigma}$ and $n_r(\boldsymbol{\sigma}) = \sum_{c\in[p]} n_r^c(\boldsymbol{\sigma})$ be the congestion of resource r in $\boldsymbol{\sigma}$, i.e., the number of its users.

The cost that a player experiences on resource r when she occupies the kth position in the schedule of r (say i is the kth user of r) is equal to $\alpha_r k + \beta_r$ (affine latency functions). Thus, the expected cost of player i in $\boldsymbol{\sigma}$ is defined as

$$\begin{aligned} cost_i(\boldsymbol{\sigma}) &= \sum_{r\in\sigma_i} \sum_{k\in[n_r(\boldsymbol{\sigma})]} ((\alpha_r k + \beta_r) \cdot \Pr[i \text{ is the } k\text{th user of } r]) \\ &= \sum_{r\in\sigma_i} \left(\alpha_r \left(n_r^{<c(i)}(\boldsymbol{\sigma}) + \frac{1}{n_r^{c(i)}(\boldsymbol{\sigma})} \sum_{k\in[n_r^{c(i)}(\boldsymbol{\sigma})]} k \right) + \beta_r \right) \\ &= \sum_{r\in\sigma_i} \left(\alpha_r \left(n_r^{<c(i)}(\boldsymbol{\sigma}) + \frac{n_r^{c(i)}(\boldsymbol{\sigma}) + 1}{2} \right) + \beta_r \right). \end{aligned}$$

The utilitarian social welfare, from now on simply the *social welfare*, of $\boldsymbol{\sigma}$ is defined as the sum of the expected costs of all players in $\boldsymbol{\sigma}$, thus equal to $\mathsf{SW}(\boldsymbol{\sigma}) = \sum_{i\in[n]} cost_i(\boldsymbol{\sigma}) = \sum_{r\in R} \sum_{k\in[n_r(\boldsymbol{\sigma})]} (\alpha_r k + \beta_r) = \sum_{r\in R} \left(\alpha_r \frac{n_r(\boldsymbol{\sigma})(n_r(\boldsymbol{\sigma})+1)}{2} + \beta_r n_r(\boldsymbol{\sigma}) \right)$, where the last equality easily follows by observing that, for each $r \in R$ with $n_r(\boldsymbol{\sigma})$ users, there is exactly one player occupying the kth position in the schedule of r for each $k \in [n_r(\boldsymbol{\sigma})]$. We shall denote by $\boldsymbol{\sigma}^*$ the *social optimum* of Γ_p^a, that is, the strategy profile minimizing the social welfare.

We shall focus on the notion of pure Nash equilibrium which is defined as follows.

Definition 1. *A strategy profile $\boldsymbol{\sigma}$ is a pure Nash equilibrium for Γ_p^a if, for each $i \in [n]$ and $S \in \mathcal{S}_i$, $cost_i(\boldsymbol{\sigma}) \leq cost_i(\boldsymbol{\sigma}_{-i}, S)$.*

By the above definition, given a pure Nash equilibrium $\boldsymbol{\sigma}$ and a social optimum $\boldsymbol{\sigma}^*$ for Γ_p^a, the following inequality holds for each $i \in [n]$:

$$\sum_{r\in\sigma_i}\left(\alpha_r\left(n_r^{<c(i)}(\boldsymbol{\sigma})+\frac{n_r^{c(i)}(\boldsymbol{\sigma})+1}{2}\right)+\beta_r\right) - \sum_{r\in\sigma_i^*}\left(\alpha_r\left(n_r^{<c(i)}(\boldsymbol{\sigma})+\frac{n_r^{c(i)}(\boldsymbol{\sigma})+2}{2}\right)+\beta_r\right)\leq 0. \quad (1)$$

Non-atomic Games. For any integer $p \geq 1$, a *priority-based affine non-atomic congestion game with p priority classes* $\Gamma_p^{na} = ([n], R, (f_i)_{i\in[n]}, (\mathcal{S}_i)_{i\in[n]}, (\alpha_r, \beta_r)_{r\in R}, (P_c)_{c\in[p]})$ has the same definition of its atomic counterpart with a different interpretation on the set of players and on how they handle their requests. For every $i \in [n]$, in fact, there is an amount of flow f_i belonging to priority class $c(i)$ that needs to be assigned to strategies in $\mathcal{S}_i$ in an arbitrarily splittable way. Let $m_i = |\mathcal{S}_i|$ denote the number of strategies available to the ith flow and set $\mathcal{S}_i = \{S_i^1, \ldots, S_i^{m_i}\}$. In this setting, a strategy profile is identified by a tuple $\boldsymbol{\sigma} = (\sigma_1^1, \ldots \sigma_1^{m_1}, \ldots, \sigma_n^1, \ldots \sigma_n^{m_n})$, where, for every $i \in [n]$ and $j \in [m_i]$, $\sigma_i^j \geq 0$ denotes the fraction of the ith flow assigned to S_i^j. We shall only consider feasible strategy profiles, i.e., such that $\sum_{j\in[m_i]} \sigma_i^j = f_i$ for each $i \in [n]$. We overload the notation of $\boldsymbol{\sigma}$ for the sake of analysing both atomic and non-atomic games under the same framework. To this aim, we also denote by $n_r^c(\boldsymbol{\sigma}) = \sum_{i\in P_c}\sum_{j\in[m_i]:r\in S_i^j}\sigma_i^j$ the total amount of flow of priority class c assigned to resource r in $\boldsymbol{\sigma}$. Similarly, we define $n_r^{<c}(\boldsymbol{\sigma}) = \sum_{c'\in[c-1]} n_r^{c'}(\boldsymbol{\sigma})$ and $n_r(\boldsymbol{\sigma}) = \sum_{c\in[p]} n_r^c(\boldsymbol{\sigma})$.

Here, the expected cost that a flow of class c experiences for each (arbitrarily small) unitary fraction assigned to resource r becomes equal to $\alpha_r\left(n_r^{<c}(\boldsymbol{\sigma}) + \frac{1}{n_r^c(\boldsymbol{\sigma})}\int_0^{n_r^c(\boldsymbol{\sigma})} t dt\right)+\beta_r = \alpha_r\left(n_r^{<c}(\boldsymbol{\sigma}) + \frac{n_r^c(\boldsymbol{\sigma})}{2}\right)+\beta_r$, while the social welfare in $\boldsymbol{\sigma}$ becomes $\mathsf{SW}(\boldsymbol{\sigma}) = \sum_{r\in R}\left(\alpha_r\frac{n_r(\boldsymbol{\sigma})^2}{2} + \beta_r n_r(\boldsymbol{\sigma})\right)$.

The notion of pure Nash equilibrium assumes the following definition.

Definition 2. *A strategy profile $\boldsymbol{\sigma}$ is a pure Nash equilibrium for Γ_p^{na} if and only if, for each $i \in [n]$, $j \in [m_i]$ such that $\sigma_i^j > 0$, and $j' \in [m_i]$,* $\sum_{r\in S_i^j}\left(\alpha_r\left(n_r^{<c(i)}(\boldsymbol{\sigma}) + \frac{n_r^{c(i)}(\boldsymbol{\sigma})}{2}\right)+\beta_r\right) \leq \sum_{r\in S_i^{j'}}\left(\alpha_r\left(n_r^{<c(i)}(\boldsymbol{\sigma}) + \frac{n_r^{c(i)}(\boldsymbol{\sigma})}{2}\right)+\beta_r\right)$.

Thus, if $\boldsymbol{\sigma}$ is a pure Nash equilibrium and $\boldsymbol{\sigma}^*$ is a social optimum for Γ_p^{na}, the following inequality holds for each $i \in [n]$:

$$\sum_{r\in\sigma_i}\left(\alpha_r\left(n_r^{<c(i)}(\boldsymbol{\sigma})+\frac{n_r^{c(i)}(\boldsymbol{\sigma})}{2}\right)+\beta_r\right) - \sum_{r\in\sigma_i^*}\left(\alpha_r\left(n_r^{<c(i)}(\boldsymbol{\sigma})+\frac{n_r^{c(i)}(\boldsymbol{\sigma})}{2}\right)+\beta_r\right)\leq 0. \quad (2)$$

Definition of PoA **and** PoS. Given a congestion game Γ, denote by $\mathsf{NE}(\Gamma)$ the set of its pure Nash equilibria. The PoA of Γ is defined as $\mathsf{PoA}(\Gamma) = \max_{\sigma \in \mathsf{NE}(\Gamma)} \frac{\mathsf{SW}(\sigma)}{\mathsf{SW}(\sigma^*)}$, while the PoS of Γ is defined as $\mathsf{PoS}(\Gamma) = \min_{\sigma \in \mathsf{NE}(\Gamma)} \frac{\mathsf{SW}(\sigma)}{\mathsf{SW}(\sigma^*)}$.

Results for $p = 1$. We end this section by recalling the known results for the priority-free case of $p = 1$. For atomic games, [39] shows that the PoA is $5/3$, while [11] proves that the PoS drops to $1+1/\sqrt{5} \approx 1.447$. For non-atomic games, it is not difficult to see that both the random and the round-robin policy induce the same set of pure Nash equilibria. Hence, by the results in [43], the PoA and the PoS are equal to $4/3$ (for instance, the classical Pigou network yields a $4/3$ lower bound on the PoS also in the random model).

3 Existence of Pure Nash Equilibria

In this section, we shall prove that priority-based congestion games always admit pure Nash equilibria. For non-atomic games, we also show that all equilibria attain the same social welfare, thus implying that the PoA and the PoS coincide within this class.

Atomic Games. A priority-based (affine) atomic congestion game with only one priority class boils down to a traditional congestion game for which existence of pure Nash equilibria (and more generally the finite improvement path property) is guaranteed by Rosenthal's Theorem [40]. However, for more priority classes, this equivalence does not hold any more and a dedicated existential proof is required. Towards this end, we need to introduce some additional notation. Given an atomic game Γ_p^a, with $p \geq 1$, and a priority class $c \in [p]$, denote by $\Gamma_{\leq c}^a$ the restriction of Γ_p^a to the players of priority class at most c; moreover, given a strategy profile $\boldsymbol{\sigma}^{<c}$ for $\Gamma_{\leq c-1}^a$, with $\boldsymbol{\sigma}^0 := \emptyset$, denote by $\overline{\Gamma}_c^a(\boldsymbol{\sigma}^{<c})$ the game obtained from $\Gamma_{\leq c}^a$ by freezing the strategic choices of all players of class $c' < c$ according to $\boldsymbol{\sigma}^{<c}$ and letting only the players of class c play. We shall denote by $\overline{\boldsymbol{\sigma}}^c$ a strategy profile for $\overline{\Gamma}_c^a(\boldsymbol{\sigma}^{<c})$, i.e., a strategy profile satisfying $\overline{\boldsymbol{\sigma}}_i^c = \boldsymbol{\sigma}_i^{<c}$ for each player i such that $c(i) < c$.

We first show that, for any strategy profile $\boldsymbol{\sigma}^{<c}$ for $\Gamma_{\leq c-1}^a$, $\overline{\Gamma}_c^a(\boldsymbol{\sigma}^{<c})$ is an exact potential game.

Lemma 1. *For any* $p \geq 2$, *affine atomic congestion game with* p *priority classes* Γ_p^a, *priority class* $c \in [p]$ *and strategy profile* $\boldsymbol{\sigma}^{<c}$ *for* $\Gamma_{\leq c-1}^a$, *game* $\overline{\Gamma}_c^a(\boldsymbol{\sigma}^{<c})$ *admits the following exact potential function:*

$$\Phi_c(\overline{\boldsymbol{\sigma}}^c) := \sum_{r \in R} \left(\alpha_r n_r^c(\overline{\boldsymbol{\sigma}}^c) \left(n_r^{<c}(\overline{\boldsymbol{\sigma}}^c) + \frac{n_r^c(\boldsymbol{\sigma}) + 3}{4} \right) + n_r^c(\overline{\boldsymbol{\sigma}}^c)\beta_r \right). \quad (3)$$

We can now prove that any priority-based affine atomic congestion game admits a pure Nash equilibrium.

Theorem 1. *For any $p \geq 1$, game Γ_p^a admits pure Nash equilibria.*

Proof. As the cost of a player of priority class c is not influenced by the choices of the players of higher classes, it follows that, given a strategy profile $\boldsymbol{\sigma}$ for Γ_p^a and a player i, it holds that σ_i is a best-response for i against $\boldsymbol{\sigma}_{-i}$ in Γ_p^a if and only if σ_i is a best-response for i against $\overline{\boldsymbol{\sigma}}_{-i}^{c(i)}$ in $\overline{\Gamma}_{c(i)}^a(\boldsymbol{\sigma}^{<c(i)})$. Thus, for each $c \in [p]$, thanks to Lemma 1, a pure Nash equilibrium for $\Gamma_{\leq c}^a$ can be constructed inductively by extending a pure Nash equilibrium for $\Gamma_{\leq c-1}^a$. □

Non-atomic Games. By following and extending [4,26,43], we show that every non-atomic priority-based affine congestion game admits pure Nash equilibria and that there is no difference between the PoA and the PoS within this class.

Theorem 2. *Every non-atomic priority-based affine congestion game admits pure Nash equilibria. Moreover, all equilibria have the same social welfare.*

4 Bounding the PoA and the PoS

In this section, we characterize the PoA and the PoS of priority-based affine congestion games for both of their versions: atomic and non-atomic. We perform our analysis by relying on the primal-dual method introduced in [7]. As the base case of $p = 1$ has been already solved, we shall focus on games with at least two priority classes.

4.1 The Primal-Dual Formulation

Fix a priority-based affine congestion game Γ_p with $p \geq 2$, a pure Nash equilibrium $\boldsymbol{\sigma}$ for Γ_p and a social optimum $\boldsymbol{\sigma}^*$ for Γ_p. For a resource $r \in R$ and a priority class $c \in [p]$, set $k_r^c = n_r^c(\boldsymbol{\sigma})$, $o_r^c = n_r^c(\boldsymbol{\sigma}^*)$, $k_r^{<c} = n_r^{<c}(\boldsymbol{\sigma})$, $o_r^{<c} = n_r^{<c}(\boldsymbol{\sigma}^*)$, $k_r = n_r(\boldsymbol{\sigma})$ and $o_r = n_r(\boldsymbol{\sigma}^*)$. Observe that, no matter whether Γ_p is an atomic or non-atomic game, all the previous quantities are well defined.

If Γ_p is an atomic game, for each $c \in [p]$, by summing inequality (1) for each $i \in [n]$ such that $c(i) = c$, we obtain

$$\sum_{r \in R} \left(\alpha_r \left(k_r^c \left(k_r^{<c} + \frac{k_r^c + 1}{2} \right) - o_r^c \left(k_r^{<c} + \frac{k_r^c + 2}{2} \right) \right) + \beta_r \left(k_r^c - o_r^c \right) \right) \leq 0.$$

Similarly, if Γ_p is a non-atomic game, for each $c \in [p]$, by summing inequality (2) for each $i \in [n]$ such that $c(i) = c$, we get

$$\sum_{r \in R} \left(\alpha_r \left(k_r^c \left(k_r^{<c} + \frac{k_r^c}{2} \right) - o_r^c \left(k_r^{<c} + \frac{k_r^c}{2} \right) \right) + \beta_r \left(k_r^c - o_r^c \right) \right) \leq 0.$$

Thus, we have that, in general, inequality

$$\sum_{r \in R} \left(\alpha_r \left(k_r^c \left(k_r^{<c} + \frac{k_r^c + \delta}{2} \right) - o_r^c \left(k_r^{<c} + \frac{k_r^c + 2\delta}{2} \right) \right) \right)$$

$$+\sum_{r\in R}\left(\beta_r\left(k_r^c - o_r^c\right)\right) \leq 0 \qquad (4)$$

holds for each $c \in [p]$, with $\delta = 1$ when dealing with atomic games and $\delta = 0$ when dealing with non-atomic ones. Moreover, also the social welfare of both $\boldsymbol{\sigma}$ and $\boldsymbol{\sigma}^*$ can be expressed in a unified manner, as we have $\mathsf{SW}(\boldsymbol{\sigma}) = \sum_{r\in R}\left(\alpha_r \frac{k_r(k_r+\delta)}{2} + \beta_r k_r\right)$ and $\mathsf{SW}(\boldsymbol{\sigma}^*) = \sum_{r\in R}\left(\alpha_r \frac{o_r(o_r+\delta)}{2} + \beta_r o_r\right)$ with the same constraints on δ.

By applying the primal-dual method to bound the PoA of Γ_p, we get the following primal linear program $PP(\Gamma_p)$:

$$\begin{aligned}
\max\ & \sum_{r\in R}\left(\alpha_r \frac{k_r(k_r+\delta)}{2} + \beta_r k_r\right)\\
s.t.\ & \sum_{r\in R}\left(\alpha_r\left(k_r^c\left(k_r^{<c} + \frac{k_r^c+\delta}{2}\right) - o_r^c\left(k_r^{<c} + \frac{k_r^c+2\delta}{2}\right)\right)\right)\\
& \quad + \sum_{r\in R}\left(\beta_r\left(k_r^c - o_r^c\right)\right) \leq 0, \quad \forall c \in [p]\\
& \sum_{r\in R}\left(\alpha_r \frac{o_r(o_r+\delta)}{2} + \beta_r o_r\right) = 1\\
& \alpha_r, \beta_r \geq 0, \quad \forall r \in R
\end{aligned}$$

The dual program $DP(\Gamma_p)$ is the following:

$$\begin{aligned}
\min\ & \gamma\\
s.t.\ & \sum_{c\in[p]}\left(x_c\left(k_r^c\left(k_r^{<c} + \frac{k_r^c+\delta}{2}\right) - o_r^c\left(k_r^{<c} + \frac{k_r^c+2\delta}{2}\right)\right)\right)\\
& \quad + \gamma\frac{o_r(o_r+\delta)}{2} - \frac{k_r(k_r+\delta)}{2} \geq 0 \qquad \forall r \in R && (5)\\
& \sum_{c\in[p]}\left(x_c(k_r^c - o_r^c)\right) + \gamma o_r - k_r \geq 0 \qquad \forall r \in R && (6)\\
& x_c \geq 0 \qquad \forall c \in [p]
\end{aligned}$$

4.2 Upper Bounds

For any $p \geq 2$, consider the following non-linear program $NLP(p)$:

$$\begin{aligned}
\min\ & \gamma\\
s.t.\ & x_1 \leq \gamma && (7)\\
& x_{c+1}^2 \leq \gamma(x_c - 1) \quad \forall c \in [p-2] && (8)\\
& x_{p-1} \leq \gamma(x_{p-1} - 1) && (9)\\
& x_p = \frac{2x_{p-1}}{x_{p-1}+1} && (10)
\end{aligned}$$

$$x_c \geq 0 \qquad \forall c \in [p] \tag{11}$$

As our main result, we show that $NLP(p)$ admits a unique optimal solution which is also feasible for $DP(\Gamma_p)$.

Theorem 3. *For every $p \geq 2$, there exists a unique optimal solution $\overline{\mathbf{s}}(p) = (\overline{x}_1(p), \ldots, \overline{x}_p(p), \overline{\gamma}(p))$ for $NLP(p)$. Moreover, $\overline{\mathbf{s}}(p)$ is a feasible solution for $DP(\Gamma_p)$ and $\overline{x}_c(p) > 1$ for each $i \in [p]$.*

Having shown that $\overline{\mathbf{s}}(p)$ is feasible for $DP(\Gamma_p)$, we can claim the following result.

Corollary 1. *For any priority-based affine congestion game Γ_p with $p \geq 2$, $\mathsf{PoA}(\Gamma_p) \leq \overline{\gamma}(p)$.*

By numerically solving $NLP(p)$, we explicitly quantify the upper bounds on the PoA for some values of p as outlined in Fig. 1 (where, for completeness, we also report the previously known bound for the case of $p = 1$, which is not covered by our analysis).

4.3 Lower Bounds

Here, we construct, given an integer $p \geq 2$, a family of singleton congestion games to obtain lower bounds on the PoS matching the upper bounds given in Corollary 1 for the PoA. These games, which cover both the atomic and non-atomic cases, are defined by relying on the optimal solution $\overline{\mathbf{s}}(p)$ for $NLP(p)$. It is important to highlight that the explicit computation of $\overline{\mathbf{s}}(p)$ is not necessary. Before presenting the promised family of games, we warm up by considering separately the cases of $p = 2, 3$ that require different constructions.

Theorem 4. *For any $\epsilon > 0$, there exist two priority-based singleton affine atomic congestion games Γ_2^a and Γ_3^a such that $\mathsf{PoS}(\Gamma_2^a) \geq 2 - \epsilon$ and $\mathsf{PoS}(\Gamma_3^a) \geq 2.3247 - \epsilon$.*

Proof. We only show here the claim for $p = 2$, the full proof is deferred to the Appendix. Game Γ_2^a is defined as follows. There are θ players of class 1 and θ players of class 2. The set of resources R is defined as follows: $R = R_1 \cup \{r_2\}$, with $R_1 = \{r_{1,1}, r_{1,2}, \ldots, r_{1,\theta}\}$. All resources in R_1 have a linear latency function with coefficient equal to $(\theta+2)/2$, while resource r_2 has a linear latency function with coefficient equal to 1. All players of class 2 have a unique strategic choice corresponding to resource r_2, while each player of class 1 can choose between two resources, called the *first* and *second* resource, respectively. More precisely, the ith player of class 1 can choose between resources $r_{1,i}$ and r_2. Observe that Γ_2^a is a singleton game. We stress that the use of dummy players with a unique strategic choice is a common technique in the literature and is not a limiting assumption, as one can always add a second choice, with an arbitrarily high cost, that can never be adopted in a pure Nash equilibrium.

It is immediate to check that the second strategy, which may cost at most $(\theta+1)/2$, is a dominant one for all players of class 1. Thus, the strategy profile $\boldsymbol{\sigma}$ in which all players of class 1 choose their second resource is the unique pure Nash equilibrium for Γ_2^a. We lower bound the PoS of Γ_2^a by comparing the social welfare of $\boldsymbol{\sigma}$ with the one yielded by the strategy profile $\boldsymbol{\sigma}^*$ in which all players of class 1 choose their first resource. In particular, we shall consider the limit of this lower bound for $\theta \to \infty$. We get $\lim_{\theta\to\infty} \mathsf{PoS}(\Gamma_2^a) \geq \lim_{\theta\to\infty} \frac{\mathsf{SW}(\boldsymbol{\sigma})}{\mathsf{SW}(\boldsymbol{\sigma}^*)} = \lim_{\theta\to\infty} \frac{\frac{1}{2}2\theta(2\theta+1)}{\theta(\frac{\theta+2}{2})+\frac{1}{2}\theta(\theta+1)} = 2$, thus showing the claim. □

The previous construction can be easily adapted to provide a lower bound for the PoA (and so also for the PoS) of non-atomic games with $p = 2, 3$.

Theorem 5. *There exists two priority-based singleton affine non-atomic congestion games Γ_2^{na} and Γ_3^{na} such that $\mathsf{PoA}(\Gamma_2^{na}) \geq 2$ and $\mathsf{PoA}(\Gamma_3^{na}) \geq 2.3247$.*

We now show how to generalize the previous constructions for any $p \geq 4$.

Theorem 6. *For any $\epsilon > 0$ and $p \geq 4$, there exists a priority-based singleton affine atomic congestion game Γ_p^a such that $\mathsf{PoS}(\Gamma_p^a) \geq \overline{\gamma}(p) - \epsilon$.*

Proof. Fix a value $\epsilon > 0$ and an integer $p \geq 4$ and consider the following singleton atomic game Γ_p^a. For every $c \in [p]$, the number of players of class c is equal to $|P_c| := \pi_c$, with

$$\pi_c = \begin{cases} \theta & \text{if } c = p, \\ \frac{\theta}{2(\overline{x}_{p-1}(p)-1)} & \text{if } c = p-1, \\ \frac{\overline{x}_{c+1}(p)}{\overline{x}_c(p)-1}\pi_{c+1} & \text{if } c \in [p-2]. \end{cases}$$

Here, the values $\overline{x}_c(p)$ for each $c \in [p]$ are the ones yielded by the optimal solution $\overline{\mathbf{s}}(p)$ for $NLP(p)$. We shall consider the case in which θ goes to infinity. Thus, as $\overline{x}_c(p) > 1$ for each $c \in [p]$ by Theorem 3, each value π_c belongs to the set of positive integers and is, so, well defined.

The set of resources is $R = R_1 \cup \{r_2, \ldots, r_p\}$, with $R_1 = \{r_{1,1}, \ldots, r_{1,|P_1|}\}$. All resources in R_1 have a linear latency function with coefficient equal to α_1, while, for $c \in [p] \setminus \{1\}$, resource r_c has a linear latency function with coefficient equal to α_c. All players of class p have a unique strategic choice corresponding to resource r_p. For each $c \in [p-1]$, instead, each player of class c can choose between two resources, called the *first* and *second* resource, respectively. For every $c \in [p-1] \setminus \{1\}$, the first and second resources of a player of class c are r_c and r_{c+1}, while the ith player of class 1 can choose between resources $r_{1,i}$ and r_2. Observe that Γ_p^a is a singleton game.

In order to maximize the PoA yielded by this instance, let us use the pair of primal-dual formulations $PP(\Gamma_p^a)$ and $DP(\Gamma_p^a)$, where we set $\boldsymbol{\sigma}$ and $\boldsymbol{\sigma}^*$ as the strategy profiles in which all players of class c, with $c \in [p-1]$, choose their second and first resource, respectively. As we consider the case of θ going to infinity, which implies that the number of players in each class grows arbitrarily large, we can get rid of small constants in the formulation, thus obtaining the following simplified primal linear program $PP(\Gamma_p^a)$:

$$\begin{aligned}
\max \ & \sum_{i=2}^{p-1} \frac{\alpha_i \pi_{i-1}^2}{2} + \frac{\alpha_p(\pi_{p-1}+\pi_p)^2}{2} \\
s.t. \ & \frac{\alpha_2 \pi_1^2}{2} - \sum_{i\in[\pi_1]} \alpha_{1,i} \leq 0, \\
& \frac{\alpha_{c+1}\pi_c^2}{2} - \alpha_c \pi_c \pi_{c-1} \leq 0, \quad \forall c \in [p-1]\setminus\{1\} \\
& \sum_{i\in[\pi_1]} \alpha_{1,i} + \sum_{i=2}^{p} \frac{\alpha_i \pi_i^2}{2} = 1 \\
& \alpha_{1,i} \geq 0, \quad \forall i \in [\pi_1] \\
& \alpha_c \geq 0, \quad \forall c \in [p]\setminus\{1\}
\end{aligned}$$

The dual program $DP(\Gamma_p^a)$ is the following:

$$\begin{aligned}
\min \ & \gamma \\
s.t. \ & -x_1 + \gamma \geq 0 \\
& \frac{\pi_{c-1}^2}{2} x_{c-1} - \pi_{c-1}\pi_c x_c + \gamma \frac{\pi_c^2}{2} - \frac{\pi_{c-1}^2}{2} \geq 0 \qquad \forall c \in [p-1]\setminus\{1\} \\
& \frac{\pi_{p-1}^2}{2} x_{p-1} + \gamma \frac{\pi_p^2}{2} - \frac{(\pi_{p-1}+\pi_p)^2}{2} \geq 0 \\
& x_c \geq 0 \qquad \forall c \in [p]
\end{aligned}$$

As $DP(\Gamma_p^a)$ is a particular instantiation of the dual program $DP(\cdot)$, it follows that its optimal solution has to be not smaller than $\overline{\gamma}(p)$. However, by substituting the values π_c and setting $x_c = \overline{x}_c(p)$ for each $c \in [p]$, $DP(\Gamma_p^a)$ rewrites as:

$$\begin{aligned}
\min \ & \gamma \\
s.t. \ & \overline{x}_1(p) \leq \gamma \\
& \overline{x}_{c+1}(p)^2 \leq \gamma(\overline{x}_c(p) - 1) \qquad \forall c \in [p-2] \\
& \overline{x}_{p-1}(p) \leq \gamma(\overline{x}_{p-1}(p) - 1) \\
& \overline{x}_p(p) = \frac{2\overline{x}_{p-1}(p)}{\overline{x}_{p-1}(p)+1}.
\end{aligned}$$

This implies that $\overline{\gamma}(p)$ is also an optimal solution for $DP(\Gamma_p^a)$.

Now consider the solution for $PP(\Gamma_p^a)$ obtained by setting $\alpha_p = 1$ and $\alpha_c = \frac{\alpha_{c+1}\pi_c}{2\pi_{c-1}}$ for each $c \in [p-1]$, where we assume $P_0 = \emptyset$, so that $\pi_0 = 0$. By the complementary slackness conditions, as this solution satisfies at equality all primal constraints which are related to a non-zero dual variable, it follows that $\mathsf{SW}(\boldsymbol{\sigma})/\mathsf{SW}(\boldsymbol{\sigma}^*) = \overline{\gamma}(p)$. This indeed shows that the PoA of Γ_p^a is at least $\overline{\gamma}(p)$.

To extend this result to the PoS, we need to show that $\boldsymbol{\sigma}$ is the unique pure Nash equilibrium for Γ_p^a. To this aim, we slightly perturb the coefficients of the latency functions by setting $\alpha_c = \frac{\alpha_{c+1}(\pi_c+1)}{2(\pi_{c-1}+1)} + \epsilon'$ for each $c \in [p-1]$, where $\epsilon' > 0$ is arbitrarily small. With this modification, we prove that, for every $c \in [p-1]$, under the assumption that all players of class $c-1$ choose their second resource, playing the second resource is a dominant strategy for all players of class c. Because of players of class $c-1$ are using their second resource, the first resource of a player of class c costs at least $(\pi_{c-1}+1)\alpha_c$, while the second one costs at most $\frac{(\pi_c+1)\alpha_{c+1}}{2}$. By the definition of α_c, the second resource always yields a strictly smaller cost, thus showing the claim. The modification decreases the ratio $\mathsf{SW}(\boldsymbol{\sigma})/\mathsf{SW}(\boldsymbol{\sigma}^*)$ of a negligible amount so that, for a suitable choice of ϵ', we have $\mathsf{PoS}(\Gamma_p^a) \geq \overline{\gamma}(p) - \epsilon$. □

The game used in proof of the previous theorem can be adapted, with some modifications, to show the same result for non-atomic games.

Theorem 7. *For any $p \geq 4$, there exists a priority-based singleton affine non-atomic congestion game Γ_p^{na} such that* $\mathsf{PoA}(\Gamma_p^{na}) \geq \overline{\gamma}(p)$.

5 Conclusions

We have given tight bounds for the PoA and the PoS of both atomic and non-atomic affine congestion games, under the assumption that the set of players is partitioned into $p \geq 2$ priority classes and the resources schedule its users according to a priority-based policy, breaking ties uniformly at random. These bounds hold even for load balancing games. Our findings outline an interesting separation between the case of $p \geq 2$ and the priority-free scenario of $p = 1$. The results are obtained by using the primal-dual method of [7]. An important consequence of this fact is that the upper bounds extend with no degradation to coarse correlated equilibria, as shown in [6].

There are several possible research directions that may be investigated. For instance, one can consider generalizations such as weighted players, polynomial latency functions, approximate Nash equilibria. Although the PoA matches the PoS even under singleton strategies, this may not be the case in presence of symmetric players or identical resources: both these restricted scenarios may hide useful properties. Moreover, as the lower bounding instances are based on a very constrained construction, it is also interesting to address special cases in which priorities classes and strategies are restricted to obey particular relationships. An orthogonal approach may be that of considering the presence of a central authority which has the power of assigning priority classes to the players so as to induce games with low PoA/PoS.

Our randomized model assumes that players are risk neutral. Different behavior may arise under alternative models of risk averseness as investigated in [39] for the priority-free case.

References

1. Aland, S., Dumrauf, D., Gairing, M., Monien, B., Schoppmann, F.: Exact price of anarchy for polynomial congestion games. SIAM J. Comput. **40**(5), 1211–1233 (2011)
2. Anshelevich, E., Dasgupta, A., Kleinberg, J., Tardos, E., Wexler, T., Roughgarden, T.: The price of stability for network design with fair cost allocation. SIAM J. Comput. **38**(4), 1602–1623 (2008)
3. Awerbuch, B., Azar, Y., Epstein, L.: The price of routing unsplittable flow. In Proceedings of STOC, pp. 57–66. ACM (2005)
4. Beckmann, M., McGuire, C.B., Winsten, C.B.: Studies in the economics of transportation. Yale University Press, New Haven (1956)
5. Bhawalkar, K., Gairing, M., Roughgarden, T.: Weighted congestion games: price of anarchy, universal worst-case examples, and tightness. ACM Trans. Econ. Comp. **2**(4), 1–23 (2014)
6. Bilò, V.: On the robustness of the approximate price of anarchy in generalized congestion games. In: Gairing, M., Savani, R. (eds.) SAGT 2016. LNCS, vol. 9928, pp. 93–104. Springer, Heidelberg (2016). https://doi.org/10.1007/978-3-662-53354-3_8
7. Bilò, V.: A unifying tool for bounding the quality of non-cooperative solutions in weighted congestion games. Theor. Comput. Syst. **62**(5), 1288–1317 (2018)
8. Bilò, V., Fanelli, A., Flammini, M., Moscardelli, L.: Performances of one-round walks in linear congestion games. Theor. Comput. Syst. **49**(1), 24–45 (2011)
9. Bilò, V., Vinci, C.: On the impact of singleton strategies in congestion games. In: Proceedings of ESA, of LIPIcs, vol. 87, pp. 17:1–17:14. Schloss Dagstuhl - Leibniz-Zentrum fuer Informatik (2017)
10. Bilò, V., Vinci, C.: The price of anarchy of affine congestion games with similar strategies. Theor. Comput. Sci. **806**, 641–654 (2020)
11. Bilò, V., Moscardelli, L., Vinci, C.: Uniform mixed equilibria in network congestion games with link failures. In: Proceedings of ICALP, of LIPIcs, vol. 107, pp. 146:1–146:14. Schloss Dagstuhl - Leibniz-Zentrum fuer Informatik (2018)
12. Caragiannis, I.: Better bounds for online load balancing on unrelated machines. In: Proceedings of SODA, pp. 972–981. ACM (2008)
13. Caragiannis, I., Flammini, M., Kaklamanis, C., Kanellopoulos, P., Moscardelli, L.: Tight bounds for selfish and greedy load balancing. Algorithmica **61**(3), 606–637 (2011)
14. Caragiannis, I., Gkatzelis, V., Vinci, C.: Coordination mechanisms, cost-sharing, and approximation algorithms for scheduling. In: Devanur, N.R., Lu, P. (eds.) WINE 2017. LNCS, vol. 10660, pp. 74–87. Springer, Cham (2017). https://doi.org/10.1007/978-3-319-71924-5_6
15. Christodoulou, G., Gairing, M.: Price of stability in polynomial congestion games. ACM Trans. Econ. Comp. **42**(2), 101–1017 (2016)
16. Christodoulou, G., Gairing, M., Giannakopoulos, Y., Spirakis, P.G.: The price of stability of weighted congestion games. SIAM J. Comput. **48**(5), 1544–1582 (2019)
17. Christodoulou, G., Koutsoupias, E.: On the price of anarchy and stability of correlated equilibria of linear congestion games. In: Brodal, G.S., Leonardi, S. (eds.) ESA 2005. LNCS, vol. 3669, pp. 59–70. Springer, Heidelberg (2005). https://doi.org/10.1007/11561071_8
18. Christodoulou, G., Koutsoupias, E.: The price of anarchy of finite congestion games. In: Proceedings of STOC, pp. 67–73. ACM (2005)

19. Christodoulou, G., Koutsoupias, E., Nanavati, A.: Coordination mechanisms. Theor. Comput. Sci. **410**(36), 3327–3336 (2009)
20. Christodoulou, G., Koutsoupias, E., Spirakis, P.G.: On the performance of approximate equilibria in congestion games. Algorithmica **61**(1), 116–140 (2011)
21. Christodoulou, G., Mirrokni, V.S., Sidiropoulos, A.: Convergence and approximation in potential games. Theor. Comput. Sci. **438**, 13–27 (2012)
22. Cole, R., Correa, J.R., Gkatzelis, V., Mirrokni, V.S., Olver, N.: Inner product spaces for minsum coordination mechanisms. In: Proceedings of STOC, pp. 539–548. ACM (2011)
23. Cominetti, R., Scarsini, M., Schröder, M., Stier-Moses, N.E.: Price of anarchy in stochastic atomic congestion games with affine costs. In: Proceedings of EC, pp. 579–580. ACM (2019)
24. Correa, J.R., Cristi, A., Oosterwijk, T.: On the price of anarchy for flows over time. In: Proceedings of EC, pp. 559–577. ACM (2019)
25. Correa, J.R., Schulz, A.S., Stier-Moses, N.E.: Selfish routing in capacitated networks. Math. Oper. Res. **29**(4), 961–976 (2004)
26. Dafermos, S.C., Sparrow, F.T.: The traffic assignment problem for a general network. J. Res. Natl. Bureau Stand. **73B**(2), 91–118 (1969)
27. de Jong, J., Kern, W., Steenhuisen, B., Uetz, M.: The asymptotic price of anarchy for k-uniform congestion games. In: Solis-Oba, R., Fleischer, R. (eds.) WAOA 2017. LNCS, vol. 10787, pp. 317–328. Springer, Cham (2018). https://doi.org/10.1007/978-3-319-89441-6_23
28. Farzad, B., Olver, N., Vetta, A.: A priority-based model of routing. Chicago J. Theor. Comput. Sci., 28 (2008)
29. Feldman, M., Immorlica, N., Lucier, B., Roughgarden, T., Syrgkanis, V.: The price of anarchy in large games. In: Proceedings of STOC, pp. 963–976. ACM (2016)
30. Gairing, M., Lücking, T., Mavronicolas, M., Monien, B.: The price of anarchy for polynomial social cost. Theor. Comput. Sci. **369**(1–3), 116–135 (2006)
31. Gairing, M., Schoppmann, F.: Total latency in singleton congestion games. In: Deng, X., Graham, F.C. (eds.) WINE 2007. LNCS, vol. 4858, pp. 381–387. Springer, Heidelberg (2007). https://doi.org/10.1007/978-3-540-77105-0_42
32. Gkatzelis, V., Kollias, K., Roughgarden, T.: Optimal cost-sharing in general resource selection games. Oper. Res. **64**(6), 1230–1238 (2016)
33. Gopalakrishnan, R., Marden, J.R., Wierman, A.: Potential games are necessary to ensure pure nash equilibria in cost sharing games. Math. Oper. Res. **39**(4), 1252–1296 (2013)
34. Ieong, S., McGrew, R., Nudelman, E., Shoham, Y., Sun, Q.: Fast and compact: a simple class of congestion games. In: Proceedings of the 20th National Conference on Artificial Intelligence, AAAI, pp. 489–494. AAAI Press (2005)
35. Klimm, M., Schmand, D., Tönnis, A.: The online best reply algorithm for resource allocation problems. In: Fotakis, D., Markakis, E. (eds.) SAGT 2019. LNCS, vol. 11801, pp. 200–215. Springer, Cham (2019). https://doi.org/10.1007/978-3-030-30473-7_14
36. Koutsoupias, E., Papadimitriou, C.: Worst-case equilibria. In: Meinel, C., Tison, S. (eds.) STACS 1999. LNCS, vol. 1563, pp. 404–413. Springer, Heidelberg (1999). https://doi.org/10.1007/3-540-49116-3_38
37. Lücking, T., Mavronicolas, M., Monien, B., Rode, M.: A new model for selfish routing. Theor. Comput. Sci. **406**(3), 187–2006 (2008)
38. Nash, J.F.: Equilibrium points in n-person games. Proc. Natl. Acad. Sci. **36**(1), 48–49 (1950)

39. Piliouras, G., Nikolova, E., Shamma, J.S.: Risk sensitivity of price of anarchy under uncertainty. ACM Trans. Econ. Comp. **5**(1), 51–527 (2016)
40. Rosenthal, R.W.: A class of games possessing pure-strategy nash equilibria. Int. J. Game Theory **2**(1), 65–67 (1973)
41. Roughgarden, T.: The price of anarchy is independent of the network topology. J. Comput. System Sci. **67**(2), 341–364 (2003)
42. Roughgarden, T.: Intrinsic robustness of the price of anarchy. J. ACM **62**(5), 321–3242 (2015)
43. Roughgarden, T., Tardos, E.: How bad is selfish routing? J. ACM **49**(2), 236–259 (2002)
44. Suri, S., Tóth, C., Zhou, Y.: Selfish load balancing and atomic congestion games. Algorithmica **47**(1), 79–96 (2007)
45. Vinci, C.: Non-atomic one-round walks in congestion games. Theor. Comput. Sci. **764**, 61–79 (2019)
46. von Falkenhausen, P., Harks, T.: Optimal cost sharing for resource selection games. Math. Oper. Res. **38**(1), 184–208 (2013)

Equilibrium Inefficiency in Resource Buying Games with Load-Dependent Costs

Eirini Georgoulaki[1], Kostas Kollias[2(✉)], and Tami Tamir[3]

[1] University of Athens, Athens, Greece
eirini.geo.98@gmail.com
[2] Google Research, Mountain View, CA, USA
kostaskollias@google.com
[3] The Interdisciplinary Center, Herzliya, Israel
tami@idc.ac.il

Abstract. We study the inefficiency of equilibria of *resource buying games*, i.e., congestion games with *arbitrary cost-sharing*. Under arbitrary cost-sharing, players do not only declare the resources they will use, they also declare and submit a payment per resource. If the total payments on a resource cover its cost, the resource is activated, otherwise it remains unavailable to the players. Equilibrium existence and inefficiency under arbitrary cost-sharing is very well understood in certain models, such as network design games, where the joint cost of every resource (edge) is constant. In the case of congestion-dependent costs the understanding is not yet complete. For increasing per player cost functions, it is known that the optimal solution can be cast as a Nash equilibrium with the appropriate selection of payments and, hence, the price of stability is 1. In this work we initially focus on the price of anarchy for linear congestion games and prove that (in the direct generalization of the arbitrary cost-sharing model to congestion-dependent costs) it grows to infinity as the number of players grows large. However, we also show that with a natural modification to the cost-sharing model, the price of anarchy becomes 17/3. Turning our attention to strong Nash equilibria, we show that the worst-case inefficiency of the best and worst stable outcomes remains the same as for Nash equilibria, with the strong price of stability staying at 1 and the strong price of anarchy staying at 17/3. These results imply arbitrary cost-sharing is comparable to fair cost-sharing as it has a better best-case scenario and a (slightly) worse worst-case scenario. We also study models with restricted strategy sets (uniform matroid congestion games) and properties of best response dynamics with arbitrary cost-sharing.

1 Introduction

The class of *unweighted congestion games* [30] includes a large collection of applications where players compete for the use of resources with congestion-dependent

T. Harks and M. Klimm (Eds.): SAGT 2020, LNCS 12283, pp. 83–98, 2020.
https://doi.org/10.1007/978-3-030-57980-7_6

costs. Players are called to select the subsets of resources they will use, with each one of them having a *strategy set* of allowable such subset selections, and these decisions induce joint costs on the resources as dictated by their respective activation-cost functions. These joint costs are split among the users of resources in a way specified by the *cost-sharing policy* of the game. Players are expected to reach a stable outcome, such as a *pure Nash equilibrium* (NE), i.e., a solution robust against unilateral deviations, or a *strong Nash equilibrium* (SE), i.e., a solution robust against group deviations.[1] Metrics of interest from the perspective of the system designer include the *price of anarchy* (PoA), i.e., the worst case ratio of the total cost in a NE divided by the optimal cost, the *price of stability* (PoS), i.e., the worst case ratio of the total cost in the best NE divided by the optimal cost, and, similarly for SE, the *strong price of anarchy* (SPoA) and the *strong price of stability* (SPoS).

A large body of work studies the above setting under the *fair cost-sharing* policy which dictates that the joint cost of a resource is split equally among its users. Among the most fundamental classes of games in these studies one finds *network design games* where a player's strategy set consists of all possible paths in a graph between the player's designated endpoints and where the joint cost of every edge is a given constant, together with *linear congestion games*, where the joint cost of a resource is quadratic in the number of players using it (with the per-player cost being linear). For network design games, [6] shows that the PoA is equal to the number of players n, whereas the PoS is $\Theta(\log n)$. For linear congestion games, the PoA was shown to be $5/2$ in [9,16] and the PoS was shown to be $1 + \sqrt{3}/3$ in [11,15]. The SPoA was shown to also be $5/2$ in [14]. Generalizations of fair cost-sharing to weighted versions of congestion games are studied in [2,10,31].

Different kinds of cost-sharing policies for congestion games have also been studied. For example, [13,28] study various types of cost-sharing methods for network design games (such as the class of *weighted Shapley values*). In the congestion-dependent costs setting, which includes linear congestion games, [21] shows that fair cost-sharing minimizes the PoA among all cost-sharing policies that dictate player costs. Other literature that studies cost-sharing in congestion games and their weighted variants includes [19,22,34].

A different flavor of cost-sharing is given by *arbitrary cost-sharing*, which induces the class of *resource buying games*. In contrast to the methods described above, which prescribe player costs on a resource, arbitrary cost-sharing allows players to declare their cost shares. Specifically, each player picks the resources that he will use and submits a different payment for each one. If the total payments for a given resource cover the cost induced by its users, the resource is activated. In the opposite case, the resource remains inaccessible. This setting has been studied comprehensively for network design games. The work in [7] shows that a NE is not guaranteed to exist under arbitrary cost-sharing and that the PoA and PoS are large (almost equal to the number of players n). For

[1] In this paper, we consider *pure* strategies, as is common in the study of resource buying games.

the special case of a common destination node, the PoS is 1 and a NE is guaranteed to exist. An SPoA of $\Theta(\log n)$ is given in [17]. Other works that study arbitrary cost-sharing in network design games include [4,5,8,12,24,26]. Summarizing the results and comparing against fair cost-sharing, we observe that, in the general network design game, arbitrary costs-sharing loses the NE existence property and increases the PoS from logarithmic to linear. The situation improves for the common destination case where NE existence is maintained and the PoS improves to 1. Interestingly, a special case of network design games where even the PoA improves has been identified in the face of *real-time scheduling games* [33] in which the PoA drops from $\Theta(\sqrt{n})$ for fair cost-sharing to 2 for arbitrary cost-sharing [20].

Less is known about resource buying games with congestion-dependent costs. The work in [25] studies classes of games with non-decreasing per player costs. Most closely related to our setting is the work in [23], which shows that for increasing per player costs, a NE always exists and that, in fact, the optimal solution can be made to be a NE with appropriate payments, thus settling the PoS to be equal to 1. In this landscape, our work sets out to further investigate the inefficiency of equilibria in linear congestion games and compare against fair cost-sharing, which achieves PoA and SPoA 5/2, and PoS $1 + \sqrt{3}/3$.

Other related work deals with selfish and greedy load balancing. In natural dynamics, a player that joins a resource needs to cover the marginal change in the resource activation cost. This property also characterizes selfish load balancing instances [11,32]. Some of our results for matroid games generalize results from these papers.

1.1 Our Results

We initially study the obvious generalization of arbitrary cost-sharing to linear costs, in which players can submit any payment for a resource. We quickly observe that the PoA can be very large with a simple and somewhat uninteresting example, the details of which are given in Sect. 3. The example relies on having one player who is restricted to a single resource and multiple others who freeload on him instead of switching to empty resources. The restricted player ends up paying an astronomical cost of n^2 even though his marginal contribution to the joint cost is much smaller (specifically $2n-1$). Given that such an instance is unreasonable from a practical perspective (a player would not tolerate paying a very large part of a resource's cost that is clearly not caused by his presence so that others may use it), we seek a minor modification to the arbitrary cost-sharing model that leads to more meaningful results. We choose to impose the *marginal contribution constraint*, which suggests that no payment larger than the marginal contribution of a player on a resource is accepted.

The marginal contribution constraint can be interpreted in two ways. In the first one, the system designer closes down resources where the constraint is violated. This is a means for the designer to reduce the PoA in a manner that is instance-oblivious, i.e., requires only local observation of the players and payments on each resource as opposed to global knowledge of the full set of players and their available strategies. In the second interpretation, players suffer a large cost when they pay more than their marginal contribution due to the perception of being exploited and they themselves deviate away from such strategies.

Some of our results refer to uniform matroid resource buying game, in which every player j is associated with a set of feasible resources, and a demand ℓ_j. The strategy space of a player includes all the subsets of size ℓ_j of his feasible resources. A singleton game is a special case of matroid game with unit demands. A prominent example of uniform matroid games, is preemptive real-time scheduling, where every player corresponds to a job of a specific length that should be processed after its release- time and before its deadline. Since preemptions are allowed, any selection of ℓ_j slots in this interval can do.

Our results on arbitrary cost-sharing with the marginal contribution constraint in linear congestion games are as follows:

- In Sect. 3 we prove that the PoA and SPoA for general games are equal to 17/3 and the SPoS is equal to 1. We also show that a NE always exists.
- In Sect. 4 we prove that the PoA and SPoA for the special case of uniform matroid games reduces to a value between 4 and 4.055. We also show that in a singleton game, the minimal size of a coalition that may benefit from a coordinated deviation from a NE profile is 3, thus a NE is stable against any coordination of two players. We also show that while the worst-case PoA is equal to the worst-case PoA, there are games for which the PoA is higher than the SPoA.
- In Sect. 5 we discuss convergence properties of best-response dynamics, showing that convergence is typically faster than fair cost-sharing. For uniform matroid games we suggest a rule for selecting the deviating player in every BR step, such that convergence is guaranteed in time lower than the players' total demand.

2 Model

In the *linear resource buying games* that we study, there is a set of n unweighted players N and a set of m resources E. Each player $j \in N$ selects a set $p_j \subseteq E$ of resources that he will use, from a set of available such profiles $S_j \subseteq 2^E$.

A profile p_j, together with payments $\xi_{e,j}$ for each $e \in p_j$ constitute the strategy (p_j, ξ_j) of player j. We write p for the complete profile and $f_e(p)$ for the load on e in p, that is, the number of players using e in p. Every resource e induces an activation cost $c_e(f_e(p)) = f_e(p)^2$ (by convention, games with such costs are called *linear*, given that the *per player* cost on a resource is linear). The players have to cover this cost with their payments. We write (p, ξ) for the complete strategies of all players. Each player j seeks to minimize his cost which is:

$$cost_j(p, \xi) = \begin{cases} \sum_{e \in p_j} \xi_{e,j}, & \text{if all } e \in p_j \text{ are open} \\ +\infty, & \text{otherwise.} \end{cases} \quad (1)$$

A resource is open if its activation cost is paid for by the players, i.e., when:

$$\sum_{j:e\in p_j} \xi_{e,j} \geq c_e(f_e(p)).$$

Given the cost structure defined above, we may describe the solution concepts we study in this paper, namely the *pure Nash equilibrium* (NE) and the *strong Nash equilibrium* (SE). The NE condition enforces that no player should be able to unilaterally change his declared payments and/or set of resources and reduce his cost. Formally, in a NE (p, ξ), we have that for every player j and every strategy (p'_j, ξ'_j) of that player:

$$cost_j(p, \xi) \leq cost_j(\{p_{-j}, p'_j\}, \{\xi_{-j}, \xi'_j\}).$$

The SE condition enforces that there should not be a set of players Γ who can coordinate to change their strategies in a way such that every one of them reduces his cost. Formally, for a SE (p, ξ) we have that, for every subset of players Γ, and for every collection of strategies (p'_Γ, ξ'_Γ) of these players, there exists some player $j \in \Gamma$ such that:

$$cost_j(p, \xi) \leq cost_j(\{p_{-\Gamma}, p'_\Gamma\}, \{\xi_{-\Gamma}, \xi'_\Gamma\}).$$

Best-response dynamics (BRD) is a natural method by which players proceed toward a pure Nash equilibrium via a local search method. Player j is said to be *sub-optimal* in (p, ξ) if he can reduce his cost by a unilateral deviation, i.e., if there exists (p'_j, ξ'_j) such that

$$cost_j(\{p_{-j}, p'_j\}, \{\xi_{-j}, \xi'_j\}) < cost_j(p, \xi).$$

In BRD, as long as the strategy profile is not a NE, a sup-optimal player is chosen to deviate to a strategy that will minimize his cost, given the profile of others.

Some of our results refer to *uniform matroid games* in which every player j is associated with a subset $M_j \subseteq E$ of the resources, and a demand ℓ_j. The strategy space of player j includes all subsets of M_j of size ℓ_j. A *singleton* game is a special case in which $\forall j, \ell_j = 1$.

The cost of a profile (p, ξ) is the total players' cost, that is, $cost(p, \xi) = \sum_j cost_j(p, \xi)$. We denote by $OPT(G)$ the cost of a social optimal solution of a game G.

We conclude the section by defining our performance metrics. We quantify the inefficiency incurred due to self-interested behavior according to the *price of anarchy* (PoA) [29] and *price of stability* (PoS) [6] measures. The PoA is the worst-case inefficiency of a pure Nash equilibrium, while the PoS measures the best-case inefficiency of a pure Nash equilibrium. Formally, Let $\mathcal{G}$ be a family of games, and let G be a game in $\mathcal{G}$. Let $\Upsilon(G)$ be the set of pure Nash equilibria of the game G. Assume that $\Upsilon(G) \neq \emptyset$.

- The *price of anarchy* of G is the ratio between the *maximal* cost of a NE and the social optimum of G. That is, $\text{PoA}(G) = \max_{p\in\Upsilon(G)} cost(p)/OPT(G)$. The *price of anarchy* of the family of games $\mathcal{G}$ is $\text{PoA}(\mathcal{G}) = sup_{G\in\mathcal{G}} \text{PoA}(G)$.

- The *price of stability* of G is the ratio between the *minimal* cost of a NE and the social optimum of G. That is, $\text{PoS}(G) = \min_{p \in \Upsilon(G)} cost(p)/OPT(G)$. The *price of stability* of the family of games $\mathcal{G}$ is $\text{PoS}(\mathcal{G}) = sup_{G \in \mathcal{G}} \text{PoS}(G)$.

The *strong price of anarchy* (SPoA) and the *strong price of stability* (SPoS) introduced in [3] are defined similarly, where $\Upsilon(G)$ refers to the set of strong equilibria.

3 General Resource Buying Games

We begin by proving that the PoA of resource buying games with linear per player costs (i.e., resource activation costs $c(x) = x^2$) grows to infinity with the number of players.

Theorem 1. *The PoA of linear resource buying games is $\Omega(n)$.*

Proof. Consider a game with n players and n resources. Player 1 can only pick resource 1, i.e., $S_1 = \{\{1\}\}$. Every other player j can pick either resource 1 or resource j, i.e., $S_j = \{\{1\}, \{j\}\}$. Let p be the profile in which every player picks resource 1 and let the declared payments be $\xi_{1,1} = n^2 - (n-1)$ and $\xi_{1,j} = 1$ for every $j > 1$.

We observe that (p, ξ) is a NE as follows. The players receive service on resource 1, since they have covered its cost. Hence, no one has an incentive to increase the payment there. Decreasing the payment will result in losing service, hence there is no incentive for that either. Each player $j > 1$ also has the option to move to the alternative resource j. There j would have to pay 1, which offers no improvement in cost.

The cost of profile p is n^2. If we let p^* be the assignment in which every player j picks resource j, we get a cost of n. This proves the PoA is at least n. □

We observe that the high PoA is given by an unrealistic and uninteresting instance. It assumes that there is one player who will effectively suffer a very large cost so others can freeload on him. To correct for such degenerate outcomes, we impose the *marginal contribution constraint* which enforces that no player may declare a cost higher than $c_e(f_e(p)) - c_e(f_e(p) - 1)$, otherwise the resource remains unavailable. Note that this expression is the highest increase that the player can cause to the joint resource cost in any ordering of the resource's users and observe that such a constraint is implicit in the large literature of arbitrary cost-sharing in network design games: when an edge has unit cost, the largest increase a player can cause to the joint cost is 1 and that is precisely that max payment seen in a NE.

Through the rest of the paper and for simplicity of exposition, when analyzing equilibria we only consider outcomes in which all players are serviced and have a finite cost, i.e., outcomes on which the payments on every used resource equal its cost. While this is automatically true for the class of SE, some extra care needs to be taken to ensure it is also true for the class of NE or otherwise players can get stuck in low payment outcomes, e.g., when every player on a resource declares a 0 payment and unilateral increases cannot cover the resource cost without violating the marginal contribution constraint. We note that imposing a cost structure that addresses this is easy to achieve with a tweak on handling underpaid resources: Resources remain closed when a player is paying more than his marginal contribution but, when players underpay, each one is charged his bid plus twice the unpaid amount. Then each player has an incentive to increase his payment up to the marginal contribution until the resource costs are covered. Hence, w.l.o.g., we may consider only outcomes in which all players are serviced. We next present our results on the inefficiency of equilibria of arbitrary cost-sharing with the marginal contribution constraint.

Let $\mathcal{G}$ be the class of linear resource buying games with the marginal contribution constraint.

Theorem 2. *$SPoS(\mathcal{G}) = 1$ (and hence also $PoS(\mathcal{G}) = 1$) and a pure NE exists for every $G \in \mathcal{G}$.*

Proof. Let p^* be an optimal profile. Assume that the players are ordered arbitrarily and every player is added greedily to his strategy in p^* and pays the marginal cost. By Theorem 6.1 in [23] this payment scheme produces a NE. We show it is also a strong NE. Assume by contradiction that p^* is not a SE and let Γ be a coalition. Let p' be the profile after the deviation of Γ. Let E^+, E^- denote the set of resources whose load increases and decreases respectively in the deviation of Γ, and let Δ_e denote the corresponding gap in the load on e.

$$\sum_e f_e(p')^2 - \sum_e f_e(p^*)^2 =$$
$$\sum_{e \in E^+} ((f_e^* + \Delta_e)^2 - (f_e^*)^2) - \sum_{e \in E^-} ((f_e^*)^2 - (f_e^* - \Delta_e)^2) < 0.$$

To see why the last expression is negative, note that the first term is exactly the added cost on E^+ that the coalition Γ has to cover and the second term is the saved cost on E^-, which is at most what is saved by the coalition. Then, the fact that the expression is negative follows from the fact that the total cost of the coalition members strictly decreases. Hence, we get a contradiction to the optimality of p^*. □

We note that the above theorem easily generalizes to cost functions of the form $c(x) = x^d$ for $d > 1$. Now that we have shown that the nice properties of arbitrary cost-sharing from [23] still hold after our modification, we proceed to analyze the PoA and SPoA. We begin with a technical lemma that captures the well known *PoA smoothness* framework [31] in our model.

Lemma 1. *Suppose λ and $\mu < 1$ are positive real numbers such that for all integers $y \geq 1$ and $x \geq 0$ it holds that*

$$(2x + 1)y \leq \lambda y^2 + \mu x^2.$$

Then we get that the PoA of linear resource buying games with the marginal contribution constraint is at most $\lambda/(1 - \mu)$.

Proof. Let p_j^* be the set of resources used by player j in the optimal solution and let p_j be the set of resources used by player j in a worst case NE. Then, if $\xi_{e,j}$ is the payment of j for resource e, we get $\sum_j \sum_{e \in p_j} \xi_{e,j}$ for the total cost. Now consider the possible deviation of each player j, in which he uses the resources in p_j^* and pays:

- $\xi_{e,j}$ for each resource e that is both in p_j and p_j^*,
- $(f_e(p)+1)^2 - f_e(p)^2$ for each resource e that is in p_j^* but not in p_j (where $f_e(p)$ is the number of players on e in the NE p).

Note that this is a valid deviation from the NE, to the set of resources used by j in the optimal solution, since all resources in p_j^* will be paid for. By the equilibrium condition, each such cost is at least $\sum_{e \in E} \xi_{e,j}$, so we get:

$$\sum_{e \in E} f_e(p)^2 = \sum_j \sum_{e \in p_j} \xi_{e,j} \leq \sum_j \sum_{e \in p_j^* \cap p_j} \xi_{e,j} + \sum_{e \in p_j^* \setminus p_j} (f_e(p)+1)^2 - f_e(p)^2$$
$$\leq \sum_j \sum_{e \in p_j^*} (f_e(p)+1)^2 - f_e(p)^2 = \sum_j \sum_{e \in p_j^*} 2f_e(p) + 1.$$

Here the last inequality follows by our marginal contribution constraint. Now if we let f_e^* be the number of players using resource e in the optimal solution and C^* be the optimal cost, we get:

$$\sum_{e \in E} f_e(p)^2 \leq \sum_j \sum_{e \in p_j^*} 2f_e(p) + 1 \leq \sum_e \sum_{j: e \in p_j^*} 2f_e(p) + 1 \leq \sum_e (2f_e(p)+1) f_e(p^*)$$
$$\leq \sum_e \lambda f_e(p^*)^2 + \mu f_e(p)^2 = \lambda \sum_{e \in E} f_e(p)^2 + \mu \sum_{e \in E} f_e(p^*)^2.$$

The last inequality follows by the assumption in the statement of the lemma. Rearranging gives

$$\frac{\sum_{e \in E} f_e(p)^2}{\sum_{e \in E} f_e(p^*)^2} \leq \frac{\lambda}{1-\mu},$$

which proves the lemma.

Lemma 2. *$PoA(\mathcal{G}) \leq 17/3$.*

Proof. Here we simply prove that values $\lambda = 3.4$ and $\mu = 0.4$ satisfy Lemma 1. Hence we focus on inequality:

$$(2x+1)y \leq 3.4y^2 + 0.4x^2.$$

The inequality trivially holds for $y = 0$. We now focus on the case with $y = 1$. It is easy to check that the inequality holds for $x = 0, 1, 2, 3$. It is similarly easy to check that is holds for every real $x > 3$ since the derivative of $(0.4x^2 - 2x + 2.4)' = 0.8x - 2$ is positive for $x > 3$ and hence $0.4x^2 - (2x+1) + 3.4$ remains positive after $x = 3$.

We now switch to $y \geq 2$. Our main inequality can be rewritten as:

$$3.4y^2 + 0.4x^2 - 2xy - y \geq 0.$$

The value of x that minimizes the left hand side is $2.5y$. It is enough to satisfy the inequality with this value of x, which is:

$$3.4y^2 + 0.4 \cdot 2.5^2 y^2 - 5y^2 - y \geq 0 \Rightarrow 0.9y^2 - y \geq 0,$$

which is true since we have assumed $y \geq 2$. □

Lemma 3. $PoA(\mathcal{G}) \geq 17/3$.

Proof. We construct the following instance with 7 players and 21 resources. The players are numbered $1, 2, \ldots, 7$ and the resources are labeled $A_1, B_1, C_1, \ldots, A_7, B_7, C_7$. Each player j wants to use one of two possible sets of resources. The first one, which will be the one used by the player in the optimal solution, is $\{A_j, B_j, C_j\}$. The second one, which will used by the player in the NE, is $\{A_{j+1}, A_{j+2}, A_{j+3}, B_{j+1}, B_{j+2}, C_{j+1}, C_{j+2}\}$. When the indices overflow (by becoming larger than 7) we assume we go back to 1 for 8, back to 2 for 9, and back to 3 for 10. In our NE we assume the players equally split the cost on every resource. Observe that every type A resource will have 3 players on it in the NE, while every type B and type C resource will have two players. Since each player uses 3 As, 2 Bs, and 2 Cs, each player's cost is: $3 \cdot 3 + 2 \cdot 2 + 2 \cdot 2 = 17$.

If a player j wishes to move to his other possible set of resources, he will have to cover the marginal increase to the costs there. He will be increasing the cost on resource A_j from 9 to 16, the cost on resource B_j from 4 to 9, and similarly the cost on resource C_j from 4 to 9. These give a total marginal payment of 17, which proves our assignment is indeed a NE. It is not hard to check that the total cost in the NE is 119 whereas the total cost in the optimal solution is 21. Taking the ratio completes the proof. □

Theorem 3. $PoA(\mathcal{G}) = 17/3$.

Proof. Follows from Lemma 2 and Lemma 3. □

Lemma 4. *For every $\epsilon > 0$, there exists a game $G \in \mathcal{G}$, such that $SPoA(G) \geq 17/3 - \epsilon$.*

Proof. We construct a game G, with n players and $3(n+3) + 18$ resources as follows. The players are numbered $1, 2, \ldots, n$, the first $3(n+3)$ resources are labeled $A_1, B_1, C_1, A_2, B_2, C_2, \ldots, A_{n+3}, B_{n+3}, C_{n+3}$, and the final 18 resources are labeled $D_1, D_2, \ldots, D_{18}$. Each player j has two strategies:

- The first one, which will be used by the player in the optimal solution, is $p_j^* = \{A_j, B_j, C_j\}$, for $j \in \{4, 5, \ldots, n\}$, $p_1^* = \{A_1, B_1, C_1, D_1, D_2, \ldots, D_9\}$, $p_2^* = \{A_2, B_2, C_2, D_{10}, D_{11}, \ldots, D_{16}\}$, and $p_3^* = \{A_3, B_3, C_3, D_{17}, D_{18}\}$.
- The second one, $p_j' = \{A_{j+1}, A_{j+2}, A_{j+3}, B_{j+1}, B_{j+2}, C_{j+1}, C_{j+2}\}$, will be used by the player in the SE.

Consider the profile (p', ξ) in which every player selects his second strategy, and the players equally split the cost on every resource. We first show that p' is a NE by examining each player separately:

- Player 1 is alone in $\{A_2, B_2, C_2\}$, shares $\{A_3, B_3, C_3\}$ with player 2, and shares $\{A_4\}$ with players 2 and 3. So $cost_1(p', \xi) = 1 \cdot 3 + 2 \cdot 3 + 3 \cdot 1 = 12$. If he switches to p_1^*, he would also be paying 12, since he would be alone on all 12 resources.
- Player 2 shares $\{A_3, B_3, C_3\}$ with player 1, two other A-resources with two other players, one B-resource with one other player and one C-resource with one other player. So $cost_2(p', \xi) = 2 \cdot 3 + 3 \cdot 2 + 2 \cdot 1 + 2 \cdot 1 = 16$. If he switches to p_2^*, he would be paying $(2^2 - 1) \cdot 3 + 1 \cdot 7 = 16$.
- Player 3 shares three A-resources with two other players, two B-resources and two C-resources with one other player. So $cost_3(p', \xi) = 3 \cdot 3 + 2 \cdot 2 + 2 \cdot 2 = 17$. If he switches to p_3^*, he would be paying $(3^2 - 2^2) \cdot 3 + 2 = 17$.
- A player $j \in \{4, 5, \ldots, n-2\}$ shares three A-resources with two other players, two B-resources and two C-resources with one other player. So $cost_j(p', \xi) = 3 \cdot 3 + 2 \cdot 2 + 2 \cdot 2 = 17$. A player that switches to p_j^*, he would be paying $(4^2 - 3^2) + (3^2 - 2^2) \cdot 2 = 17$.
- Player $n - 1$ shares two A-resources with two other players, one A-resource, two B-resources and two C-resources with one other player. So $cost_{n-1}(p', \xi) = 3 \cdot 2 + 2 \cdot 1 + 2 \cdot 2 + 2 \cdot 2 = 16$. If he switches to p_{n-1}^*, he would be paying $(4^2 - 3^2) + (3^2 - 2^2) \cdot 2 = 17$.
- Player n shares one A-resource with two other players, one A-resource, one B-resource and one C-resource with one other player, and is alone on one A-resource, one B-resource and one C-resource. So $cost_n(p', \xi) = 3 \cdot 1 + 2 \cdot 1 + 2 \cdot 1 + 2 \cdot 1 + 1 \cdot 1 + 1 \cdot 1 + 1 \cdot 1 = 12$. If he switches to p_n^*, he would be paying $(4^2 - 3^2) + (3^2 - 2^2) \cdot 2 = 17$.

To see that (p', ξ) is a SE, suppose that an arbitrary subset of the players switch to their other strategy. Then the lowest-numbered player j in the subset experiences no improvement, since the resources that j would occupy if he switches are still occupied by the same players as in p'.

From the latter and from the fact that no resources are being shared in the optimal solution, we get that

$$SPoA(G) \geq \frac{17n - 12}{3n + 18} \rightarrow \frac{17}{3}, \text{ as } n \rightarrow \infty$$

This means that for every $\epsilon > 0$, the exists a game such that $SPoA(G) \geq 17/3 - \epsilon$, which completes the proof. □

Theorem 4. $SPoA(\mathcal{G}) = 17/3$.

Proof. Follows from Lemma 2 and Lemma 4. □

4 Uniform Matroid Resource Buying Games

In a uniform matroid resource buying game, every player j is associated with a subset $M_j \subseteq E$ of the resources, and a demand ℓ_j. The strategy space of player j includes all subsets of M_j of size ℓ_j. A singleton game is a special case of matroid games in which $\forall j, \ell_j = 1$. Let $\mathcal{G}_{UM}$ be the class of resource buying games with the marginal contribution constraint, and uniform matroid strategies.

For any resource buying game instance, a possible algorithm for computing a NE is to index the players arbitrarily, and then assign them in that order. If a player is added to a resource e with current load f_e, then $\xi_{e,j} = 2f_e + 1$. Every player selects a strategy that minimizes his total payment. It is easy to verify that the resulting profile is a NE, as the load on the resources can only increase after a player j is assigned, and every profitable deviation of j contradicts his greedy choice at the assignment time.

By the above, for the case of singleton games, we conclude that the PoA of our game is at least the approximation ratio of greedy load balancing with the objective of minimizing the loads' L_2-norm. This problem is studied in [11,32]. In fact, the lower bound of $4-\epsilon$ presented in [11], can be adapted for our game. We present a simpler lower bound that exploits the payment distribution flexibility in our game, and also handles coordinated deviations.

Theorem 5. *For every $\epsilon > 0$, there exists a game $G \in \mathcal{G}_{UM}$, with $SPoA(G) \geq 4 - \epsilon$. The lower bound is achieved already by a singleton game.*

We defer the proof to our full version. For the upper bound, we show that the elegant analysis in [11] for bounding the approximation ratio of greedy load balancing, can also be used for our game. The main challenge is to show that the 2-neighborhood property they define for the load balancing problem, is also valid in games with arbitrary payment distribution and uniform matroid strategies.

Theorem 6. $PoA(\mathcal{G}_{UM}) \leq 2\sqrt{21}/3 + 1 \approx 4.055$.

Proof. Let G be a uniform matroid game achieving the highest PoA, let (p, ξ) be a NE of G, and let p^* be an optimal solution such that the ratio of the total cost in p to the total cost in p^* is maximal. As shown in the proof of Lemma 2, for every profile (p, ξ) where ξ obeys the marginal contribution constraint, the total cost of p is at most $\sum_e (2f_e + 1) f^*_e$. In addition, in every matroid game, the total load on the resources is fixed. Specifically, $\sum_{e \in E} f_e = \sum_{e \in E} f^*_e = \sum_j \ell_j$.

We can assume that the strategy space of player j is exactly $p_j \cup p^*_j$. That is, player j needs to select ℓ_j resources from $p_j \cup p^*_j$. If the set of feasible resources for j includes more resources, then they can be removed without hurting the stability of p.

Define a directed graph Δ as follows. The vertex set of Δ consists of one vertex for every resource. The edge set reflects the difference between p and p^* and is defined in the following way. For every player j, since G is a matroid game, $|p_j| = |p^*_j|$. Define a mapping $H_j : p^*_j \rightarrow p_j$. If $e \in p_j \cap p^*_j$ then $H_j(e) = e$, else, the mapping is arbitrary as long as it is 1-to-1 and onto. Player j contributes ℓ_j

edges to Δ, one edge for every pair $(e, H_j(e))$. Thus, a directed edge may be a self loop (e, e) if player j uses e in both p and p^*, or an edge (e_1, e_2) if j uses resource e_1 only in the optimal solution, and resource e_2 only in the NE. We say that resource e is of type f_e/f_e^*. Note that f_e and f_e^* correspond, respectively, to the in-degree and out-degree of e in the graph Δ. We show that for any game instance we can construct another game instance that has at least the same PoA and satisfies the following 2-neighborhood property, defined in [11]: *the incoming edge of any resource of type* $1/1$*originates from a resource of type* $0/1$. Formally (an extension of the definition in [11]),

Claim. Let j be a resource for which $f_e = f_e^* = 1$. Assume $e \in p_a$ and $e \in p_b^*$. That is, player a is the only player using e in p, and player b is the only player using e in p^*. Then, (i) $a \neq b$, (ii) Let e' be the resource such that $H_a(e) = e'$, then $f_{e'} = 0$ and $f_{e'}^* = 1$.

Proof. (i) Assume by contradiction that $a = b$, that is, the same player is the only player that uses e in both profiles. Construct a new game instance by excluding resource e and reducing by one the demand of player a. If $\ell_j = 1$ in G, then player a is totally excluded from G. In the resulting game, both the optimal cost and the cost of p are decreased by 1, and, therefore, the PoA increases.

(ii) Given that $a \neq b$, the two players define a path $\langle e' - e - e'' \rangle$ in Δ, such that $H_a(e') = e$ and $H_b(e) = e''$. That is, $e' \in p_a^*, e \in p_a, e \in p_b^*$ and $e'' \in p_b$. Assume by contradiction that $f_{e'} > 0$, that is, e' is not empty in p. Construct a new game instance by (i) excluding resource e and reducing by one ℓ_a and ℓ_b. If a demand is reduced to 0, then exclude the corresponding player from G, (ii) introducing a new player c whose demand is $\ell_c = 1$ and whose strategy space is $\{e', e''\}$. Set $f_c^* = \{e'\}$ and $f_c = \{e''\}$. Also, set $\xi_{e'',c} = \xi_{e'',b}$, that is, the payment of the new player for using e'' is exactly the payment of b for using e''. We show that the resulting game has a higher PoA, by showing that the resulting profile is a NE. Since p is a NE, we know that b cannot benefit from replacing e'' by e. Since $f_e = 1$, this implies that $\xi_{e'',b} \leq 3$. In the new instance, the cost of c for using e'' is therefore at most 3. Our assumption that $f_{e'} > 0$ implies that replacing e'' by e' would result in cost at least 3 for c, thus, it is not beneficial, and the strategy of c is stable. No other player can benefit from changing his strategy, since all the loads are as in p. In the modified game, both the optimal cost and the cost of p are decreased by 1, and therefore the PoA increases.

We turn to show that $f_{e'}^* = 1$, that is, a is the only player that uses e' in p^*. Assume by contradiction that $f_{e'}^* > 1$. Thus, some other player, c, is together with a on e' in p^*. Construct a new game by introducing a new resource that is only feasible for a. In an optimal solution of the modified instance, a is the only player on the new resource, thus, the optimal cost is reduced by at least 3. On the other hand, p remains a NE, as also in p, a is using a resource with load 1. Again, we get a modified game with an increased PoA.

Summing up, the following three conditions hold: $\sum_e f_e^2 \leq \sum_e (2f_e + 1) f_e^*$, $\sum_{e \in E} f_e = \sum_{e \in E} f_e^*$, and the 2-neighborhood property is valid. Therefore, we have the three building blocks required for the analysis of [11] to get the PoA bound. □

In our full version, we also prove the following results on coordinated deviations in uniform matroid games.

Theorem 7. *The minimal size of a coalition that has a profitable deviation from a NE profile of a singleton game is* 3.

Theorem 8. *There exists a symmetric singleton game G and a NE profile (p, ξ) such that (p, ξ) is a NE, and there exists a set of* 3 *players that have a profitable coordinated deviation from (p, ξ).*

5 Convergence Rate of BRD

Given a strategy profile, the best-response (BR) of player j is the set of strategies that minimize his cost after fixing the strategies and payments of all other players. A player is sub-optimal in (p, ξ) if his current strategy is not in his BR set. If no player is sub-optimal, then (p, ξ) is a NE.

We analyze the convergence time of BRD by letting a player deviate to his BR and updating the payments of resources that the player departs from. We assume that these updates are not counted as a change of strategy and that only a change in the set of resources selected by a player counts as such. This fits analysis of BR convergence in other models – in which players costs are modified when other players change strategies.

It is well known that BRD converges to a NE in congestion games with fair cost-sharing. However, the BR-sequence may be exponentially long [1,18]. We first bound the number of steps in every BR sequence in a general resource buying game. The bound we achieve is identical to the bound for singleton games with fair cost-sharing [27].

Theorem 9. *For every resource buying game, and every initial profile (p_0, ξ_0), every BRD staring from (p_0, ξ_0) converges to a NE within less than $n^2 m$ steps.*

We defer the proof to our full version. For uniform matroid games, we suggest a rule for selecting in every BR step the deviating sub-optimal player, such that, if the initial profile is based on fair cost-sharing, then BRD converges within less than $\sum_j \ell_j$ steps. In particular, for singleton games, we get a bound of less than n steps on the convergence time, starting from an arbitrary profile with fair cost-sharing.

The intuition is that, unlike regular congestion games, the payment of a player does not increase if other players join resources he is using. Thus, every migration sets an upper bound on the cost of a player in the final NE.

Consider any BR sequence performed in a uniform matroid game. Denote by (p^t, ξ^t) the profile after t BR steps. In particular (p^0, ξ^0) is the initial profile. Observe that in a BR move of player j, he exchanges $k \leq \ell_j$ resources. Without loss of generality, every exchange is associated with a reduced cost. That is,

$$\text{For all } e_{out} \in p_j^t \setminus p_j^{t+1} \text{ and } e_{in} \in p_j^{t+1} \setminus p_j^t, \text{ it holds that } \xi_{e_{in},j}^{t+1} < \xi_{e_{out},j}^t. \quad (2)$$

This holds since otherwise, $p_j^{t+1} \cup \{e_{out}\} \setminus \{e_{in}\}$ is a better or not worse deviation.

For every profile (p, ξ), and every sub-optimal player j, let $z_j(p, \xi)$ be the minimal payment of j for a resource that he wishes to exchange in a BR move. Finally, let m_0 be the number of resources with positive load in the initial profile.

Theorem 10. *In uniform matroid games, if ξ_0 is based on fair cost-sharing, and in every BR step a sub-optimal player with minimal $z_j(p, \xi)$ is activated, then a NE is reached after at most $\sum_j \ell_j - m_0$ steps.*

We defer the proof to our full version. We note that for every n, m_0, the above analysis is tight for a symmetric singleton game with n resources. If in the initial profile the players are assigned on $m_0 < n$ resources, then in turn, each activated player will select an empty resource.

References

1. Ackermann, H., Röglin, H., Vöcking, B.: On the impact of combinatorial structure on congestion games. J. ACM **55**(6), 251–2522 (2008)
2. Aland, S., Dumrauf, D., Gairing, M., Monien, B., Schoppmann, F.: Exact price of anarchy for polynomial congestion games. SIAM J. Comput. **40**(5), 1211–1233 (2011)
3. Andelman, N., Feldman, M., Mansour, Y.: Strong price of anarchy. Games Econ. Behav. **65**(2), 289–317 (2009)
4. Anshelevich, E., Caskurlu, B.: Exact and approximate equilibria for optimal group network formation. Theor. Comput. Sci. **412**(39), 5298–5314 (2011)
5. Anshelevich, E., Caskurlu, B.: Price of stability in survivable network design. Theory Comput. Syst. **49**(1), 98–138 (2011)
6. Anshelevich, E., Dasgupta, A., Kleinberg, J.M., Tardos, É., Wexler, T., Roughgarden, T.: The price of stability for network design with fair cost allocation. SIAM J. Comput. **38**(4), 1602–1623 (2008)
7. Anshelevich, E., Dasgupta, A., Tardos, É., Wexler, T.: Near-optimal network design with selfish agents. Theory Comput. **4**(1), 77–109 (2008)
8. Anshelevich, E., Karagiozova, A.: Terminal backup, 3D matching, and covering cubic graphs. SIAM J. Comput. **40**(3), 678–708 (2011)
9. Awerbuch, B., Azar, Y., Epstein, A.: The price of routing unsplittable flow. In: Proceedings of the 37th Annual ACM Symposium on Theory of Computing, Baltimore, MD, USA, 22–24 May 2005, pp. 57–66 (2005)
10. Bhawalkar, K., Gairing, M., Roughgarden, T.: Weighted congestion games: the price of anarchy, universal worst-case examples, and tightness. ACM Trans. Econ. Comput., 2(4), 141–1423 (2014)
11. Caragiannis, I., Flammini, M., Kaklamanis, C., Kanellopoulos, P., Moscardelli, L.: Tight bounds for selfish and greedy load balancing. Algorithmica **61**(3), 606–637 (2011)
12. Cardinal, J., Hoefer, M.: Non-cooperative facility location and covering games. Theor. Comput. Sci. **411**(16–18), 1855–1876 (2010)
13. Chen, H., Roughgarden, T., Valiant, G.: Designing network protocols for good equilibria. SIAM J. Comput. **39**(5), 1799–1832 (2010)

14. Chien, S., Sinclair, A.: Strong and pareto price of anarchy in congestion games. In: Albers, S., Marchetti-Spaccamela, A., Matias, Y., Nikoletseas, S., Thomas, W. (eds.) ICALP 2009. LNCS, vol. 5555, pp. 279–291. Springer, Heidelberg (2009). https://doi.org/10.1007/978-3-642-02927-1_24
15. Christodoulou, G., Koutsoupias, E.: On the price of anarchy and stability of correlated equilibria of linear congestion games. In: Algorithms - ESA 2005, 13th Annual European Symposium, Palma de Mallorca, Spain, 3–6 October 2005, Proceedings, pp. 59–70 (2005)
16. Christodoulou, G., Koutsoupias, E., Nanavati, A.: Coordination mechanisms. In: Díaz, J., Karhumäki, J., Lepistö, A., Sannella, D. (eds.) ICALP 2004. LNCS, vol. 3142, pp. 345–357. Springer, Heidelberg (2004). https://doi.org/10.1007/978-3-540-27836-8_31
17. Epstein, A., Feldman, M., Mansour, Y.: Strong equilibrium in cost sharing connection games. Games Econ. Behav. **67**(1), 51–68 (2009)
18. Fabrikant, A., Papadimitriou, C., Talwar, K.: The complexity of pure nash equilibria. In Proceedings 36th ACM Symposium on Theory of Computing, pp. 604–612 (2004)
19. Gairing, M., Kollias, K., Kotsialou, G.: Tight bounds for cost-sharing in weighted congestion games. In: Halldórsson, M.M., Iwama, K., Kobayashi, N., Speckmann, B. (eds.) ICALP 2015. LNCS, vol. 9135, pp. 626–637. Springer, Heidelberg (2015). https://doi.org/10.1007/978-3-662-47666-6_50
20. Georgoulaki, E., Kollias, K.: On the price of anarchy of cost-sharing in real-time scheduling systems. In: Caragiannis, I., Mirrokni, V., Nikolova, E. (eds.) WINE 2019. LNCS, vol. 11920, pp. 200–213. Springer, Cham (2019). https://doi.org/10.1007/978-3-030-35389-6_15
21. Gkatzelis, V., Kollias, K., Roughgarden, T.: Optimal cost-sharing in weighted congestion games. In: Liu, T.-Y., Qi, Q., Ye, Y. (eds.) WINE 2014. LNCS, vol. 8877, pp. 72–88. Springer, Cham (2014). https://doi.org/10.1007/978-3-319-13129-0_6
22. Gopalakrishnan, R., Marden, J.R., Wierman, A.: Potential games are Necessary to ensure pure nash equilibria in cost sharing games. Math. Oper. Res. **39**(4), 1252–1296 (2014)
23. Harks, T., Peis, B.: Resource buying games. Algorithmica **70**(3), 493–512 (2014)
24. Hoefer, M.: Non-cooperative tree creation. Algorithmica **53**(1), 104–131 (2009)
25. Hoefer, M.: Competitive cost sharing with economies of scale. Algorithmica **60**(4), 743–765 (2011)
26. Hoefer, M.: Strategic cooperation in cost sharing games. Int. J. Game Theory **42**(1), 29–53 (2013)
27. Ieong, S., McGrew, R., Nudelman, E., Shoham, Y., Sun, Q.: Fast and compact: a simple class of congestion games. In: Proceedings of the 20th National Conference on Artificial Intelligence, AAAI 2005, vol. 2 (2005)
28. Kollias, K., Roughgarden, T.: Restoring pure equilibria to weighted congestion games. ACM Trans. Econ. Comput. **3**(4), 211–2124 (2015)
29. Koutsoupias, E., Papadimitriou, C.H.: Worst-case equilibria. Comput. Sci. Rev. **3**(2), 65–69 (2009)
30. Rosenthal, R.W.: A class of games possessing pure-strategy nash equilibria. Int. J. Game Theory **2**(1), 65–67 (1973)
31. Roughgarden, T.: Intrinsic robustness of the price of anarchy. J. ACM **62**(5), 321–3242 (2015)
32. Suri, S., Toth, C., Zhou, Y.: Selfish load balancing and atomic congestion games. Algorithmica **47**, 79–96 (2007)

33. Tamir, T.: Cost-sharing games in real-time scheduling systems. In: Christodoulou, G., Harks, T. (eds.) WINE 2018. LNCS, vol. 11316, pp. 423–437. Springer, Cham (2018). https://doi.org/10.1007/978-3-030-04612-5_28
34. von Falkenhausen, P., Harks, T.: Optimal cost sharing for resource selection games. Math. Oper. Res. **38**(1), 184–208 (2013)

A Unifying Approximate Potential for Weighted Congestion Games

Yiannis Giannakopoulos(✉) and Diogo Poças

TU Munich, Munich, Germany
{yiannis.giannakopoulos,diogo.pocas}@tum.de

Abstract. We provide a unifying, black-box tool for establishing existence of approximate equilibria in weighted congestion games and, at the same time, bounding their Price of Stability. Our framework can handle resources with general costs—including, in particular, decreasing ones—and is formulated in terms of a set of parameters which are determined via elementary analytic properties of the cost functions.

We demonstrate the power of our tool by applying it to recover the recent result of Caragiannis and Fanelli [ICALP'19] for polynomial congestion games; improve upon the bounds for fair cost sharing games by Chen and Roughgarden [Theory Comput. Syst., 2009]; and derive new bounds for nondecreasing concave costs. An interesting feature of our framework is that it can be readily applied to *mixtures* of different families of cost functions; for example, we provide bounds for games whose resources are conical combinations of polynomial and concave costs.

In the core of our analysis lies the use of a unifying approximate potential function which is simple and general enough to be applicable to arbitrary congestion games, but at the same time powerful enough to produce state-of-the-art bounds across a range of different cost functions.

Keywords: Atomic congestion games · Potential games · Approximate equilibria · Price of stability

1 Introduction

Atomic congestion games are one of the most well-studied topics in *algorithmic game theory* [24,30]. In their most general form, players have weights and compete over a common set of resources; the cost of each resource is a function of the total weight of the players that end up using it. As a result, they can model a wide range of interesting applications including, e.g., network routing [29] and

Supported by the Alexander von Humboldt Foundation with funds from the German Federal Ministry of Education and Research (BMBF). Y. Giannakopoulos is an associated researcher with the Research Training Group GRK 2201 "Advanced Optimization in a Networked Economy", funded by the German Research Foundation (DFG). A full version of this paper is available at [16]: https://arxiv.org/abs/2005.10101.

T. Harks and M. Klimm (Eds.): SAGT 2020, LNCS 12283, pp. 99–113, 2020.
https://doi.org/10.1007/978-3-030-57980-7_7

load balancing [32], but also even cost-sharing games (via the use of decreasing cost functions) like fair network design [31].

An important special case is that of *unweighted* congestion games, where the costs depend only on the *number* of players that use each edge. In a seminal paper, Rosenthal [27] proved that unweighted congestion games always have (pure Nash) equilibria. A key tool in his derivation was the novel use of a *potential function*, which is able to capture the different players' deviations in a very elegant and concise way. Then, the desired equilibrium is derived as the minimizer of that function (over all feasible outcomes of the game). This technique can also be viewed as an *equilibrium refinement*, and has been a very influential idea in game theory [23]. It allows us not only to establish the existence of equilibria, but in many cases, this special potential-minimizer equilibrium has additional desired properties.

Of particular importance to us in this paper, is that it has been the de facto method for proving *Price of Stability (PoS)* bounds in congestion games (see, e.g., [24, Ch. 18, 19]). The PoS notion [1,10] captures the minimum approximation ratio of the social cost, among all equilibria, to the socially optimum outcome of the game (that might not be an equilibrium). In other words, the PoS is the best-case counterpart of the notorious Price of Anarchy (PoA) notion introduced by Koutsoupias and Papadimitriou [21,26]

Unfortunately, though, it is a well-known fact that general *weighted* congestion games do *not* always have equilibria [28] and thus, do not admit a potential function. To alleviate this, a line of work has focused on designing *approximate* potential functions (see, e.g., [4,6,8,9,18]): the minimizer of such functions is guaranteed to be an approximate equilibrium (as opposed to an *exact* one that is given by Rosenthal's potential in the unweighted case), while at the same time it can achieve a good approximation ratio to the optimal social cost (providing, thus, an upper bound for the approximate-equilibrium extension of the PoS notion). However, most of those prior works use different approximate potentials, designed specially for the particular cost-function model that each one studies.

Our goal in this paper is to provide a simple, high-level framework whose interface is agnostic to the underlying potential function technicalities and which can readily be instantiated for all resource costs at hand to derive meaningful bounds.

1.1 Related Work

Following the seminal work of [27,28], a long line of results has been devoted to the (non)existence of equilibria in weighted congestion games. [14,17,22] demonstrated that equilibria might not exist even in very simple classes of games, including network congestion games with quadratic cost functions and games where player weights are either 1 or 2. On the other hand, [14,19,25] showed that equilibria do exist in games with affine or exponential cost functions; [13,20] proved the same for singleton games (where players can only occupy single resources). Dunkel and Schulz [11] were able to extend the nonexistence instance of Fotakis et al. [14] to a hardness gadget, in order to show that, deciding whether a congestion game with step cost functions has an equilibrium, is a (strongly) NP-complete problem.

Regarding the existence of *approximate* equilibria in general weighted congestion games, [7] showed that games with n players always have n-approximate equilibria, and this guarantee is tight (up to logarithmic factors); they also proved that the corresponding decision problem, i.e., of the existence of $\tilde{\Theta}(n)$-approximate equilibria, is NP-complete.

A lot of work has been focused on the important special case of *polynomial* congestions games, parameterized by the maximum degree d of the cost functions. Although, due to [14] we already know that exact equilibria do not in general exist in such games, Caragiannis et al. [5] were the first to show that α-approximate equilibria do exist for $\alpha = d!$; this factor was later improved to $\alpha = d + 1$ [8,18] and $\alpha = d$ [4]. As a matter of fact, Caragiannis and Fanelli [4] provide an even more comprehensive result that, for any choice of a parameter $\delta \in [0, 1]$, simultaneously establishes the existence of $(d+\delta)$-approximate equilibria and gives an upper bound of $\frac{d+1}{\delta+1}$ on their PoS. They achieve this by designing an appropriate approximate potential function, tailored to polynomial costs. On the nonexistence front, [18] first gave instances of very simple, two-player polynomial congestion games that do not have α-approximate equilibria, for $\alpha \approx 1.153$. This was recently improved to $\alpha = \Omega(\frac{\sqrt{d}}{\log d})$ by Christodoulou et al. [7], who also established NP-hardness of the corresponding existence decision problem.

The work of Hansknecht et al. [18] is very relevant for our approach in this paper, since they also propose a "generic" approximate potential function that can, in principle, be applied to general cost functions. They instantiate it for polynomial costs to derive their aforementioned existence of $(d + 1)$-approximate equilibria. Additionally, they also state a result about the existence of $\frac{3}{2}$-approximate equilibria in games with nondecreasing concave costs; however, this proof in their paper is not complete. Furthermore, [18] focuses just on the existence of approximate equilibria, and thus it does not provide any PoS bounds.

Another well-studied class of congestion games is that of fair cost sharing, where each resource has a constant initial cost which is split equally among the players that use it. Thus, such games have *decreasing* cost functions. Finding the PoS for the special, undirected network version of such games is a notorious open problem in the field (see, e.g., [1–3,12,15]). Very relevant for us is the work of Chen and Roughgarden [6] who showed that general weighted fair cost sharing games always have α-approximate equilibria whose PoS is at most $O\left(\frac{\log W}{\alpha}\right)$, for any choice of parameter $\alpha = \Omega(\log w_{\max})$, where $w_{\max}$ is the maximum weight of any player and W the maximum possible load in any resource. They achieve this by designing a special approximate potential function, tailored to the specific form of the cost functions.

1.2 Our Results and Techniques

We propose a new approximate potential function (see (7)) for weighted congestion games with general cost functions. In particular, our potential can be instantiated beyond the standard model of polynomial cost functions and the common assumption of non-decreasing monotonicity. However, this potential is only used in the analysis part of our paper: we hide away its specific form by hard-coding it within the proof of a unifying tool (Theorem 1). Then, this tool can be used in a black-box way to readily derive both existence of approximate equilibria, and bounds on their PoS. This proof makes also use of a general, high-level lemma that can capture the essence of the *potential method* as a technique for deriving existence and PoS bounds for approximate equilibria (Lemma 1); we believe this might be of independent interest, since in future work it could be used for alternative potential functions, beyond our choice of (7) in this paper.

Our framework effectively works in two steps. Given a congestion game, first one has to determine how *good* its cost functions are with respect to two simple, analytic properties (Definition 2). Then, the resulting "goodness" parameters can be plugged straight into our master theorem (Theorem 1) to deduce the existence of an (α, β)-equilibrium; that is, an α-approximate (pure Nash) equilibrium whose social cost is at most a factor of β away from the optimum.

We demonstrate the power of our tool by applying it to recover and improve prior bounds on the existence of (α, β)-equilibria for well-studied classes of congestion games, as well as to derive novel results. The simplicity and the algebraic nature of our tool allows us to produce fine-grained bounds in the form of a parametric trade-off curve that describes the relation between the α and β parameters of the (α, β)-equilibrium; in other words, all our results give a *continuum* of existence bounds. Our bounds are summarized in Table 1.

Table 1. Our main results on the existence of (α, β)-equilibria for different cost models. For polynomials of degree d we recover the result of [4]. For fair costs our results improve those of [6] and for concave costs we extend those of [18]. For mixtures of different cost functions, namely polynomial and concave, our results are novel.

Cost functions	Previous work	Our results	
		General	Extreme points
Polynomials of degree $\leq d$	$\left(\lambda, \frac{d+1}{\lambda}\right)$, for $\lambda \in [d, d+1]$ [4]	$\left(\lambda, \frac{d+1}{\lambda}\right)$, for $\lambda \in [d, d+1]$ [Theorem 2]	$(d, 1+\frac{1}{d})$, $(d+1, 1)$
Concave	$\left(\frac{3}{2}, \infty\right)$ [18]	$\left(\lambda, \frac{\lambda}{\lambda-1}\right)$, for $\lambda \in \left[\frac{3}{2}, 2\right]$ [Theorem 3]	$\left(\frac{3}{2}, 3\right)$, $(2, 2)$
Polynomials + Concave	N/A	$\left(\lambda, 1+\frac{d+1}{\lambda}\right)$, for $\lambda \in [d, d+1]$ [Theorem 5]	$(d, 2+\frac{1}{d})$, $(d+1, 2)$
Fair cost sharing	$\left(\lambda, 1+\frac{2\log_2(1+W)}{\lambda}\right)$, for $\lambda = \Omega(\ln w_{\max})$ [6]	$\left(\Theta(\ln w_{\max})+\lambda, 1+\frac{\ln W}{\lambda}\right)$, for $\lambda \geq 1$ [Theorem 4]	$(\Theta(\ln w_{\max}), 1+\ln W)$, $(\Theta(\ln W), \Theta(1))(\Theta(\ln W), \Theta(1))$

More specifically, first (Theorem 2) we rederive the recent bounds of [4] for polynomial congestion games, in a more "clean", high-level way. Then (Theorem 4), we improve the α, β parameters on the (α, β)-equilibrium existence results of [6] for fair cost-sharing games (a more detailed comparison can be seen in Fig. 1). Furthermore, we derive new results for (nondecreasing) concave costs: we show that $(\lambda, \frac{\lambda}{\lambda-1})$-equilibria always exist, for all $\lambda \in [\frac{3}{2}, 2]$ (Theorem 3). The special corner case of a $(\frac{3}{2}, 3)$-equilibrium is compatible, thus, with the $\frac{3}{2}$-approximate equilibrium existence stated in [18].

Another interesting characteristic of our tool is its *modularity*: it can readily combine different cost functions to give bounds for more complex congestion games (see Definition 3). For example, we prove that games with cost functions that are conical combinations of d-degree polynomials and concave costs, always have $\left(\lambda, 1 + \frac{d+1}{\lambda}\right)$-equilibria, where λ ranges in $[d, d+1]$ (Theorem 5).

Finally, an added advantage of our black-box method is that it also results in arguably simpler and more streamlined proofs for the existence and PoS bounds.

Before concluding the overview of our results, we want to elaborate a bit more on the comparison to the potential approach of Hansknecht et al. [18]. Although [18] does not deal with PoS bounds, as far as existence of approximate equilibria is concerned, their paper is rather similar in principle to ours. They propose a general potential function which is based on a discrete interpretation of the cost function's integral, which corresponds to the first component of our potential in (7). We take a different approach by using directly the *actual* integral, and also adding an extra term that corresponds to a weighted average of the costs of the players' weights. In that way, we avoid a lot of the intricate technicalities that are involved with the discrete arguments (e.g., orderings of the weights) in [18], making the application of our potential (via our high-level tool of Theorem 1) more "tractable" for a wider range of cost functions.

Due to space constraints, all omitted proofs can be found at the full version of the paper [16].

2 Model and Notation

We use $\mathbb{R}_+$ to denote the set of nonnegative real numbers.

In a *(weighted) congestion game* $\mathcal{G}$ there are finite, nonempty sets of *players* N and *resources* E. Let $n = |N|$. Each player $i \in N$ has a *weight* $w_i \in \mathbb{R}_+$ and a *strategy set* $S_i \subseteq 2^E$. We use $w_{\min} = \min_{i \in N} w_i$ and $w_{\max} = \max_{i \in N} w_i$ for the minimum and maximum player weights, respectively, and for a subset of players $I \subseteq N$, we use $w_I = \sum_{i \in I} w_i$ to denote the sum of their weights. For the special case of $w_{\min} = w_{\max} = 1$, that is, if all weights are 1, we say that $\mathcal{G}$ is *unweighted*.

Associated with each resource $e \in E$ is a *cost function* $c_e : \mathbb{R}_+ \longrightarrow \mathbb{R}_+$. In general, we will make no extra assumptions on the cost functions. However, important special cases, that we will also study as applications of the main tool of our paper, include *polynomial congestion games* of degree d, for $d \geq 1$ integer, and *fair cost sharing games*. In the former, the cost functions are polynomials with nonnegative coefficients and degree at most d; in the latter, cost functions are (decreasing) of the form $c_e(x) = \frac{a_e}{x}$ where a_e is a positive real.

A (pure) *strategy profile* (or *outcome*) is a choice of strategies $\boldsymbol{s} = (s_1, s_2, ..., s_n) \in \boldsymbol{S} = S_1 \times \cdots \times S_n$. We use the standard game-theoretic notation $\boldsymbol{s}_{-i} = (s_1, \ldots, s_{i-1}, s_{i+1}, \ldots s_n)$, $\boldsymbol{S}_{-i} = S_1 \times \cdots \times S_{i-1} \times S_{i+1} \times \cdots \times S_n$. In that way, for example, we can denote $\boldsymbol{s} = (s_i, \boldsymbol{s}_{-i})$. Given a profile $\boldsymbol{s} \in \boldsymbol{S}$, we define the *load* $x_e(\boldsymbol{s})$ of resource e as the total weight of players that use resource e at outcome $\boldsymbol{s}$, i.e., $x_e(\boldsymbol{s}) = w_{N_e(\boldsymbol{s})} = \sum_{i \in N: e \in s_i} w_i$, where $N_e(\boldsymbol{s})$ is the set of players using e. We will use $W = \sum_{i \in N} w_i$ to denote the maximum possible load of any resource. The *cost* of player i is defined by $C_i(\boldsymbol{s}) = \sum_{e \in s_i} c_e(x_e(\boldsymbol{s}))$. The *social cost* of a strategy profile $\boldsymbol{s}$ is the weighted sum of the players' costs

$$C(\boldsymbol{s}) = \sum_{i \in N} w_i \cdot C_i(\boldsymbol{s}) = \sum_{e \in E} x_e(\boldsymbol{s}) \cdot c_e(x_e(\boldsymbol{s})).$$

We use $\mathrm{OPT}(\mathcal{G}) = \min_{\boldsymbol{s} \in \boldsymbol{S}} C(\boldsymbol{s})$ to denote the *optimum social cost* over all outcomes.

An outcome $\boldsymbol{s}$ is an *α-approximate* (pure Nash) *equilibrium*, for $\alpha \geq 1$, if

$$C_i(\boldsymbol{s}) \leq \alpha \cdot C_i(s_i', \boldsymbol{s}_{-i}) \qquad \text{for all } i \in N,\ s_i' \in S_i \tag{1}$$

That is, no player can unilaterally deviate from $\boldsymbol{s}$ and improve her cost by more than a factor of α. Notice that for the special case of $\alpha = 1$ we get the definition of the standard, *exact* pure Nash equilibrium. We denote the set of all α-equilibria of $\mathcal{G}$ by $\mathrm{NE}_\alpha(\mathcal{G})$ Then, the *α-approximate Price of Stability (α-PoS)* of $\mathcal{G}$ is the social cost of the best-case Nash equilibrium over the optimum social cost:

$$\mathrm{PoS}_\alpha(\mathcal{G}) = \min_{\boldsymbol{s} \in \mathrm{NE}_\alpha(\mathcal{G})} \frac{C(\boldsymbol{s})}{\mathrm{OPT}(\mathcal{G})}. \tag{2}$$

For $\alpha = 1$ we get the standard definition of the Price of Stability (PoS) for exact equilibria [1]. We combine the notions of an approximate equilibrium with approximating the optimum social cost in the following definition:

Definition 1 ((α, β)-equilibrium). *Fix a congestion game $\mathcal{G}$. A strategy profile $\boldsymbol{s}$ is an (α, β)-equilibrium if it is an α-approximate equilibrium of $\mathcal{G}$ (see (1)) and its social cost is at most β times the optimal cost of $\mathcal{G}$, i.e., $C(\boldsymbol{s}) \leq \beta \cdot \mathrm{OPT}(\mathcal{G})$.*

Notice that if a game has an (α, β)-equilibrium then, due to (2), its α-PoS is at most β.

2.1 Equivalent Cost Functions

It is not difficult to see that, in any weighted congestion game, the cost functions of each resource are actually evaluated on finitely many points: although our model assumes c_e to be defined over the entire $\mathbb{R}_+$, its values outside the domain $\{x_e(\boldsymbol{s}) \mid \boldsymbol{s} \in \boldsymbol{S}\}$ are irrelevant. In particular, this domain is included within the set of different sums of weights

$$\mathcal{W} = \left\{ \sum_{i \in N} y_i \cdot w_i \;\middle|\; y_i \in \{0, 1\},\ i \in N \right\}.$$

This means that one only needs to define costs on at most $|\mathcal{W}| = 2^n$ different values: any two games whose costs coincide on $\mathcal{W}$ are equivalent.

However, it is still convenient to treat our costs as functions over $\mathbb{R}_+$. First, because this allows for simple and succinct representations. But of particular importance to us, is also the fact that our main tool (Theorem 1) can be applied to all *integrable* cost functions (so that Definition 2 can be utilized). From the above discussion, it should be obvious that any congestion game has (infinitely) many equivalent representations, that is, different extensions from $\mathcal{W}$ to $\mathbb{R}_+$. Such an extension can always be done in a way that c_e is an integrable function (since $\mathcal{W}$ is finite).

It is interesting to point out here that different representations can potentially give different existence and PoS bounds via our tool. Although we do not deal with this feature for most of the paper, it is important for our fair cost sharing results (Sect. 4.3); since function $x \mapsto 1/x$ is not integrable over the interval $[0, w_{\min})$ (and as a matter of fact, not even defined on $x = 0$) we have the freedom, according to the discussion above, to redefine it in any way we want on $[0, w_{\min})$, so that it is a well-defined, integrable function over $\mathbb{R}_+$.

3 The Main Tool

In this section we present our framework for establishing existence of (α, β)-equilibria in weighted congestion games with general cost functions. We begin with the following lemma, that tries to distil and abstract the potential method technique in congestion games. Specialized or restricted forms of it have essentially been used, even if not explicitly stated, in multiple works in the past (see, e.g., [4,6,18]). It can be seen as a more fine-grained version of [8, Lemma 4.1], although some extra care is needed to adapt it to the more abstract setting of our paper and utilize its full power.

Lemma 1 (Potential Method). *Fix a congestion game. Assume that, for each resource e, there exist positive reals $\alpha_{1,e}, \alpha_{2,e}, \beta_{1,e}, \beta_{2,e}$, and a function $\phi_e : 2^N \longrightarrow \mathbb{R}$ such that $\phi_e(\emptyset) = 0$ and*

$$\alpha_{1,e} \leq \frac{\phi_e(I \cup \{i\}) - \phi_e(I)}{w_i \cdot c_e(w_I + w_i)} \leq \alpha_{2,e} \qquad \textit{for all } i \in N,\ I \subseteq N \setminus \{i\}; \tag{3}$$

$$\beta_{1,e} \leq \frac{\phi_e(I)}{w_I \cdot c_e(w_I)} \leq \beta_{2,e} \qquad \textit{for all } \emptyset \neq I \subseteq N. \tag{4}$$

Then the game has an (α, β)-equilibrium with

$$\alpha = \max_{e \in E} \frac{\alpha_{2,e}}{\alpha_{1,e}} \quad \textit{and} \quad \beta = \frac{\max_{e \in E} \beta_{2,e}/\alpha_{1,e}}{\min_{e \in E} \beta_{1,e}/\alpha_{1,e}}.$$

We continue with defining a critical notion that will act as the medium to utilize our main black-box tool in Theorem 1. It involves a set of parameters, that determine how "well" a given cost function behaves with respect to two specific, simple analytic properties (namely (5) and (6)). These properties can be interpreted as bounds on the average of the cost function over continuous intervals.

Definition 2 (Good Cost Functions). *Fix a congestion game $\mathcal{G}$. A function $c : \mathbb{R}_+ \longrightarrow \mathbb{R}_+$ will be called* $(\alpha_1, \alpha_2, \beta_1, \beta_2)$-good *(with respect to $\mathcal{G}$), for $\alpha_1, \alpha_2, \beta_1, \beta_2 > 0$, if there exists a nonnegative constant ξ such that, for all $x \in \{0\} \cup [w_{\min}, W]$, $w \in [w_{\min}, w_{\max}]$:*

$$\alpha_1 \cdot c(x+w) - \xi \cdot c(w) \le \frac{1}{w} \int_x^{x+w} c(t)\,dt \le \alpha_2 \cdot c(x+w) - \xi \cdot c(w) \tag{5}$$

and for all $x \in [w_{\min}, W]$:

$$\beta_1 \cdot c(x) - \xi \cdot c_{\min}(x) \le \frac{1}{x} \int_0^{x} c(t)\,dt \le \beta_2 \cdot c(x) - \xi \cdot c_{\max}(x), \tag{6}$$

where $c_{\min}(x) = \min_{y \in [w_{\min}, x]} c(y)$, $c_{\max}(x) = \max_{y \in [w_{\min}, x]} c(y)$.

Definition 3 (Good Games). *A congestion game will be called $\{(\alpha_{1,j}, \alpha_{2,j}, \beta_{1,j}, \beta_{2,j})\}_{j \in J}$-good if any cost function is a conical combination of such good functions. Formally, for any $e \in E$ there exists a nonempty $J_e \subseteq J$ and nonnegative constants $\{\lambda_{e,j}\}_{j \in J_e}$, such that*

$$c_e(t) = \sum_{j \in J_e} \lambda_{e,j} c_j(t)$$

where, for all $j \in J$, c_j is a $(\alpha_{1,j}, \alpha_{2,j}, \beta_{1,j}, \beta_{2,j})$-good function (see Definition 2).

Remark 1. Notice that an important special case of Definition 3 is when $J = E$, $J_e = \{e\}$, and $\lambda_{e,e} = 1$, meaning that the actual cost functions of the game are good themselves. As a matter of fact, it is not hard to see that any good game $\mathcal{G}$ can be transformed to a strategically equivalent one $\mathcal{G}'$ that has that property. First, replace each resource e of $\mathcal{G}$ with a gadget of "parallel" resources $\{(e, j) | j \in J_e\}$, each having a cost function of $c_{(e,j)}(t) = \lambda_{e,j} c_j(t)$; this results in a strategically equivalent game $\mathcal{G}'$ with resources $E' = \{(e, j) | e \in E, j \in J_e\}$. Next, just observe that Definition 2 is invariant under nonnegative scalar multiplication: since functions c_j satisfy conditions (5) and (6), so do functions $\lambda_{e,j} \cdot c_j$ that are exactly the cost functions of the new game $\mathcal{G}'$.

Remark 2 (Increasing Good Functions). If a cost function is nondecreasing, then (6) can be replaced by the (weaker, sufficient) condition:

$$\beta_1 c(x) \le \frac{1}{x} \int_0^{x} c(t)\,dt \le (\beta_2 - \xi) c(x), \tag{6'}$$

since $0 \le c(y) \le c(x)$ for any $y \in [w_{\min}, x]$.

Now we are ready to state our main tool. This is essentially the interface of our entire framework: under the hood it uses a specific potential function form (see (7)), but its statement involves only the goodness parameters of the cost functions, as defined above. In that way, one can readily derive meaningful bounds about the existence of (α, β)-equilibria in a black-box way, just by studying the simple analytic properties given in (2) and the plugging the parameters in the theorem below:

Theorem 1. *Any* $\{(\alpha_{1,j}, \alpha_{2,j}, \beta_{1,j}, \beta_{2,j})\}_{j \in J}$*-good congestion game has an* (α, β)*-equilibrium with*

$$\alpha = \max_{j \in J} \frac{\alpha_{2,j}}{\alpha_{1,j}} \qquad \textit{and} \qquad \beta = \frac{\max_{j \in J} \beta_{2,j}/\alpha_{1,j}}{\min_{j \in J} \beta_{1,j}/\alpha_{1,j}}.$$

Proof. First notice that, by Remark 1, it is without loss to assume that $J = E$ and that any cost function c_e, $e \in E$, is $(\alpha_{1,e}, \alpha_{2,e}, \beta_{1,e}, \beta_{2,e})$-good. Denote by ξ_e (a choice of) the parameter ξ for which resource e satisfies Definition 2.

We will then show that functions

$$\phi_e(I) = \int_0^{w_I} c_e(t)\, dt + \xi_e \sum_{i \in I} w_i c_e(w_i) \tag{7}$$

satisfy the conditions of Lemma 1,

Fix some resource $e \in E$, a player i and a subset $I \subseteq N \setminus \{i\}$ of remaining players. For simplicity, from now on we drop the e subscripts and also denote $w = w_i$ and $x = w_I$. Then,

$$\begin{aligned}
\phi(I \cup \{i\}) - \phi(I) &= \int_0^{x+w} c_e(t)\, dt - \int_0^{x} c_e(t)\, dt \\
&\quad + \xi_e \left(\sum_{j \in I} w_j c_e(w_j) - \sum_{j \in I \cup \{i\}} w_j c_e(w_j) \right) \\
&= \int_x^{x+w} c(t)\, dt + \xi w c(w).
\end{aligned}$$

So, by deploying (5), it is not difficult to see that

$$\alpha_1 c(x+w) \leq \frac{1}{w} \left[\phi(I \cup \{i\}) - \phi(I) \right] \leq \alpha_2 c(x+w),$$

and thus condition (3) of Lemma 1 is indeed satisfied.

Next, observe that since $w_j \in [w_{\min}, w_{\max}]$ for all $j \in I$, and $x = \sum_{j \in I} w_j$, we have the bounds

$$c_{\min}(x) \leq \min_{j \in I} c(w_j) \leq \frac{1}{x} \sum_{j \in I} w_j c(w_j) \leq \max_{j \in I} c(w_j) \leq c_{\max}(x), \tag{8}$$

where the first and the last inequalities hold due to the fact that $\{w_j \mid j \in I\} \subseteq [w_{\min}, x]$. Assuming $I \neq \emptyset$, we have that $x \in [w_{\min}, W]$ and so we can use (8) and (6) to bound $\frac{1}{x}\phi(I)$ from below and above by:

$$\beta_1 c(x) \leq \frac{1}{x}\phi(I) = \frac{1}{x}\int_0^x c(t)\,dt + \xi\frac{1}{x}\sum_{j\in I} w_j c(w_j) \leq \beta_2 c(x).$$

Thus, condition (4) of Lemma 1 is also satisfied.

4 Applications

In this section we present several applications of our black-box Theorem 1, that demonstrate both its power and simplicity. In accordance to the nature of that tool, they all share a common structure: first, we prove lemmas describing the right goodness parameters (according to Definition 2) for each special cost function of interest (see Lemmas 2 to 4); then, we plug them in Theorem 1 to derive our bounds (see Theorems 2 to 4).

4.1 Polynomial Costs

We start with polynomial cost functions, arguably the most studied setting in congestion games. We recover the result from Caragiannis and Fanelli [4] that, for polynomials of degree at most d with nonnegative coefficients, there exist $(d+\delta)$-approximate equilibria with social cost at most $\frac{d+1}{d+\delta}$ times the optimum, for any $\delta \in [0, 1]$. This is the currently best known guarantee of (α, β)-equilibria for polynomial cost functions. Let us begin by analysing the goodness parameters of each monomial.

Lemma 2. *Any monomial of degree $d \geq 1$ is $\left(\mu, 1, \frac{1}{d+1}, \mu\right)$-good, for any $\mu \in [\frac{1}{d+1}, \frac{1}{d}]$.*

For the special case of constant cost functions, i.e., 0-degree monomials, it is not difficult to get the following:

Lemma 3. *Any constant function is $(1, 1, 1, 1)$-good.*

Theorem 2. *Any weighted polynomial congestion game of degree $d \geq 1$ has an $(\lambda, \frac{d+1}{\lambda})$-equilibrium, for any $\lambda \in [d, d+1]$.*

The parameter λ quantifies the trade-off curve between the approximation guarantee on the existence of α-approximate equilibria and their PoS. At one extreme case $\lambda = d+1$, we get that $\alpha = d+1$ and $\beta = 1$; in other words, there always exist $(d+1)$-approximate equilibria with an optimal PoS of 1 (as a matter of fact, from [8] we already know that every social optimum is itself a $(d+1)$-approximate equilibrium). At the other extreme case $\lambda = d$, we get that

$$\alpha = d, \qquad \beta = \frac{d+1}{d} = 1 + \frac{1}{d};$$

in other words, there always exist d-approximate equilibria with PoS at most $1 + \frac{1}{d}$.

4.2 Concave Costs

We now look at nondecreasing concave cost functions. The best known result in this setting is due to Hansknecht et al. [18], who state that 3/2-approximate equilibria exist. However, the proof in their paper is not complete. Moreover, the PoS of the existing approximate equilibria is not discussed. In this section, not only we provide a simpler proof of this result, but we also extend it for a range of λ-approximate equilibria with $\lambda \in [3/2, 2]$, and for a guarantee on the PoS.

Lemma 4. *Any nondecreasing concave function is* $(\mu, \mu + \frac{1}{2}, \frac{1}{2}, \mu + \frac{1}{2})$*-good, for all* $\mu \in [\frac{1}{2}, 1]$.

Theorem 3. *Any weighted congestion game with nondecreasing concave cost functions has a* $(\lambda, \frac{\lambda}{\lambda-1})$*-equilibrium, for any* $\lambda \in [\frac{3}{2}, 2]$.

4.3 Fair Cost Sharing

In this section, we focus on the fair cost sharing model in which $c_e(x) = \frac{a_e}{x}$, where a_e is a positive, resource-dependent value. We assume that $w_{\min} = 1$; this is without loss, since we can just rescale the player weights. This setting was studied by Chen and Roughgarden [6]. Here we improve on their results (see Fig. 1), with a simpler proof.

We must notice that the function $x \mapsto a_e/x$ is not integrable in an interval starting at 0, and hence we cannot immediately apply our Definition 2. However, based on our discussion in Sect. 2.1 we can modify the game in order to overcome this. First, we assume for our analysis that $a_e = 1$ since any other choice of a_e can be seen as a trivial conical combination of the function $1/x$ (see Definition 3). Next, we change the cost function $c_e(x)$ to be constant and equal to λ in the interval $[0, 1)$, for some $\lambda \geq 1$.

Lemma 5. *Fix a weighted congestion game with* $w_{\min} = 1$*. For any* $\lambda \geq 1$*, the cost function*

$$c(x) = \begin{cases} 1/x, & x \geq 1, \\ \lambda, & 0 \leq x < 1 \end{cases}$$

is $(\alpha_1, \alpha_2, \beta_1, \beta_2)$*-good with*

$$\alpha_1 = 1, \qquad \alpha_2 = \max\left(\left(1 + \frac{1}{w_{\max}}\right)\ln(1 + w_{\max}), \ln(w_{\max}) + \lambda\right),$$

$$\beta_1 = \lambda, \qquad \beta_2 = \ln W + \lambda.$$

Theorem 4. *Fix a fair cost sharing game with unit minimum weight* ($w_{\min} = 1$)*, and let* $w_{\max}, W$ *be the maximum weight and the maximum total load. Then, for any* $\lambda \geq 1$*, our game has an* (α, β)*-equilibrium where*

$$\alpha = \max\left(\left(1 + \frac{1}{w_{\max}}\right)\ln(1 + w_{\max}), \ln(w_{\max}) + \lambda\right), \qquad \beta = 1 + \frac{\ln W}{\lambda}.$$

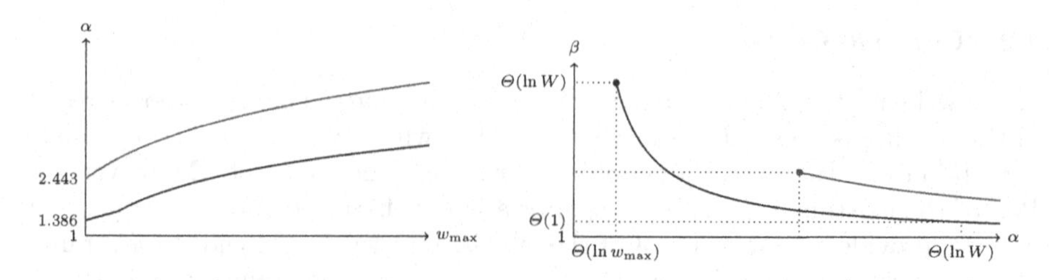

Fig. 1. Fair cost sharing games. Left: guarantee on the existence of α-approximate equilibria, as a function of $w_{\max}$, given by Theorem 4 (setting $\lambda = 1$). Right: trade-off curve for the existence of (α, β)-equilibria, given by Theorem 4; here we choose $w_{\max} = 3$, $W = 50$. For comparison, the previously best bounds [6, Theorem 5.1 and Lemma 5.3] are plotted in red, while our results are in blue. The fact that the blue line of the right plot starts earlier is a direct consequence of our results providing a strictly better (smaller) absolute existence guarantee α (see left plot).

Proof. Combining Theorem 1 with Lemma 5 we conclude that, for $\lambda \geq 1$, our game has an (α, β)-equilibrium with

$$\alpha = \frac{\alpha_2}{\alpha_1} = \max\left(\left(1 + \frac{1}{w_{\max}}\right) \ln(1 + w_{\max}), \ln(w_{\max}) + \lambda\right),$$
$$\beta = \frac{\beta_2/\alpha_1}{\beta_1/\alpha_1} = \frac{\ln W + \lambda}{\lambda} = 1 + \frac{\ln W}{\lambda}.$$

□

The parameter λ quantifies the trade-off curve between the approximation guarantee on equilibria and their price of stability. At one extreme case $\lambda = 1$, we get that

$$\alpha = \max\left(\left(1 + \frac{1}{w_{\max}}\right) \ln(1 + w_{\max}), \ln(w_{\max}) + 1\right) = \Theta(\ln w_{\max}),$$
$$\beta = 1 + \ln W.$$

In other words, there exist $\Theta(\ln w_{\max})$-approximate equilibria with price of stability $\Theta(\ln W)$. At the other extreme case $\lambda = \Theta(\ln W)$, we get that

$$\alpha = \max\left(\left(1 + \frac{1}{w_{\max}}\right) \ln(1 + w_{\max}), \ln(w_{\max}) + \Theta(\ln W)\right) = \Theta(\ln W),$$
$$\beta = 1 + \frac{\ln W}{\Theta(\ln W)} = \Theta(1)$$

in other words, there exist $\Theta(\ln W)$-approximate equilibria with *constant* price of stability $\Theta(1)$. The complete trade-off curve can be seen in Fig. 1 (right). We can also compare our results with the best known upper bounds. In [6, Lemma 5.3], it was shown that α-approximate equilibria exist for $\alpha \geq \log_2[e(1 + w_{\max})]$; and in [6, Theorem 5.1], it was shown that $\left(f, 1 + \frac{2\log_2(1+W)}{f}\right)$-equilibria exist for any $f \geq 2\log_2[e(1 + w_{\max})]$. As Fig. 1 shows, we improve on both results.

4.4 Mixtures of Cost Functions

A big advantage of our approach is that we can study the existence of (α, β)-equilibria for games that merge cost functions of two or more different types. For example, in this section we look at congestion games that have *both* concave costs and polynomial costs (as well as any conical combination). Interestingly, we show that this results in only a small increase in the PoS guarantee of Theorem 2, while the existence guarantee stays the same. For the following theorem we consider polynomials of degree at least 2, since affine functions are themselves concave and would be already captured by Theorem 3.

Theorem 5. *Any weighted congestion game with cost functions that are conical combinations of concave and polynomial costs of maximum degree $d \geq 2$ has an $(\lambda, 1 + \frac{d+1}{\lambda})$-equilibrium, for any $\lambda \in [d, d+1]$.*

Proof. Fix a maximum degree $d \geq 2$ for the polynomial costs and a parameter $\lambda \in [d, d+1]$. By defining $\mu = \frac{d+1}{2\lambda}$ we have that $\frac{1}{2} \leq \mu \leq \frac{1}{2}\left(1 + \frac{1}{d}\right) \leq 1$, and so by applying Lemma 4 we can derive that any concave cost is $(\mu, \mu + \frac{1}{2}, \frac{1}{2}, \mu + \frac{1}{2})$-good. Next, by Lemmas 2 and 3 we can derive that all monomials of degree $k = 0, \dots, d-1$ are $(\frac{1}{k+1}, 1, \frac{1}{k+1}, \frac{1}{k+1})$-good and the monomial of degree d is $(\frac{1}{\lambda}, 1, \frac{1}{d+1}, \frac{1}{\lambda})$-good.

Deploying our black-box tool Theorem 1 (and shortcutting some calculations that we have already performed in the proof of Theorem 2) we can guarantee the existence of an (α, β)-equilibrium with

$$\alpha = \max\left\{1 + \frac{1}{2\mu}, \lambda\right\} = \max\left\{1 + \frac{\lambda}{d+1}, \lambda\right\} = \lambda,$$

since $2 \leq d \leq \lambda \leq d+1$, and

$$\beta = \frac{\max\left\{\frac{\mu+1/2}{\mu}, 1\right\}}{\min\left\{\frac{1/2}{\mu}, \frac{\lambda}{d+1}\right\}} = \frac{1 + \frac{1}{2\mu}}{\min\left\{\frac{1}{2\mu}, \frac{\lambda}{d+1}\right\}} = 1 + 2\mu = 1 + \frac{d+1}{\lambda},$$

where for the third equality we used that, from the definition of μ, $\frac{1}{2\mu} = \frac{\lambda}{d+1}$. □

Acknowledgements. We thank Martin Gairing for interesting discussions.

References

1. Anshelevich, E., Dasgupta, A., Kleinberg, J., Tardos, É., Wexler, T., Roughgarden, T.: The price of stability for network design with fair cost allocation. SIAM J. Comput. **38**(4), 1602–1623 (2008). https://doi.org/10.1137/070680096
2. Bilò, V., Caragiannis, I., Fanelli, A., Monaco, G.: Improved lower bounds on the price of stability of undirected network design games. Theory Comput. Syst. **52**(4), 668–686 (2013). https://doi.org/10.1007/s00224-012-9411-6

3. Bilò, V., Flammini, M., Moscardelli, L.: The price of stability for undirected broadcast network design with fair cost allocation is constant. Games Econ. Behav. (2014). https://doi.org/10.1016/j.geb.2014.09.010
4. Caragiannis, I., Fanelli, A.: On approximate pure Nash equilibria in weighted congestion games with polynomial latencies. In: Proceedings of the 46th International Colloquium on Automata, Languages, and Programming (ICALP), pp. 133:1–133:12 (2019). https://doi.org/10.4230/LIPIcs.ICALP.2019.133
5. Caragiannis, I., Fanelli, A., Gravin, N., Skopalik, A.: Efficient computation of approximate pure Nash equilibria in congestion games. In: Proceedings of the 52nd IEEE Annual Symposium on Foundations of Computer Science (FOCS), pp. 532–541 (2011). https://doi.org/10.1109/focs.2011.50
6. Chen, H.L., Roughgarden, T.: Network design with weighted players. Theory Comput. Syst. **45**(2), 302–324 (2009). https://doi.org/10.1007/s00224-008-9128-8
7. Christodoulou, G., Gairing, M., Giannakopoulos, Y., Poças, D., Waldmann, C.: Existence and complexity of approximate equilibria in weighted congestion games. In: Proceedings of the 47th International Colloquium on Automata, Languages, and Programming (ICALP), pp. 32:1–32:18 (2020). https://doi.org/10.4230/LIPIcs.ICALP.2020.32
8. Christodoulou, G., Gairing, M., Giannakopoulos, Y., Spirakis, P.G.: The price of stability of weighted congestion games. SIAM J. Comput. **48**(5), 1544–1582 (2019). https://doi.org/10.1137/18M1207880
9. Christodoulou, G., Koutsoupias, E., Spirakis, P.G.: On the performance of approximate equilibria in congestion games. Algorithmica **61**(1), 116–140 (2011). https://doi.org/10.1007/s00453-010-9449-2
10. Correa, J.R., Schulz, A.S., Stier-Moses, N.E.: Selfish routing in capacitated networks. Math. Oper. Res. **29**(4), 961–976 (2004). https://doi.org/10.1287/moor.1040.0098
11. Dunkel, J., Schulz, A.S.: On the complexity of pure-strategy Nash equilibria in congestion and local-effect games. Math. Oper. Res. **33**(4), 851–868 (2008). https://doi.org/10.1287/moor.1080.0322
12. Fiat, A., Kaplan, H., Levy, M., Olonetsky, S., Shabo, R.: On the price of stability for designing undirected networks with fair cost allocations. In: Proceedings of the 33rd International ColloquiumAutomata, Languages and Programming (ICALP), pp. 608–618 (2006). https://doi.org/10.1007/11786986_53
13. Fotakis, D., Kontogiannis, S., Koutsoupias, E., Mavronicolas, M., Spirakis, P.: The structure and complexity of Nash equilibria for a selfish routing game. Theoret. Comput. Sci. **410**(36), 3305–3326 (2009). https://doi.org/10.1016/j.tcs.2008.01.004
14. Fotakis, D., Kontogiannis, S., Spirakis, P.: Selfish unsplittable flows. Theoret. Comput. Sci. **348**(2), 226–239 (2005). https://doi.org/10.1016/j.tcs.2005.09.024
15. Freeman, R., Haney, S., Panigrahi, D.: On the price of stability of undirected multicast games. In: Cai, Y., Vetta, A. (eds.) WINE 2016. LNCS, vol. 10123, pp. 354–368. Springer, Heidelberg (2016). https://doi.org/10.1007/978-3-662-54110-4_25
16. Giannakopoulos, Y., Poças, D.: A unifying approximate potential for weighted congestion games. CoRR abs/2005.10101, May 2020. https://arxiv.org/abs/2005.10101
17. Goemans, M., Mirrokni, V., Vetta, A.: Sink equilibria and convergence. In: Proceedings of the 46th Annual IEEE Symposium on Foundations of Computer Science (FOCS), pp. 142–151 (2005). https://doi.org/10.1109/SFCS.2005.68

18. Hansknecht, C., Klimm, M., Skopalik, A.: Approximate pure Nash equilibria in weighted congestion games. In: Proceedings of the 17th International Workshop on Approximation Algorithms for Combinatorial Optimization Problems (APPROX), pp. 242–257 (2014). https://doi.org/10.4230/LIPIcs.APPROX-RANDOM.2014.242
19. Harks, T., Klimm, M.: On the existence of pure Nash equilibria in weighted congestion games. Math. Oper. Res. **37**(3), 419–436 (2012). https://doi.org/10.1287/moor.1120.0543
20. Harks, T., Klimm, M., Möhring, R.H.: Strong equilibria in games with the lexicographical improvement property. Int. J. Game Theory **42**(2), 461–482 (2012). https://doi.org/10.1007/s00182-012-0322-1
21. Koutsoupias, E., Papadimitriou, C.: Worst-case equilibria. In: Meinel, C., Tison, S. (eds.) STACS 1999. LNCS, vol. 1563, pp. 404–413. Springer, Heidelberg (1999). https://doi.org/10.1007/3-540-49116-3_38
22. Libman, L., Orda, A.: Atomic resource sharing in noncooperative networks. Telecommun. Syst. **17**(4), 385–409 (2001). https://doi.org/10.1023/A:1016770831869
23. Monderer, D., Shapley, L.S.: Potential games. Games Econ. Behav. **14**(1), 124–143 (1996). https://doi.org/10.1006/game.1996.0044
24. Nisan, N., Roughgarden, T., Tardos, É., Vazirani, V. (eds.): Algorithmic Game Theory. Cambridge University Press (2007). https://doi.org/10.1017/CBO9780511800481
25. Panagopoulou, P.N., Spirakis, P.G.: Algorithms for pure Nash equilibria in weighted congestion games. J. Exp. Algorithmics **11**, 27 (2007). https://doi.org/10.1145/1187436.1216584
26. Papadimitriou, C.: Algorithms, games, and the internet. In: Proceedings of the 33rd Annual ACM Symposium on Theory of Computing (STOC), pp. 749–753 (2001). https://doi.org/10.1145/380752.380883
27. Rosenthal, R.W.: A class of games possessing pure-strategy Nash equilibria. Int. J. Game Theory **2**(1), 65–67 (1973). https://doi.org/10.1007/BF01737559
28. Rosenthal, R.W.: The network equilibrium problem in integers. Networks **3**(1), 53–59 (1973). https://doi.org/10.1002/net.3230030104
29. Roughgarden, T.: Routing games. In: Nisan, N., Roughgarden, T., Tardos, É., Vazirani, V. (eds.) Algorithmic Game Theory, Chap. 18. Cambridge University Press (2007). https://doi.org/10.1017/CBO9780511800481.020
30. Roughgarden, T.: Twenty Lectures on Algorithmic Game Theory. Cambridge University Press (2016). https://doi.org/10.1017/cbo9781316779309
31. Tardos, É., Wexler, T.: Network formation games and the potential function method. In: Nisan, N., Roughgarden, T., Tardos, É., Vazirani, V. (eds.) Algorithmic Game Theory, Chap. 19. Cambridge University Press (2007). https://doi.org/10.1017/cbo9780511800481.021
32. Vöcking, B.: Selfish load balancing. In: Nisan, N., Roughgarden, T., Tardos, É., Vazirani, V. (eds.) Algorithmic Game Theory, Chap. 20. Cambridge University Press (2007). https://doi.org/10.1017/cbo9780511800481.022

The Impact of Spillback on the Price of Anarchy for Flows over Time

Jonas Israel and Leon Sering(✉)

Technische Universität Berlin, Berlin, Germany
j.israel@tu-berlin.de, sering@math.tu-berlin.de

Abstract. Flows over time enable a mathematical modeling of traffic that changes as time progresses. In order to evaluate these dynamic flows from a game theoretical perspective we consider the *price of anarchy* (PoA). In this paper we study the impact of spillback effects on the PoA, which turn out to be substantial. It is known that, in general, the PoA is unbounded in the spillback setting. We extend this by showing that it is still unbounded even when considering networks with unit edge capacities and that the *Braess ratio* can be arbitrarily large.

In contrast to that, we show that on a fixed network the PoA as a function of the flow amount is bounded by a constant and also upper bound the PoA for the set of networks where the outflow capacities satisfy certain constraints depending on the quickest flow. This upper bound only depends on the worst spillback factor of the Nash flows over time of the given network. It therefore provides a way to quantify the impact of spillback to the quality of the dynamic equilibria.

In addition, we show the surprising fact that the introduction of spillback behavior can actually speed up dynamic equilibria in some networks.

Keywords: Nash flow over time · Dynamic equilibria · Deterministic queuing · Price of anarchy · Spillback · Traffic

Related Version: A full version of this paper including all proofs is available at https://arxiv.org/abs/2007.04218.

1 Introduction

Road traffic is an integral part of modern societies, which consists of many users with individual behaviors and goals. For this reason traffic dynamics are very hard to predict and can barely be controlled. However, through recent technologies such as intelligent navigation systems it might be possible to positively

Funded by the Deutsche Forschungsgemeinschaft (DFG, German Research Foundation) under Grant BR 4744/2-1 and Germany's Excellence Strategy – The Berlin Mathematics Research Center MATH+ (EXC-2046/1, project ID: 390685689).

T. Harks and M. Klimm (Eds.): SAGT 2020, LNCS 12283, pp. 114–129, 2020.
https://doi.org/10.1007/978-3-030-57980-7_8

affect the behavior of traffic, steering it towards shorter travel times leading to less pollution and an overall improved quality of life.

In the following research work we focus on a mathematical traffic flow model called *flows over time with spillback*. Here, the network is depicted as a graph with a source and a sink, and the traffic flow can progress in continuous time from one vertex over an edge to the next vertex. We consider flow as a continuous stream affected by two types of temporal factors. First, flow does not travel instantaneously through the network but needs actual time to traverse an edge, and second, flow on an edge may change over time. Compared to static network flows these temporal components enable us to model traffic realistically through different congestion levels. To model road constraints within the network, we equip each edge with an inflow and an outflow capacity governing with which rate flow can enter and leave the edge, a length characterizing the time it takes a flow-particle to travel from the tail to the head of the edge, and finally, a storage capacity which describes how much flow volume fits on the edge. If the desired outflow exceeds the outflow capacity of an edge the excess flow queues up in front of the bottle-neck at the head of the edge. If at any point in time the queue of an edge is so large that the amount of flow traversing the edge plus the amount of flow in the queue equals the storage capacity, the edge is considered full and new flow can only enter if at least as much flow leaves at the same time. With this mechanic it is possible to model *spillback*, i.e., the phenomenon that traffic congestion at one street can block exits or intersections further upstream. The ability to model spillback within the framework of flows over time is a very recent discovery [22], which has not been studied much yet.

As we experience in our everyday lives traffic is not performing optimal most of the time, but rather consists of agents that behave egoistically. Thus, we are interested in game theoretic aspects of this flow model, particularly in the *price of anarchy* (PoA), the ratio of the worst uncoordinated behavior described via a dynamic equilibrium, and the optimal flow behavior measured by some social cost function. In real-world scenarios that ratio could give us an idea of how much one can possibly improve traffic through optimized traffic control, for example through modern navigation systems or autonomous driving. Even though it has been shown in [22] that the PoA in networks with spillback is unbounded in general we investigate the dependency of the PoA on several parameters, for example, the minimal spillback factor, which measures how much the capacities of an edge are reduced due to spillback. Another interesting phenomenon of selfish road users we study is the well known Braess paradox [2]. It states that the overall travel time of all users might decrease if a frequently used road segment gets closed. In reverse, this means that building new roads between heavily used section of the network might cause more congestion and longer travel times.

Related Work. Flows over time were first introduced by Ford and Fulkerson [8] in the context of an optimization problem to route as much flow as possible in a given time horizon. Gale [9] proved the existence of earliest arrival flows which optimize the amount of flow routed to the sink simultaneously for all points in time and Wilkinson [26] later presented an algorithm to compute these flows. For an overview on flows over time from an optimization point of view we refer to the survey by Skutella [23]. From a game theoretic point of view, flows over time were first considered by Vickrey [24] in the setting of transportation research. In the last years the theory of Nash equilibria for flow models has been advanced significantly. From the introduction of the price of anarchy by Koutsoupias and Papadimitriou [13,16] and the congestion games studied by Roughgarden and Tardos [18,19] (both for static flows), over existence results concerning the dynamic (i.e., time dependent) model by Meunier and Wagner [15], to the constructive approach to dynamic equilibria by Koch and Skutella [12]. Here, the authors present a novel notion of dynamic equilibria, called *Nash flows over time*, which enabled a whole set of proceeding research. This new research includes the study of existence, uniqueness and the long-term behavior of Nash flows over time by Cominetti et al. [4–6], the work by Macko et al. [14] about the Braess paradox for flows over time as well as the extension to multi-terminal settings [21]. Of special interest to the paper at hand are the results by Bhaskar et al. [1] and very recently by Correa et al. [7] about the PoA for flows over time. Since it was already shown that the evacuation-PoA (maximizing the flow amount within some time horizon) is unbounded [12], they focus on the time-PoA (minimizing the completion time for a given flow amount) for which they establish an upper bound of $\frac{e}{e-1}$ under some constraints on the capacities of the network. Sering and Vargas Koch [22] generalized the flows over time model in order to represent spillback and transferred the results about dynamic equilibria to this extension. Very recently, Graf et al. [10] characterized an alternative equilibrium concept for flows over time, where particles do not predict the future evolution of the flow but instead reconsider their route choice on every node. In addition, there is a active research line on packet routing models, where traffic is represented by atomic vehicles that traverses the network in discrete time steps. Recent progress in this area is due to Cao et al. [3], Scarsini et al. [20], Harks et al. [11] and Peis et al. [17].

Contribution and Outline. We study the price of anarchy of flows over time with spillback introduced in [22], which is known to be unbounded in general. After introducing the model in Sect. 2, we show in Sect. 3 that the PoA stays unbounded even if we restrict the set of networks to a specific topology but allow arbitrary capacity, or in reverse if we only allow unit capacities but therefore more complex graph structures. Furthermore, we show that the Braess ratio can be arbitrarily large depending only on the minimum edge capacity. Even though it seems that the addition of full edges and spillback only increases completion times this is not a general rule, as we show that there are examples where the completion time of Nash flows over time is larger when disabling spillback. In contrast to the above lower bounds we show in Sect. 4 that if we consider the case of temporal routing games on a fixed network, i.e., only the flow amount that gets routed through the network varies, the PoA is bounded by a constant. In the end we translate the ideas of [1] to the model with spillback and prove an upper bound of $\frac{\mathsf{c}e}{\mathsf{c}e-1}$ on the PoA in networks with specific conditions on the capacities in dependency of a maximal flow over time. This upper bound only depends on the worst spillback factor c of the Nash flows over time of the given network, and therefore provides a way to quantify the impact of spillback to the PoA (note that e denotes the Euler constant here). Finally, we give a brief conclusion and outlook for further research in Sect. 5.

2 The Model

In the following we want to recall the essential definitions of the flow over time model with deterministic queuing. We consider the extended version that handles spillback effects, as introduced in [22] and mainly stick to the same notation.

Flow Dynamics. We consider a *network* $\Gamma = (G, s, t, r_0, \tau, \nu^+, \nu^-, \sigma)$ given by a directed graph $G = (V, E)$ with a single *source* s and a single *sink* t, such that every vertex is reachable from s. We have a *network inflow rate* of $r_0 > 0$ determining the constant rate of flow entering the network from time 0 onward. Furthermore, every edge $e \in E$ is equipped with a *transit time* $\tau_e \geq 0$, an *in- and outflow capacity* $\nu_e^+ > 0$ and $\nu_e^- > 0$ as well as a *storage capacity* $\sigma_e > 0$. In order to avoid undefined flow behavior, we require that traversing flow alone can never fill up an edge, i.e., $\sigma_e > \nu_e^+ \cdot \tau_e$ and that the total transit time of every directed cycle is strictly positive. For technical reason we furthermore assume that all properties are rational numbers.

A *flow over time* is given by a family of locally integrable and bounded functions $f = (f_e^+, f_e^-)_{e \in E}$, where $f_e^+, f_e^- : \mathbb{R}_{\geq 0} \to \mathbb{R}_{\geq 0}$ denote the in- and outflow rate of edge e at every point in time. The *cumulative in-* and *outflow* and the *queue size* are given by

$$F_e^+(\theta) := \int_0^\theta f_e^+(\xi)\,\mathrm{d}\xi, \quad F_e^-(\theta) := \int_0^\theta f_e^-(\xi) \quad \text{and} \quad z_e(\theta) := F_e^+(\theta - \tau_e) - F_e^-(\theta).$$

We require that flow is preserved at every edge e (*non-deficit constraint*) and at every vertex $v \in V \setminus \{t\}$ (*conservation constraint*), which means, for every point in time θ we have

$$z_e(\theta) \geq 0 \qquad \text{and} \qquad \sum_{e \in \delta_v^+} f_e^+(\theta) - \sum_{e \in \delta_v^-} f_e^-(\theta) = \begin{cases} 0 & \text{for } v \in V \setminus \{s, t\}, \\ r_0 & \text{for } v = s. \end{cases}$$

Here, δ_v^- is the set of all *incoming* and δ_v^+ the set of all *outgoing* edges of node v. An edge e is *full* at time θ if the total amount of flow on e, called *edge load*, $d_e(\theta) := F_e^+(\theta) - F_e^-(\theta)$ reaches the storage capacity σ_e. The *inflow bound* $b_e^+(\theta)$ denotes that current inflow capacity, which might be smaller than ν_e^+ due to spillback, and the *push rate* $b_e^-(\theta)$ specify the current *desired* outflow rate, which is reached whenever there are no restrictions of following links. Formally, we have

$$b_e^+(\theta) := \begin{cases} \nu_e^+ & \text{if } d_e(\theta) < \sigma_e \\ \min\{f_e^-(\theta), \nu_e^+\} & \text{else,} \end{cases} \quad \text{and} \quad b_e^-(\theta) := \begin{cases} \nu_e^- & \text{if } z_e(\theta) > 0, \\ \min\{f_e^+(\theta - \tau_e), \nu_e^-\} & \text{else.} \end{cases}$$

A flow over time f is *feasible* if for all edges e and all times θ it satisfies $f_e^+(\theta) \leq b_e^+(\theta)$ (*inflow condition*) and if there exists a $c_v \in (0,1]$ for every $v \in V$ such that for every $e \in \delta^-(v)$ $f_e^-(\theta) = \min\{b_e^-(\theta), \nu_e^- \cdot c_v\}$ (*fair allocation condition*). Furthermore, we require for all time θ that every vertex v with an incoming edge $e_1 \in \delta_v^-$ with $f_{e_1}^-(\theta) < b_{e_1}^-(\theta)$ (called *throttled edge*) there exists an outgoing edge $e_2 \in \delta_v^+$ with $f_{e_2}^+(\theta) = b_{e_2}^+(\theta)$ (*no-slack condition*). Finally, the set of full edges should be cycle free at every point in time (*no-deadlock condition*).

For a given $v \in V$ the maximal value c that satisfies the fair allocation condition at a given point in time θ is called *spillback factor* denoted by $c_v(\theta)$. This value denotes the reduction of the outflow capacity due to spillback leading to the *effective outflow capacity* of $\nu_e^- \cdot c_v(\theta)$. If the outflow rate $f_{uv}^-(\theta)$ of an incoming edge is strictly smaller than the push rate $b_{uv}^-(\theta)$, this edge is throttled implying $c_v(\theta) < 1$, which means that there is spillback at v. In this case the no-slack condition ensures that there is a reason for the spillback in form of an outgoing exhausted edge vw: $f_{vw}^+(\theta) = b_{vw}^+(\theta)$. The spillback factor will play an important role throughout this paper. For more details and further intuition on the definitions of a feasible flow over time in this setting we refer to [22].

Nash Flows Over Time. In order to define Nash flows over time we need to define the arrival time of every particle of the flow. To simplify the notation we identify every particle with the point in time θ when it enters the network at the source. For every edge e we define the *waiting time function* $q_e : \mathbb{R}_{\geq 0} \to \mathbb{R}_{\geq 0}$ by $q_e(\theta) := \min\left\{ q \geq 0 \;\middle|\; \int_{\theta+\tau_e}^{\theta+\tau_e+q} f_e^-(\xi)\,\mathrm{d}\xi = z_e(\theta+\tau_e) \right\}$, i.e., $q_e(\theta)$ denotes the time a particle entering e at time θ waits in the queue. For every vertex v the *earliest arrival time function* $\ell_v : \mathbb{R}_{\geq 0} \to \mathbb{R}_{\geq 0}$ denotes the earliest point in time the particle θ (which enters the network at time θ) can reach v:

$$\ell_v(\theta) := \begin{cases} \theta & \text{if } v = s, \\ \min\limits_{e=uv \in E} \ell_u(\theta) + \tau_e + q_e(\ell_u(\theta)) & \text{else.} \end{cases}$$

For a given particle θ the *current shortest path network* $G'_\theta = (V, E'_\theta)$ is the network of all edges $e = uv$ that are *active for* θ, i.e., for which $\ell_v(\theta) = \ell_u(\theta) + \tau_e + q_e(\ell_u(\theta))$. It contains all s-v-paths that particle θ can use to be at v at the earliest possible point in time. Furthermore, we denote the *resetting edges* E_θ^* as the set of edges for which particle θ encounters a queue when taking a current shortest path and $\bar{E}_\theta$ denotes the set of edges which are full when particle θ reaches its tail. More precisely, $E_\theta^* := \{e = uv \in E | q_e(\ell_u(\theta)) > 0\}$ and $\bar{E}_\theta := \{e = uv \in E | d_e(\ell_u(\theta)) = \sigma_e\}$.

Definition 1 (Nash flow over time). *We call a feasible flow over time f a* Nash flow over time, *or* dynamic equilibrium, *if almost every particle uses a current shortest s-t-path, i.e., if $f_e^+(\theta) > 0$ implies $\theta \in \ell_u(\Theta_e)$ for all $e = uv \in E$*

and almost all $\theta \in \mathbb{R}_{\geq 0}$*, where* $\Theta_e := \{\theta \in \mathbb{R}_{\geq 0} | e \in E'_\theta\}$ *denotes all particles for which* e *is active.*

Equivalently, it has been shown [22, Lemma 4.1] that a feasible flow over time is a dynamic equilibrium if and only if $F_e^+(\ell_u(\theta)) = F_e^-(\ell_v(\theta))$ for all $e = uv \in E$ and all $\theta \in \mathbb{R}_{\geq 0}$. By setting $x_e(\theta) := F_e^+(\ell_u(\theta)) = F_e^-(\ell_v(\theta))$ we observe that $(x_e(\theta))_{e \in E}$ form a static s-t-flow of value $r_0 \cdot \theta$. Since the x_e are absolute continuous, their derivatives

$$x'_e(\theta) = f_e^+(\ell_u(\theta)) \cdot \ell'_u(\theta) = f_e^-(\ell_v(\theta)) \cdot \ell'_v(\theta) \tag{1}$$

exist almost everywhere and can be seen as the strategy of particle θ (as for every θ it is a static s-t-flow of value r_0). For a fixed θ the derivatives $x'_e := x'_e(\theta)$ and $\ell'_v := \ell'_v(\theta)$ together with the spillback factors $c_v := c_v(\ell_v(\theta))$ are called *spillback thin flows* and with $b_e^+ := b_e^+(\ell_u(\theta))$ for all $e = uv$ satisfy the following equations:

$$\begin{aligned}
\ell'_s &= 1, \\
\ell'_v &= \min_{e=uv \in E'_\theta} \rho_e(\ell'_u, x'_e, c_v) && \text{for } v \in V \setminus \{s\}, \\
\ell'_v &= \rho_e(\ell'_u, x'_e, c_v) && \text{for } e = uv \in E'_\theta \text{ with } x'_e > 0, \\
\ell'_v &\geq \max_{e=vw \in E'_\theta} \frac{x'_e}{b_e^+} && \text{for } v \in V, \\
\ell'_v &= \max_{e=vw \in E'_\theta} \frac{x'_e}{b_e^+} && \text{for } v \in V \text{ with } c_v < 1,
\end{aligned}$$

where

$$\rho_e(\ell'_u, x'_e, c_v) := \begin{cases} \frac{x'_e}{c_v \cdot \nu_e^-} & \text{if } e = uv \in E^*_\theta, \\ \max\left\{ \ell'_u, \frac{x'_e}{c_v \cdot \nu_e^-} \right\} & \text{if } e = uv \in E'_\theta \setminus E^*_\theta. \end{cases}$$

It turns out that the particles of a Nash flow over time f can be divided into intervals, so called *phases*, for which the derivatives (and thus the inflow and outflow rates) stay constant. We denote the set of phases by $\mathcal{I}_f$. The transition points between two phases correspond to one or multiple *events*: A new edge (and therefore new s-t-paths) can become active, a queue can deplete, an edge can become full or the outflow rate (and hence the inflow bound) of a full edge might change. Note however, that an event at edge $e = uv$ for a particle θ does not happen at time θ itself but rather at time $\ell_u(\theta)$ when the particle entering the network at time θ reaches vertex u (while taking a shortest s-u path).

Games, Optimal Flows and the Price of Anarchy. For a *temporal routing game* we consider a finite volume of flow $M \in (0, \infty)$ entering the network. For a Nash flow over time f the last particle enters the network at time $\frac{M}{r_0}$ and leaves the network at time $\ell_t(\frac{M}{r_0})$. As the network satisfy the first-in-first-out-principle (FIFO), ℓ_t is non-decreasing, which means that $T_f := \ell_t(\frac{M}{r_0})$ denotes the *completion time* when the entire flow of volume M has reached t. Most of the time we identify a network Γ with its corresponding temporal routing game (i.e., Γ and M). In contrast to dynamic equilibria, optimal *quickest flows* can be computed by determining a time horizon T with $\ell_t(\frac{M}{r_0}) = T$ by applying a binary search framework to the maximum flow over time problem. Hereby, a maximum flow over time for time horizon T can be constructed via a feasible static flow y maximizing $T \cdot |y| - \sum_{e \in E} \tau_e \cdot y_e$. This *underlying static flow* y is then temporally repeated, which means a rate of y_p is sent into every s-t path $p \in \mathcal{P}_{st}$ over time $[0, T - \tau_p]$. The arrival time of the last particle at t (i.e., the optimal completion time) is denoted by $T_{\text{opt}}(M)$. For more details on optimal flows over time we refer to Skutella's survey [23].

In this paper we consider the *time price of anarchy* (which we simply refer to as "price of anarchy"). For a given temporal routing game Γ it measures the worst ratio between the arrival time at t for the last particle in a Nash flow over time and the arrival time in an optimal flow: $\text{PoA}(\Gamma) := \frac{T_{EQ}(\Gamma)}{T_{\text{opt}}(\Gamma)}$.[1] As it is unknown whether the arrival time functions ℓ_t are unique over all Nash flows over time, we need to consider the worst dynamic equilibrium, i.e., $T_{EQ}(\Gamma) := \sup_{f \in \mathcal{F}(\Gamma)} T_f$, where $\mathcal{F}(\Gamma)$ denotes the set of Nash flows over time in Γ.

Further Notation. We enumerate the event points by the order of their occurrence seen by particles at the source, i.e., $\theta_i < \theta_{i+1}$ and say phase i is given by (θ_{i-1}, θ_i) (using $\theta_0 = 0$).[2] In addition, we consider the point in time $\frac{M}{r_0}$ when the last particle enters the network as the last event r, i.e., $\theta_r := \frac{M}{r_0}$. Since the edge sets E'_θ, E^*_θ, $\bar{E}_\theta$, the inflow bound, and hence, the spillback thin flow stay constant within each phase i we use the following notation for $\theta \in (\theta_{i-1}, \theta_i)$

$$G'_i := G'_\theta, E'_i := E'_\theta, E^*_i := E^*_\theta, \bar{E}_i := \bar{E}_\theta, x'_i := x'(\theta), \ell'_{i,v} := \ell'_v(\theta), c_{i,v} := c_v(\ell_v(\theta)).$$

The inflow at the sink is also constant in a phase. We denote this by the *capacity* $\kappa_i := f_t^+(\ell_t(\theta))$ for some $\theta \in (\theta_{i-1}, \theta_i)$ where we use $f_t^+(\theta) := \sum_{vt \in \delta^-(t)} f_{vt}^-(\theta)$. Finally, the derivatives of the waiting times $(q_e(\ell_u(\theta)))'$ stay constant within a phase as they are either $\ell'_v(\theta) - \ell'_u(\theta)$ if $e = uv$ is active or 0 otherwise. For $\theta \in (\theta_{i-1}, \theta_i)$ we write $q'_{i,e} := (q_e(\ell_u(\theta)))'$ and $q'_{i,p} := \sum_{e \in p} q'_{i,e}$ for an s-t path p.

3 Lower Bounds on the Price of Anarchy

We first show in 3 that the PoA can be unbounded even on very simple graphs (an observation first made in [22]) and that the same is true for graphs with unit capacities. Afterwards, in 3 and 3, we use similar constructions to investigate the Braess paradox for flows over time with spillback and to show that

[1] All results from Sect. 4 can also be translated to the *total delay price of anarchy* measuring the arrival times of all particles combined, similarly as is done in [1].

[2] We imagine i as a natural number. But since it is an open question whether the event point converges to a finite limit, it is possible to expand the index set to the ordinal numbers up to ω^ω. In this case the i-th phase should be defined as (θ_i, θ_{i+1}) as it is not possible to determine a predecessor of an ordinal number. For the sake of simplicity however, we stick to the definition where $(0, \theta_1)$ is the first phase.

there exist networks on which Nash flows over time with spillback are faster than their respective counterparts without spillback.

PoA Depending on Graph Structure or Capacities. Consider the network Γ given in Fig. 1 and a Nash flow over time f of it. Since in the first phase the shortest path is (e_1, e_2), edge e_2 fills up quickly. Once this happens flow already queues up at the end of edge e_1, and thus, e_3 is never used by f. An optimal flow can use e_3 and therefore routes flow to the sink much faster, resulting in an unbounded PoA. This construction can easily be generalized to all graphs that have the graph given in Fig. 1 as a minor.

Theorem 1. *(cf. [22, introductary example]) Let G be any graph that has the graph given in Fig. 1 as a minor, then there exists a temporal routing game Γ on G with $PoA(\Gamma) \in \Omega(\frac{1}{\nu^-_{min}})$ where $\nu^-_{min} := \min_{e \in E}\{\nu^-_e : \nu^-_e > 0\}$.*

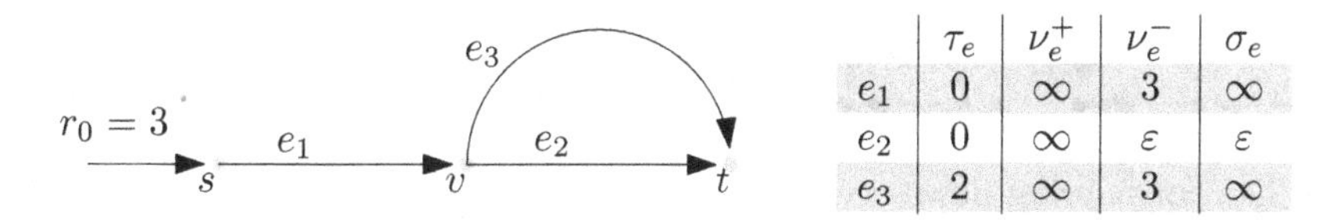

	τ_e	ν^+_e	ν^-_e	σ_e
e_1	0	∞	3	∞
e_2	0	∞	ε	ε
e_3	2	∞	3	∞

Fig. 1. This is a network on which Nash flows over time with spillback have unbounded price of anarchy (see Theorem 1). A similar example was first given in [22, Fig. 2].

To avoid the above unboundedness one could ask for the PoA for temporal routing games on graphs with restricted edge capacities. By constrictions of the model we have to set the inflow and storage capacities of all edges $e \in \delta^+(s)$ to $\nu^+_e > r$ and $\sigma_e = \infty$, respectively. We say a network has *unit edge capacities* if for all edges $e \notin \delta^+(s)$ it holds that $\nu^+_e = \nu^-_e = \sigma_e = 1$ and further for all edges $e \in \delta^+(s)$ also $\nu^-_e = 1$. Unfortunately, even when restricting to networks with unit edge capacities the PoA is unbounded.

Theorem 2. *capacities There exists a family of networks with unit edge capacities and $\tau_e \in \{0, 1\}$ for all edges for which the PoA is linear in the number of edges.*

This can be seen by considering the network given in Fig. 1 and exchanging e_1 and e_3 with bunches of unit-capacity parallel edges and setting $\nu^-_{e_2} = \sigma_{e_2} = 1$. If we use enough parallel edges we can generate a similar flow behavior as we encountered when lowering the capacity of edge e_2 in the proof of Theorem 1.

Nevertheless, we show another way of constraining edge capacities to achive an interesting upper bound on the PoA in Sect. 4.2.

Braess Ratio. In his work on selfish routing with static flows [2] Braess showed that there are networks where adding an edge can paradoxically increase congestion leading to a worse equilibrium. In line with the paper of Macko et al. [14]

we define the *Braess ratio* for flows over time with spillback as follows. Let Γ be a temporal routing game on a graph G and let $\Gamma(H)$ be the same instance restricted to some subgraph $H \subseteq G$. Then the Braess ratio of Γ is

$$\mathrm{BR}(\Gamma) = \max_{H \subseteq G} \frac{T_{\mathrm{EQ}}(\Gamma)}{T_{\mathrm{EQ}}(\Gamma(H))}.$$

We say graph G *admits a Braess paradox* if there is a temporal routing game Γ on G with $\mathrm{BR}(\Gamma) > 1$. In [14] it is shown that the Braess ratio for flows over time without spillback (for a slightly different cost function instead of the last completion time) is arbitrarily large depending linearly on the number of edges of the underlying graph. The authors furthermore show that a graph G or its transpose (the graph where every edge uv is replaced by the edge vu and s and t are swapped) admit a Braess paradox if and only if G contains at least one of the following graphs as a topological minor.

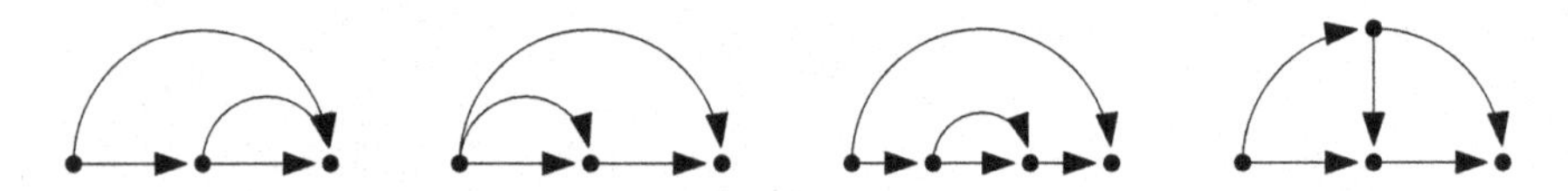

When considering flows over time with spillback and the graph in Fig. 1 it is easy to see that this graph admits a Braess paradox with arbitrarily large Braess ratio even though it does not have one of the graphs above as a topological minor (and neither does its transpose). To see this choose H to be the subgraph where from the graph in Fig. 1 we delete edge e_2.

Corollary 1. *For any $a \in \mathbb{R}$ there exists a temporal routing game Γ on the graph given in Fig. 1 such that the Braess ratio satisfies $BR(\Gamma) > a$.*

Spillback Can Improve Completion Time. The following proposition shows that there are temporal routing games where Nash flows with spillback perform better than Nash flows without spillback. This might at first be surprising, as spillback seems to only be obstructive to routing flow fast. But it is indeed possible to construct networks where spillback leads to shorter completion times. In the network depicted in Fig. 2 there are two parallel edges, namely e_3 and $e_{3'}$, for which it holds that the completion time of a Nash flow is worse if the edges are present compared to the same network without those edges. We exploit this in our construction: In the spillback model one of these 'bad' edges becomes full nearly instantaneously yielding the other 'bad' edge to never get active. Thus, the spillback Nash flow routes flow only over one of those 'bad' edges. Since in the Koch-Skutella model without spillback both of these parallel edges get active at some point, the Nash flow over time here uses both of them resulting in a worse completion time.

Proposition 1. *In the network Γ given in Fig. 2 the completion time of any Nash flow over time with spillback is less than the completion time of the Nash flow over time without spillback on the same network using $\nu_e := \min\{\nu_e^+, \nu_e^-\}$.*

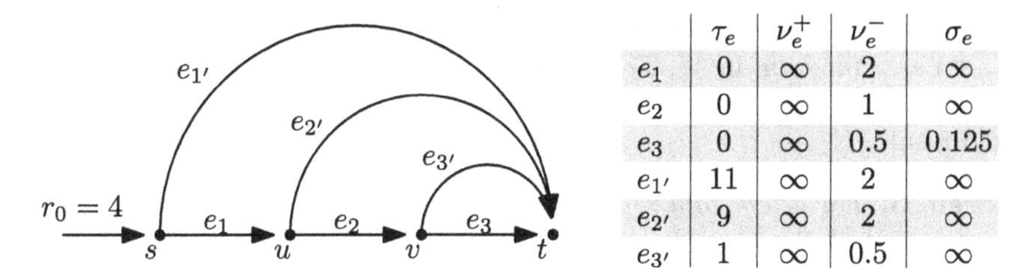

	τ_e	ν_e^+	ν_e^-	σ_e
e_1	0	∞	2	∞
e_2	0	∞	1	∞
e_3	0	∞	0.5	0.125
$e_{1'}$	11	∞	2	∞
$e_{2'}$	9	∞	2	∞
$e_{3'}$	1	∞	0.5	∞

Fig. 2. This example shows that Nash flows over time with spillback can be faster than Nash flows over time without spillback, see Proposition 1.

4 Upper Bounds on the Price of Anarchy

In the following we prove two upper bounds on the price of anarchy. First, we show that for a single, fixed network the PoA is bounded by a constant in the long run. After that we show that if for a given network we are allowed to decrease the outflow capacities by a certain amount then the PoA only depends on the worst spillback factor of the Nash flows over time.

4.1 Price of Anarchy for a Fixed Network

Until now we have studied the PoA depending on the structure of the underlying graph or its capacities. For both questions we constructed games satisfying strong constraints that still have unbounded PoA. Now we are interested in the PoA of a network where every parameter is fixed except for the target amount M, i.e., we ask the question of how the PoA behaves in the long run on a single network.

Lemma 1. *For a temporal routing game Γ on a fixed network the completion time of the optimal flow depending on the target amount M is bounded by $T_{opt}(M) \in \Theta(M)$.*

This result is mainly due to the fact that the optimal flow does not build up any queues. Therefore its completion time depends mainly on M and the minimum edge-capacity, which we consider to be fixed.

For the classification of the asymptotic long term behavior of $T_{\text{EQ}}(M)$ we use the following auxiliary lemma that gives us a lower bound on the spillback factors of a Nash flow. The lemma follows by an application of [22, Lemma 3].

Lemma 2. *For a temporal routing game Γ there exists an $\varepsilon > 0$ such that for any Nash flow over time $f \in \mathcal{F}(\Gamma)$ the spillback factors satisfy* $\min\{c_v(\theta) : v \in V, \theta \in \mathbb{R}_{\geq 0}\} > \varepsilon$.

To get an asymptotic bound on $T_{\mathrm{EQ}}(M) = \sup_{f \in \mathcal{F}(\Gamma)} T_f(M)$ we first argue that seen as a function in M, T_f is a piece-wise linear and non-decreasing function. We can then use Lemma 2 to bound its derivative and with that obtain the desired result.

Theorem 3. *For a temporal routing game Γ on a fixed network the completion time of any Nash flow over time $f \in \mathcal{F}(\Gamma)$ is bounded by $T_f(M) \in \Theta(M)$.*

We can now use Lemma 1 and Theorem 3 and the fact that $\mathrm{PoA}(M) = \frac{T_{\mathrm{EQ}}(M)}{T_{\mathrm{opt}}(M)}$ to bound the PoA for a fixed network. In order to do so we consider the PoA as a function of the target flow amount M.

Theorem 4. *For a temporal game Γ on a fixed network, i.e. when treating everything except the amount of flow M as a constant, the price of anarchy is bounded by a constant, $PoA(M) \in \Theta(1)$.*

4.2 Bound on the Price of Anarchy for Saturated Graphs

In this section we focus on networks with an additional constraint on the edge capacities. Given a game Γ we know that the quickest flow of Γ is also a temporally repeated flow, i.e., it has an underlying static flow y. We say that y *saturates every edge* of the given graph if for each edge the outflow capacity is exhausted by y, i.e., for each $e \in E$ we have $\nu_e^- = y_e$ and additionally it holds that $|y| = \sum_{sv \in \delta^+(s)} y_{sv} = r_0$. We call the underlying graph of such a game a *saturated graph.* Even though restricting attention to saturated graphs may seem harsh, note, that every network can be made saturated by lowering the edge capacities. This can be imagined to be done by a system operator in a Stackelberg strategy-like scenario [25] and is applicable in many real-world examples. For one, streets can be narrowed down by a city administration in practice.

For temporal routing games on saturated graphs we will show that the PoA can be bounded by a value that is only dependent on the worst spillback factor of all Nash flows over time. In order to do that we adapt the idea of the proofs given by Bhaskar et al. [1] for the Koch-Skutella model to the spillback model. Note, however, that the proofs given in [1] implicitly assume only finitely many phases, which has not been proven for any of the two models. Our generalization also holds for the case of an infinite number of phases in both models.[3]

In principle the proof works as follows. For a given game Γ the relation of the completion time of any Nash flow over time of Γ to the optimal completion time can be determined by examining the capacity of the current shortest path network and the derivatives of the waiting times for a single phase of the Nash flow. One can then bound the derivatives of the waiting times and use the fact that the PoA is the maximum over the relation of the optimal completion time to the completion times of all Nash flows. This achieves the desired bound.

[3] Note, that in [7] an even more general result is shown for the Koch-Skutella model.

Bound on the Derivatives of the Waiting Times. We start by proving a relation between the derivative of the label-function at the sink and the inflow into the sink. Our proof of this result uses a different idea than the one given in [1] and is considerably shorter.

Lemma 3. *(cf. [1, Lemma 15]) Let Γ be a temporal routing game and let $f \in \mathcal{F}(\Gamma)$ with corresponding labels ℓ. Then for any $\theta \leq \frac{M}{r_0}$ we have*

$$\ell_t'(\theta) = \frac{r_0}{f_t^+(\ell_t(\theta))}.$$

Proof. Let $\theta \leq \frac{M}{r_0}$ be arbitrary. Using that $x'(\theta)$ is a static s-t flow of value r_0 and $x'_{vt}(\theta) = f^-_{vt}(\ell_t(\theta)) \cdot \ell'_t(\theta)$ from Eq. (1) we obtain

$$r_0 = \sum_{vt \in \delta^-(t)} x'_{vt}(\theta) = \sum_{vt \in \delta^-(t)} f^-_{vt}(\ell_t(\theta)) \cdot \ell'_t(\theta) = \ell'_t(\theta) \cdot f_t^+(\ell_t(\theta)).$$

Since $f_t^+(\ell_t(\theta)) > 0$ for all θ, rearranging terms give the desired result. □

We now proceed with a path-wise bound on the derivatives of the waiting times $q'_{i,p}$ for a single phase of the Nash flow over time i using the capacities κ_i.

Lemma 4. *(cf. [1, Lemma 18]) Let Γ be a temporal routing game where the static flow underlying the quickest flow saturates every edge and let $f \in \mathcal{F}(\Gamma)$. For any s-t path p, the travel time is bounded by*

$$\tau_p \geq \ell_t(\theta_r) - \sum_{i \in \mathcal{I}_f} (1 + q'_{i,p}) \cdot \frac{\kappa_i}{r_0} \cdot (\ell_t(\theta_i) - \ell_t(\theta_{i-1})).$$

In the proof we first establish a dependence of the length of a phase as it is experienced at the source and at the sink, respectively. Then we express τ_p in terms of the label functions ℓ and the waiting times q and their derivatives. The result then follows from applying Lemma 3.

Relation of the Completion Times of Nash Flow and Quickest Flow. The following lemma enables us to give a first relation of the completion times of the optimal quickest flow and a Nash flow over time.

Lemma 5. *(cf. [1, Lemma 19]) Let Γ be a temporal routing game where the static flow y underlying the quickest flow saturates every edge and let $f \in \mathcal{F}(\Gamma)$. Then, the completion time T_{opt} of the optimal flow and the completion time T_f of the Nash flow f are related as*

$$r_0 \cdot T_{opt} = \sum_{p \in \mathcal{P}_{s,t}} y_p \tau_p + \sum_{i \in \mathcal{I}_f} \kappa_i \cdot (\ell_t(\theta_i) - \ell_t(\theta_{i-1})),$$

where $\mathcal{P}_{s,t}$ is the set of all simple s-t paths in G and $\ell_t(\theta_r) = T_f$.

The proof idea is to compare the arrival rates of both flows at the sink t where we use a flow decomposition along paths for the optimal flow and a decomposition by phases for the Nash flow over time.

By combining the previous two lemmas we can now derive a lower bound on the inverse of the PoA that we will afterwards use to achieve an upper bound on the actual PoA. But in order to proof that we first need the following.

Lemma 6. *Let* $\lambda_i := \frac{\kappa_i}{r_0} \cdot \sum_{p \in \mathcal{P}_{s,t}} y_p q'_{i,p}$ *for each phase* $i \in \mathcal{I}_f$. *Then,*

$$\sum_{i \in \mathcal{I}_f} \lambda_i \cdot (\ell_t(\theta_i) - \ell_t(\theta_{i-1})) \leq (\ell_t(\theta_r) - \ell_t(\theta_0)) \cdot \sup_{i \in \mathcal{I}_f} \lambda_i.$$

In the proof we first establish that the set $\{\lambda_i : i \in \mathcal{I}_f\}$ is bounded and then use this and the telescoping principle to bound the left hand side.

The next lemma establishes the aforementioned bound on the inverse of the PoA. It is in this proof that the number of α-extension phases comes into play. If we assume that the supremum in the statement of Lemma 6 is attained by some phase $i \in \mathcal{I}_f$, which is in particular true if there are only finitely many phases, then we can prove Lemma 7 without the ε error and the proofs go through similar to [1]. But since it is still an open problem whether the number of those phases is always finite (in the Koch-Skutella model as well as the spillback model), we prove it here for the case of infinitely many α-extension phases.

Lemma 7. *Let Γ be a temporal routing game where the static flow y underlying the quickest flow saturates every edge and let $f \in \mathcal{F}(\Gamma)$. Then for every $\varepsilon > 0$ there exists a phase i of f such that*

$$\frac{T_{opt}}{T_f} + \varepsilon \geq 1 - \frac{\kappa_i}{{r_0}^2} \sum_{e \in E} \nu_e^- q'_{i,e}.$$

The proof idea is to sum $y_p \tau_p$ over all paths $p \in \mathcal{P}_{s,t}$ and using Lemma 4 to bound this from below. Afterwards, we use Lemmas 5 and 6 to obtain a lower bound on $\frac{T_{\text{opt}}}{T_f}$ in terms of a supremum of the capacities and derivatives of the queuing delay over all phases. Since we do not know whether this supremum is attained we have to inject the ε error and after rearranging terms we obtain the desired result.

Upper Bound for Saturated Graphs. We can now turn the lower bound in Lemma 7 into an upper bound on the price of anarchy by proving a bound on the sum of the right-hand side of the expression given in Lemma 7. Here for the first time the spillback factors of the Nash flow over time play an important role.

Lemma 8. *Let Γ be a temporal routing game and $f \in \mathcal{F}(\Gamma)$. In any phase i of f where $\frac{r_0}{\kappa_i} \geq 1$ we have*

$$\sum_{e \in E} \nu_e^- q'_{i,e} \leq \frac{r_0}{\mathsf{c}_i^f} \ln\left(\frac{r_0}{\kappa_i}\right),$$

where $\mathsf{c}_i^f := \min\{c_v(\theta) : v \in V, \theta \in (\theta_{i-1}, \theta_i)\}$ *is the minimal* c_v *of* f *in phase* i.

The proof utilizes [7, Claim 12] and follows the line of argumentation in [1] but incorporates the added complexity of the spillback model. We obtain that $c_v \nu_e^- q_e' = x_e' \cdot (1 - \frac{\ell_u'}{\ell_v'})$ for every edge $e = uv$ and then sum this expression over all edges in the graph. Rearranging and plugging in the above expression then yields the desired result.

We can now obtain the desired upper bound on the price of anarchy.

Theorem 5. *Let Γ be a temporal routing game where the static flow y underlying the quickest flow saturates every edge of the graph. If the minimal spillback factor satisfies $\mathsf{c} := \min_{f \in \mathcal{F}(\Gamma)} \min\{c_v(\theta) : v \in V, \theta \in \mathbb{R}_{\geq 0}\} > \frac{1}{e}$, then the price of anarchy is bounded by $\frac{T_{EQ}}{T_{opt}} \leq \frac{\mathsf{c}e}{\mathsf{c}e - 1}$.*

Proof. For any $f \in \mathcal{F}(\Gamma)$ with completion time T_f we know that $f_t^+(\theta) = \sum_{vt \in \delta^-(t)} f_{vt}^-(\theta) \leq r_0$ for all $\theta \in \mathbb{R}_{\geq 0}$ since we only consider saturated graphs. Thus, we have $\frac{r_0}{\kappa_i} \geq 1$ in all phases of f. From Lemmas 7 and 8 we obtain that for every $\varepsilon > 0$ there exists a phase i of f such that

$$\frac{T_{\text{opt}}}{T_f} + \varepsilon \geq 1 - \frac{\kappa_i}{{r_0}^2} \sum_{e \in E} \nu_e^- q_{i,e}' \geq 1 - \frac{\kappa_i}{{r_0}^2} \frac{r_0}{\mathsf{c}_i^f} \cdot \ln\left(\frac{r_0}{\kappa_i}\right) = 1 - \frac{a_i}{\mathsf{c}} \cdot \ln\left(\frac{1}{a_i}\right),$$

where $\mathsf{c} := \min_{f \in \mathcal{F}(\Gamma)} \min\{c_v(\theta) : v \in V, \theta \in \mathbb{R}_{\geq 0}\} \leq \mathsf{c}_i^f$ and $a_i := \frac{\kappa_i}{r_0}$.

Simple calculus shows that the term $\frac{a_i}{\mathsf{c}} \cdot \ln\left(\frac{1}{a_i}\right)$ is maximized for $a_i = \frac{1}{e}$. Using the above inequality, derived from some phase i, for any $\varepsilon > 0$ we obtain

$$\frac{T_{\text{opt}}}{T_f} + \varepsilon \geq 1 - \frac{1}{\mathsf{c}e} = \frac{\mathsf{c}e - 1}{\mathsf{c}e}.$$

Since by assumption we have $\mathsf{c} > \frac{1}{e}$ we can take the inverse of the inequality to obtain $\frac{T_f}{T_{\text{opt}}} \leq \frac{\mathsf{c}e}{\mathsf{c}e-1}$. We finish by noting that $T_{\text{EQ}} = \sup_{f \in \mathcal{F}(\Gamma)} T_f$. □

5 Conclusions

Our work shows that the PoA is highly dependent on spillback effects. Although, even in restricted network classes the completion times of dynamic equilibria can be arbitrarily bad compared to a quickest flow, the PoA can still be bounded in terms of the spillback factors under some constraints on the edge capacities. Transferred to real-world traffic this means the interplay between selfish traffic users is critical in particular in high congested areas.

Even though we give a substantial analysis of the PoA in the flow over time model with spillback, there are still some open problems remaining. Is the bound we establish in Theorem 5 tight? Are there any bounds in the case of $\mathsf{c} \leq \frac{1}{e}$ or is it possible to enforce $\mathsf{c} > \frac{1}{e}$ through some Stackelberg-like strategy? Do the results of the recent work of Correa et al. [7] also transfer to the spillback setting? On the more applied side of the research it would also be very interesting to algorithmically identify street segments (edges) which are especially vulnerable for spillback. In the long run this could help road administrations to decide which roads should be expanded (increasing the storage capacity) or which roads should be narrowed or closed (due to the Braess effect).

References

1. Bhaskar, U., Fleischer, L., Anshelevich, E.: A stackelberg strategy for routing flow over time. Games Econ. Behav. **92**, 232–247 (2015)
2. Braess, D.: Über ein paradoxon aus der verkehrsplanung. Unternehmensforschung **12**(1), 258–268 (1968). https://doi.org/10.1007/BF01918335
3. Cao, Z., Chen, B., Chen, X., Wang, C.: A network game of dynamic traffic. In: Proceedings of the 2017 ACM Conference on Economic and Computation, pp. 695–696 (2017)
4. Cominetti, R., Correa, J.R., Larré, O.: Existence and uniqueness of equilibria for flows over time. In: Aceto, L., Henzinger, M., Sgall, J. (eds.) ICALP 2011. LNCS, vol. 6756, pp. 552–563. Springer, Heidelberg (2011). https://doi.org/10.1007/978-3-642-22012-8_44
5. Cominetti, R., Correa, J., Larré, O.: Dynamic equilibria in fluid queueing networks. Oper. Res. **63**(1), 21–34 (2015)
6. Cominetti, R., Correa, J., Olver, N.: Long term behavior of dynamic equilibria in fluid queuing networks. In: Eisenbrand, F., Koenemann, J. (eds.) IPCO 2017. LNCS, vol. 10328, pp. 161–172. Springer, Cham (2017). https://doi.org/10.1007/978-3-319-59250-3_14
7. Correa, J., Cristi, A., Oosterwijk, T.: On the price of anarchy for flows over time. In: Proceedings of the 2019 ACM Conference on Economic and Computation, pp. 559–577. ACM (2019)
8. Ford Jr., L.R., Fulkerson, D.R.: Flows in Networks. Princeton University Press, Princeton (2015)
9. Gale, D.: Transient flows in networks. Mich. Math. J. **6**(1), 59–63 (1959)
10. Graf, L., Harks, T., Sering, L.: Dynamic flows with adaptive route choice. Math Program. 1–27 (2020). https://doi.org/10.1007/s10107-020-01504-2
11. Harks, T., Peis, B., Schmand, D., Tauer, B., Vargas Koch, L.: Competitive packet routing with priority lists. ACM Trans. Econ. Comput. **6**(1), 4 (2018)
12. Koch, R., Skutella, M.: Nash equilibria and the price of anarchy for flows over time. Theory Comput. Syst. **49**(1), 71–97 (2011). https://doi.org/10.1007/s00224-010-9299-y
13. Koutsoupias, E., Papadimitriou, C.: Worst-case equilibria. In: Meinel, C., Tison, S. (eds.) STACS 1999. LNCS, vol. 1563, pp. 404–413. Springer, Heidelberg (1999). https://doi.org/10.1007/3-540-49116-3_38
14. Macko, M., Larson, K., Steskal, Ľ.: Braess's paradox for flows over time. In: Kontogiannis, S., Koutsoupias, E., Spirakis, P.G. (eds.) SAGT 2010. LNCS, vol. 6386, pp. 262–275. Springer, Heidelberg (2010). https://doi.org/10.1007/978-3-642-16170-4_23
15. Meunier, F., Wagner, N.: Equilibrium results for dynamic congestion games. Transp. Sci. **44**(4), 524–536 (2010)
16. Papadimitriou, C.H.: Algorithms, games, and the internet. In: Orejas, F., Spirakis, P.G., van Leeuwen, J. (eds.) ICALP 2001. LNCS, vol. 2076, pp. 1–3. Springer, Heidelberg (2001). https://doi.org/10.1007/3-540-48224-5_1

17. Peis, B., Tauer, B., Timmermans, V., Vargas Koch, L.: Oligopolistic competitive packet routing. In: 18th Workshop on Algorithmic Approaches for Transportation Modelling, Optimization, and Systems (2018)
18. Roughgarden, T.: Selfish Routing and the Price of Anarchy, vol. 174. MIT Press, Cambridge (2005)
19. Roughgarden, T., Tardos, É.: How bad is selfish routing? J. ACM (JACM) **49**(2), 236–259 (2002)
20. Scarsini, M., Schröder, M., Tomala, T.: Dynamic atomic congestion games with seasonal flows. Oper. Res. **66**(2), 327–339 (2018)
21. Sering, L., Skutella, M.: Multi-source multi-sink Nash flows over time. In: 18th Workshop on Algorithmic Approaches for Transportation Modelling, Optimization, and Systems, vol. 65, pp. 12:1–12:20 (2018)
22. Sering, L., Vargas Koch, L.: Nash flows over time with spillback. In: Proceedings of the Thirtieth Annual ACM-SIAM Symposium on Discrete Algorithms, pp. 935–945. SIAM (2019)
23. Skutella, M.: An introduction to network flows over time. In: Cook, W., Lovász, L., Vygen, J. (eds.) Research Trends in Combinatorial Optimization, pp. 451–482. Springer, Heidelberg (2009). https://doi.org/10.1007/978-3-540-76796-1_21
24. Vickrey, W.S.: Congestion theory and transport investment. Am. Econ. Rev. **59**(2), 251–260 (1969)
25. Von Stackelberg, H.: Marktform und Gleichgewicht. J. Springer, Heidelberg (1934)
26. Wilkinson, W.L.: An algorithm for universal maximal dynamic flows in a network. Oper. Res. **19**(7), 1602–1612 (1971)

Dynamic Equilibria in Time-Varying Networks

Hoang Minh Pham and Leon Sering(✉)

Technische Universität Berlin, Berlin, Germany
hoang.m.pham@campus.tu-berlin.de, sering@math.tu-berlin.de

Abstract. Predicting selfish behavior in public environments by considering Nash equilibria is a central concept of game theory. For the dynamic traffic assignment problem modeled by a flow over time game, in which every particle tries to reach its destination as fast as possible, the dynamic equilibria are called Nash flows over time. So far, this model has only been considered for networks in which each arc is equipped with a constant capacity, limiting the outflow rate, and with a transit time, determining the time it takes for a particle to traverse the arc. However, real-world traffic networks can be affected by temporal changes, for example, caused by construction works or special speed zones during some time period. To model these traffic scenarios appropriately, we extend the flow over time model by time-dependent capacities and time-dependent transit times. Our first main result is the characterization of the structure of Nash flows over time. Similar to the static-network model, the strategies of the particles in dynamic equilibria can be characterized by specific static flows, called thin flows with resetting. The second main result is the existence of Nash flows over time, which we show in a constructive manner by extending a flow over time step by step by these thin flows.

Keywords: Nash flows over time · Dynamic equilibria · Deterministic queuing · Time-varying networks · Dynamic traffic assignment

Related Version: A full version of this paper including all proofs is available at https://arxiv.org/abs/2007.01525.

1 Introduction

In the last decade the technological advances in the mobility and communication sector have grown rapidly enabling access to real-time traffic data and autonomous driving vehicles in the foreseeable future. One of the major advantages of self-driving and communicating vehicles is the ability to directly use

Funded by the Deutsche Forschungsgemeinschaft (DFG, German Research Foundation) under Germany's Excellence Strategy – The Berlin Mathematics Research Center MATH+ (EXC-2046/1, project ID: 390685689).

T. Harks and M. Klimm (Eds.): SAGT 2020, LNCS 12283, pp. 130–145, 2020.
https://doi.org/10.1007/978-3-030-57980-7_9

information about the traffic network including the route-choice of other road users. This holistic view of the network can be used to decrease travel times and distribute the traffic volume more evenly over the network. As users will still expect to travel along a fastest route it is important to incorporate game theoretical aspects when analyzing the dynamic traffic assignment. The results can then be used by network designers to identify bottlenecks beforehand, forecast air pollution in dense urban areas and give feedback on network structures. In order to obtain a better understanding of the complicated interplay between traffic users it is important to develop strong mathematical models which represent as many real-world traffic features as possible. Even though the more realistic models consider a time-component, the network properties are considered to stay constant in most cases. Surely, this is a serious drawback as real road networks often have properties that vary over time. For example, the speed limit in school zones is often reduced during school hours, roads might be completely or partially blocked due to construction work and the direction of reversible lanes can be switched, causing a change in the capacity in both directions. A more exotic, but nonetheless important setting are evacuation scenarios. Consider an inhabited region of low altitude with a high risk of flooding. As soon as there is a flood warning everyone needs to be evacuated to some high-altitude-shelter. But, due to the nature of rising water levels, roads with low altitude will be impassable much sooner than roads of higher altitude. In order to plan an optimal evacuation or simulate a chaotic equilibrium scenario it is essential to use a model with time-varying properties. This research work is dedicated to providing a better understanding of the impact of dynamic road properties on the traffic dynamics in the Nash flow over time model. We will transfer all essential properties of Nash flows over time in static networks to networks with time-varying properties.

1.1 Related Work

The fundamental concept for the model considered in this paper are *flows over time* or *dynamic flows*, which were introduced back in 1956 by Ford and Fulkerson [8,9] in the context of optimization problems. The key idea is to add a time-component to classical network flows, which means that the flow particles need time to travel through the network. In 1959 Gale [10] showed the existence of so called *earliest arrival flows*, which solve several optimization problems at once, as they maximize the amount of flow reaching the sink at all points in time simultaneously. Further work on these optimal flows is due to Wilkinson [25], Fleischer and Tardos [7], Minieka [17] and many others. For formal definitions and a good overview of optimization problems in flow over time settings we refer to the survey of Skutella [22].

In order to use flows over time for traffic modeling it is important to consider game theoretic aspects. Some pioneer work goes back to Vickrey [23] and Yagar [26]. In the context of classical (static) network flows, equilibria were introduced by Wardrop [24] in 1952. In 2009 Koch and Skutella [14] (see also [15] and Koch's PhD thesis [13]) started a fruitful research line by introducing dynamic equilibria, also called *Nash flows over time*, which will be the central concept in this paper. In a Nash flow over time every particle chooses a quickest path from the origin to the destination, anticipating the route choice of all other flow particles. Cominetti et al. showed the existence of Nash flows over time [3,4] and studied the long term behavior [5]. Macko et al. [16] studied the Braess paradox in this model and Bhaskar et al. [1] and Correa et al. [6] bounded the price of anarchy under certain conditions. In 2018 Sering and Skutella [20] transferred Nash flows over time to a model with multiple sources and multiple sinks and in the following year Sering and Vargas Koch [21] considered Nash flows over time in a model with spillback.

A different equilibrium concept in the same model was considered by Graf et al. [11] by introducing instantaneous dynamic equilibria. In these flows over time the particles do not anticipate the further evolution of the flow, but instead reevaluate their route choice at every node and continue their travel on a current quickest path. In addition to that, there is an active research line on packet routing games. Here, the traffic agents are modeled by atomic packets (vehicles) of a specific size. This is often combined with discrete time steps. Some of the recent work on this topic is due to Cao et al. [2], Harks et al. [12], Peis et al. [18] and Scarsini et al. [19].

1.2 Overview and Contribution

In the *base model*, which was considered by Koch and Skutella [15] and by the follow up research [1,3–6,16,20], the network is constant and each arc has a constant capacity and constant transit time. In real-world traffic, however, temporary changes of the infrastructure are omnipresent. In order to represent this, we extend the base model to networks with time-varying capacities (including the network inflow rate) and time-varying transit times.

We start in Sect. 2 by defining the flow dynamics of the deterministic queuing model with time-varying arc properties and proving some first auxiliary results. In particular, we describe how to turn time-dependent speed limits into time-dependent transit times. In Sect. 3 we introduce some essential properties, such as the earliest arrival times, which enable us to define Nash flows over time. As in the base model, it is still possible to characterize such a dynamic equilibrium by the underlying static flow. Taking the derivatives of these parametrized static flows provides thin flows with resetting, which are defined in Sect. 4. We show that the central results of the base model transfer to time-varying networks, and in particular, that the derivatives of every Nash flow over time form a thin flow with resetting. In Sect. 5 we show the reverse of this statement: Nash flows over time can be constructed by a sequence of thin flows with resetting, which, in the end, proves the existence of dynamic equilibria. We close this section with a detailed example. Finally, in Sect. 6 we present a conclusion and give a brief outlook on further research directions.

Due to space restrictions we omit most technical proofs. They can be found in the appendix of the full version available at https://arxiv.org/abs/2007.01525.

2 Flow Dynamics

We consider a directed graph $G = (V, E)$ with a source s and a sink t such that each node is reachable by s. In contrast to the Koch-Skutella model, which we will call *base model* from now on, this time each arc e is equipped with a time-dependent capacity $\nu_e \colon [0, \infty) \to (0, \infty)$ and a time-dependent speed limit $\lambda_e \colon [0, \infty) \to (0, \infty)$, which is inversely proportional to the transit time. We consider a time-dependent network inflow rate $r \colon [0, \infty) \to [0, \infty)$ denoting the flow rate at which particles enter the network at s. We assume that the amount of flow an arc can support is unbounded and that the network inflow is unbounded as well, i.e., for all $e \in E$ we require that

$$\int_0^\theta \nu_e(\xi)\,\mathrm{d}\xi \to \infty, \quad \int_0^\theta \lambda_e(\xi)\,\mathrm{d}\xi \to \infty \quad \text{and} \quad \int_0^\theta r(\xi)\,\mathrm{d}\xi \to \infty \quad \text{for } \theta \to \infty.$$

Later on, in order to be able to construct Nash flows over time, we will additionally assume that all these functions are right-constant, i.e., for every $\theta \in [0, \infty)$ there exists an $\varepsilon > 0$ such that the function is constant on $[\theta, \theta + \varepsilon)$.

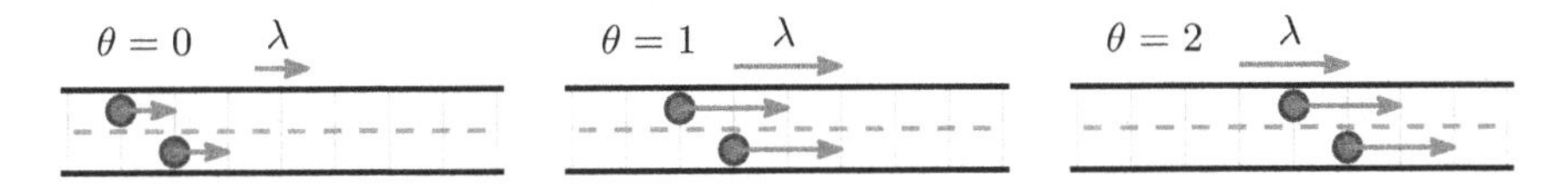

Fig. 1. Consider a road segment with time-dependent speed limit that is low in the time interval $[0, 1)$ and large afterwards. All vehicles, independent of their position, first traverse the link slowly and immediately speed up to the new speed limit at time 1.

Speed Limits. Let us focus on the transit times first. We have to be careful how to model the transit time changes, since we do not want to lose the following two properties of the base model:

(i) We want to have the first-in-first-out (FIFO) property for arcs, which leads to FIFO property of the network for Nash flows over time [15, Theorem 1].
(ii) Particles should never have the incentive to wait on a node.

In other words, we cannot simply allow piecewise-constant transit times, since this could lead to the following case: If the transit time of an arc is high at the beginning and gets reduced to a lower value at some later point in time, then particles might overtake other particles on that arc. Thus, particles might arrive earlier at the sink if they wait right in front of the arc until its transit time drops. Hence, we let the speed limit change over time instead. In order to keep the number of parameters of the network as small as possible, we assume that the lengths of all arcs equal 1 and, instead of a transit time, we equip every arc $e \in E$ with a time-dependent speed limit $\lambda_e \colon [0, \infty) \to (0, \infty)$. Thus, a particle might traverse the first part of an arc at a different speed than the remaining distance if the maximal speed changes midway; see Fig. 1.

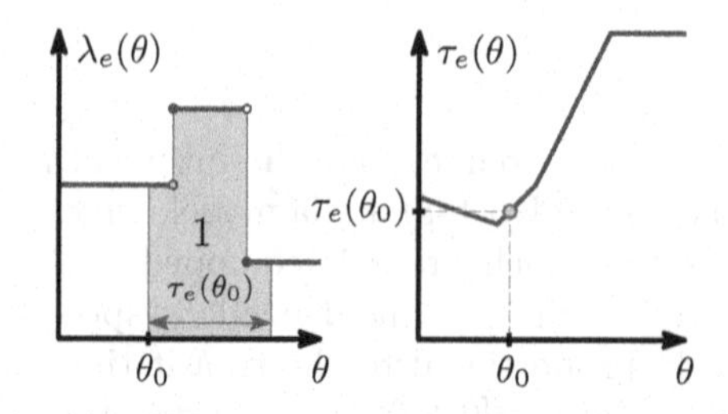

Fig. 2. From speed limits (*left side*) to transit times (*right side*). The transit time $\tau_e(\theta)$ denotes the time a particle needs to traverse the arc when entering at time θ. We normalize the speed limits by assuming that all arcs have length 1, and hence, the transit time $\tau_e(\theta)$ equals the length of an interval starting at θ such that the area under the speed limit graph within this interval is 1.

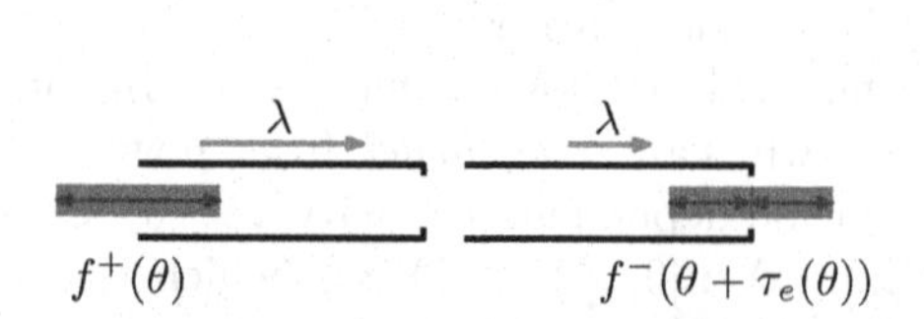

Fig. 3. An illustration of how the flow rate changes depending on the speed limits. *On the left:* As the speed limit λ is high, the flow volume entering the arc per time unit is represented by the area of the long rectangle. *On the right:* The speed limit is halved, and therefore, the same amount of flow needs twice as much time to leave the arc (or enter the queue if there is one). Hence, if there is no queue, the outflow rate at time $\tau + \tau_e(\theta)$ is only half the size of the inflow rate at time θ.

Transit Times. Note that we assume the point queue of an arc to always right in front of the exit. Hence, a particle entering arc e at time θ immediately traverses the arc of length 1 with a time-dependent speed of λ_e. The *transit time* $\tau\colon [0,\infty) \to [0,\infty)$ is therefore given by

$$\tau_e(\theta) := \min \left\{ \tau \geq 0 \;\middle|\; \int_\theta^{\theta+\tau} \lambda_e(\xi)\, \mathrm{d}\xi = 1 \right\}.$$

Since we required $\int_0^\theta \lambda_e(\xi)\, \mathrm{d}\xi$ to be unbounded for $\theta \to \infty$, we always have a finite transit time. For an illustrative example see Fig. 2.

The following lemma shows some basic properties of the transit times.

Lemma 1. *For all $e \in E$ and almost all $\theta \in [0,\infty)$ we have:*

(i) The function $\theta \mapsto \theta + \tau_e(\theta)$ is strictly increasing.
(ii) The function τ_e is continuous and almost everywhere differentiable.
(iii) For almost all $\theta \in [0,\infty)$ we have $1 + \tau_e'(\theta) = \frac{\lambda_e(\theta)}{\lambda_e(\theta+\tau_e(\theta))}$.

These statement follow by simple computation and some basic Lebesgue integral theorems.

Speed Ratios. The ratio in Lemma 1 (iii) will be important to measure the outflow of an arc depending on the inflow. We call $\gamma_e\colon [0,\infty) \to [0,\infty)$ the *speed ratio* of e and it is defined by $\gamma_e(\theta) := \frac{\lambda_e(\theta)}{\lambda_e(\theta+\tau_e(\theta))} = 1 + \tau_e'(\theta)$. Figuratively speaking, this ratio describes how much the flow rate changes under different speed limits. If, for example, $\gamma_e(\theta) = 2$, as depicted in Fig. 3, this means that the speed limit was twice as high when the particle entered the arc as it is at the moment the particle enters the queue. In this case the flow rate is halved on its way, since the same amount of flow that entered within one time unit, needs two time units to leave it. With the same intuition the flow rate is increased whenever $\gamma_e(\theta) < 1$. Note that in figures of other publications on flows over time the flow rate is often pictured by the width of the flow. But for time-varying networks this is not accurate anymore as the transit speed can vary. Hence, in this paper the flow rates are given by the width of the flow multiplied by the current speed limit.

A *flow over time* is specified by a family of locally integrable and bounded functions $f = (f_e^+, f_e^-)_{e \in E}$ denoting the in- and outflow rates. The *cumulative in- and outflows* are given by

$$F_e^+(\theta) := \int_0^\theta f_e^+(\xi)\,\mathrm{d}\xi \qquad \text{and} \qquad F_e^-(\theta) := \int_0^\theta f_e^-(\xi)\,\mathrm{d}\xi.$$

A flow over time *conserves flow on all arcs* e:

$$F_e^-(\theta + \tau_e(\theta)) \le F_e^+(\theta) \qquad \text{for all } \theta \in [0, \infty], \tag{1}$$

and *conserves flow at every node* $v \in V \setminus \{t\}$ for almost all $\theta \in [0, \infty)$:

$$\sum_{e \in \delta_v^+} f_e^+(\theta) - \sum_{e \in \delta_v^-} f_e^-(\theta) = \begin{cases} 0 & \text{if } v \in V \setminus \{t\}, \\ r(\theta) & \text{if } v = s. \end{cases} \tag{2}$$

A particle entering an arc e at time θ reaches the head of the arc at time $\theta + \tau_e(\theta)$ where it lines up at the point queue. Thereby, the *queue size* $z_e \colon [0, \infty) \to [0, \infty)$ at time $\theta + \tau_e(\theta)$ is defined by $z_e(\theta + \tau_e(\theta)) := F_e^+(\theta) - F_e^-(\theta + \tau_e(\theta))$.

We call a flow over time in a time-varying network *feasible* if we have for almost all $\theta \in [0, \infty)$ that

$$f_e^-(\theta + \tau_e(\theta)) = \begin{cases} \nu_e(\theta + \tau_e(\theta)) & \text{if } z_e(\theta + \tau_e(\theta)) > 0, \\ \min\left\{ \frac{f_e^+(\theta)}{\gamma_e(\theta)}, \nu_e(\theta + \tau_e(\theta)) \right\} & \text{else,} \end{cases} \tag{3}$$

and $f_e^-(\theta) = 0$ for almost all $\theta < \tau_e(0)$.

Note that the outflow rate depends on the speed ratio $\gamma_e(\theta)$ if the queue is empty (see Fig. 3). Otherwise, the particles enter the queue, and therefore, the outflow rate equals the capacity independent of the speed ratio. Furthermore, we observe that every arc with a positive queue always has a positive outflow, since the capacities are required to be strictly positive. And finally, (3) implies (1), which can easily be seen by considering the derivatives of the cumulative flows whenever we have an empty queue, i.e., $F_e^-(\theta + \tau_e(\theta)) = F_e^+(\theta)$. By (3) we have that $f_e^-(\theta + \tau_e(\theta)) \cdot (1 + \tau_e'(\theta)) \le f_e^+(\theta)$. Hence, (2) and (3) are sufficient for a family of functions $f = (f_e^+, f_e^-)_{e \in E}$ to be a feasible flow over time.

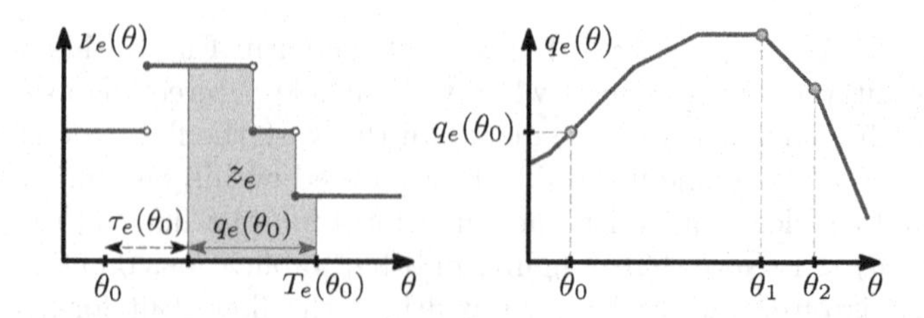

Fig. 4. Waiting times for time-dependent capacities. The waiting time of a particle θ_0 (*right side*) is given by the length of the interval starting at $\theta_0 + \tau_e(\theta_0)$ such that the area underneath the capacity graph equals the queue size at time $\theta_0 + \tau_e(\theta_0)$ (*left side*). The right boundary of the interval equals the exit time $T_e(\theta_0)$. The waiting time does not only depend on the capacity but also on the inflow rate and the transit times. For example, if the capacity and the speed limit are constant but the inflow rate is 0, the waiting time will decrease with a slope of 1 (*right side* within $[\theta_1, \theta_2]$).

The *waiting time* $q_e\colon [0,\infty) \to [0,\infty)$ of a particle that enters the arc at time θ is defined by

$$q_e(\theta) := \min\left\{ q \geq 0 \;\middle|\; \int_{\theta+\tau_e(\theta)}^{\theta+\tau_e(\theta)+q} \nu_e(\xi)\,\mathrm{d}\xi = z_e(\theta + \tau_e(\theta)) \right\}.$$

As we required $\int_0^\theta \nu_e(\xi)\,\mathrm{d}\xi$ to be unbounded for $\theta \to \infty$ the set on the right side is never empty. Hence, $q_e(\theta)$ is well-defined and has a finite value. In addition, q_e is continuous since ν_e is always strictly positive. The *exit time* $T_e\colon [0,\infty) \to [0,\infty)$ denotes the time at which the particles that have entered the arc at time θ finally leave the queue. Hence, we define $T_e(\theta) := \theta + \tau_e(\theta) + q_e(\theta)$. In Fig. 4 we display an illustrative example for the definition of waiting and exit times.

With these definitions we can show the following lemma.

Lemma 2. *For a feasible flow over time f it holds for all $e \in E$, $v \in V$ and $\theta \in [0,\infty)$ that:*

(i) $q_e(\theta) > 0 \;\Leftrightarrow\; z_e(\theta + \tau_e(\theta)) > 0$.
(ii) $z_e(\theta + \tau_e(\theta) + \xi) > 0$ *for all* $\xi \in [0, q_e(\theta))$.
(iii) $F_e^+(\theta) = F_e^-(T_e(\theta))$.
(iv) For $\theta_1 < \theta_2$ *with* $F_e^+(\theta_2) - F_e^+(\theta_1) = 0$ *and* $z_e(\theta_2 + \tau_e(\theta_2)) > 0$ *we have* $T_e(\theta_1) = T_e(\theta_2)$.
(v) The functions T_e are monotonically increasing.
(vi) The functions q_e and T_e are continuous and almost everywhere differentiable.
(vii) For almost all $\theta \in [0,\infty)$ we have

$$T_e'(\theta) = \begin{cases} \frac{f_e^+(\theta)}{\nu_e(T_e(\theta))} & \text{if } q_e(\theta) > 0, \\ \max\left\{ \gamma_e(\theta), \frac{f_e^+(\theta)}{\nu_e(T_e(\theta))} \right\} & \text{else.} \end{cases}$$

Most of the statements follow immediately from the definitions and some involve minor calculations. For (vi) we use Lebesgue's differentiation theorem.

3 Nash Flows over Time

In order to define a dynamic equilibrium we consider the particles as players in a dynamic game. For this the set of particles is identified by the non-negative reals denoted by $\mathbb{R}_{\geq 0}$. The flow volume is hereby given by the Lebesgue-measure, which means that $[a, b] \subseteq \mathbb{R}_{\geq 0}$ with $a < b$ contains a flow volume of $b - a$. The flow particles enter the network according to the ordering of the reals beginning with particle 0. It is worth noting that a particle $\phi \in \mathbb{R}_{\geq 0}$ can be split up further so that for example one half takes a different route than the other half. As characterized by Koch and Skutella, a dynamic equilibrium is a feasible flow over time, where almost all particles only use current shortest paths from s to t. Note that we assume a game with full information. Consequently, all particles know all speed limit and capacity functions in advance and have the ability to perfectly predict the future evolution of the flow over time. Hence, each particle perfectly knows all travel times and can choose its route accordingly. We start by defining the earliest arrival times for a particle $\phi \in \mathbb{R}_{\geq 0}$.

The *earliest arrival time functions* $\ell_v \colon \mathbb{R}_{\geq 0} \to [0, \infty)$ map each particle ϕ to the earliest time $\ell_v(\phi)$ it can possibly reach node v. Hence, it is the solution to

$$\ell_v(\phi) = \begin{cases} \min \left\{ \theta \geq 0 \,\middle|\, \int_0^\theta r(\xi)\,\mathrm{d}\xi = \phi \right\} & \text{for } v = s, \\ \min\limits_{e = uv \in \delta_v^-} T_e(\ell_u(\phi)) & \text{else.} \end{cases} \tag{4}$$

Note that for all $v \in V$ the earliest arrival time function ℓ_v is non-decreasing, continuous and almost everywhere differentiable. This holds directly for ℓ_s and for $v \neq s$ it follows inductively, since these properties are preserved by the concatenation $T_e \circ \ell_u$ and by the minimum of finitely many functions.

For a particle ϕ we call an arc $e = uv$ *active* if $\ell_v(\phi) = T_e(\ell_u(\phi))$. The set of all these arcs are denoted by E'_ϕ and these are exactly the arcs that form the current shortest paths from s to some node v. For this reason we call the subgraph $G'_\phi = (V, E'_\phi)$ the *current shortest paths network* for particle ϕ. Note that G'_ϕ is acyclic and that every node is reachable by s within this graph. The arcs where particle ϕ experiences a waiting time when traveling along shortest paths only are called *resetting arcs* denoted by $E^*_\phi := \{ e = uv \in E \mid q_e(\ell_u(\phi)) > 0 \}$.

Nash flows over time in time-varying networks are defined in the exact same way as Cominetti et al. defined them in the base model [3, Definition 1].

Definition 1 (Nash flow over time). *We call a feasible flow over time f a* Nash flow over time *if the following* Nash flow condition *holds:*

$$f_e^+(\theta) > 0 \;\Rightarrow\; \theta \in \ell_u(\Phi_e) \quad \textit{for all } e = uv \in E \textit{ and almost all } \theta \in [0, \infty), \tag{N}$$

where $\Phi_e := \{ \phi \in \mathbb{R}_{\geq 0} \mid e \in E'_\phi \}$ is the set of particles for which arc e is active.

As Cominetti et al. showed in [4, Theorem 1] these Nash flows over time can be characterized as follows.

Lemma 3. *A feasible flow over time f is a Nash flow over time if, and only if, for all $e = uv \in E$ and all $\phi \in \mathbb{R}_{\geq 0}$ we have $F_e^+(\ell_u(\phi)) = F_e^-(\ell_v(\phi))$.*

Since the exit and the earliest arrival times have the same properties in time-varying networks as in the base model, this lemma follows with the exact same proof that was given by Cominetti et al. for the base model [4, Theorem 1]. The same is true for the following lemma; see [4, Proposition 2].

Lemma 4. *Given a Nash flow over time the following holds for all particles ϕ:*

(i) $E_\phi^* \subseteq E_\phi'$.
(ii) $E_\phi' = \{\, e = uv \mid \ell_v(\phi) \geq \ell_u(\phi) + \tau_e(\theta) \,\}$.
(iii) $E_\phi^* = \{\, e = uv \mid \ell_v(\phi) > \ell_u(\phi) + \tau_e(\theta) \,\}$.

Motivated by Lemma 3 we define the *underlying static flow* for $\phi \in \mathbb{R}_{\geq 0}$ by

$$x_e(\phi) := F_e^+(\ell_u(\phi)) = F_e^-(\ell_v(\phi)) \quad \text{for all } e = uv \in E.$$

By the definition of ℓ_s and the integration of (2) we have $\int_0^{\ell_s(\phi)} r(\xi)\,\mathrm{d}\xi = \phi$, and hence, $x_e(\phi)$ is a static s-t-flow (classical network flow) of value ϕ, whereas the derivatives $(x_e'(\phi))_{e \in E}$ form a static s-t-flow of value 1.

4 Thin Flows

Thin flows with resetting, introduced by Koch and Skutella [15], characterize the derivatives $(x_e')_{e \in E}$ and $(\ell_v')_{v \in V}$ of Nash flows over time in the base model. In the following we will transfer this concept to time-varying networks.

Consider an acyclic network $G' = (V, E')$ with a source s and a sink t, such that every node is reachable by s. Each arc is equipped with a capacity $\nu_e > 0$ and a speed ratio $\gamma_e > 0$. Furthermore, we have a network inflow rate of $r > 0$ and an arc set $E^* \subseteq E'$. We obtain the following definition.

Definition 2 (Thin flow with resetting in a time-varying network). *A static s-t flow $(x_e')_{e \in E}$ of value 1 together with a node labeling $(\ell_v')_{v \in V}$ is a* thin flow with resetting *on E^* if:*

$$\ell_s' = \frac{1}{r} \tag{TF1}$$

$$\ell_v' = \min_{e = uv \in E'} \rho_e(\ell_u', x_e') \quad \textit{for all } v \in V \setminus \{\, s \,\}, \tag{TF2}$$

$$\ell_v' = \rho_e(\ell_u', x_e') \quad \textit{for all } e = uv \in E' \textit{ with } x_e' > 0, \tag{TF3}$$

$$\textit{where} \qquad \rho_e(\ell_u', x_e') := \begin{cases} \frac{x_e'}{\nu_e} & \textit{if } e = uv \in E^*, \\ \max\left\{\, \gamma_e \cdot \ell_u', \frac{x_e'}{\nu_e} \,\right\} & \textit{if } e = uv \in E' \backslash E^*. \end{cases}$$

The derivatives of a Nash flow over time in time-varying networks do indeed form a thin flow with resetting as the following theorem shows.

Theorem 1. *For almost all* $\phi \in \mathbb{R}_{\geq 0}$ *the derivatives* $(x'_e(\phi))_{e \in E'_\phi}$ *and* $(\ell'_v(\phi))_{v \in V}$ *of a Nash flow over time* $f = (f_e^+, f_e^-)_{e \in E}$ *form a thin flow with resetting on* E^*_ϕ *in the current shortest paths network* $G'_\phi = (V, E'_\phi)$ *with network inflow rate* $r(\ell_s(\phi))$ *as well as capacities* $\nu_e(\ell_v(\phi))$ *and speed ratios* $\gamma_e(\ell_u(\phi))$ *for each arc* $e = uv \in E$.

Proof. Let $\phi \in \mathbb{R}_{\geq 0}$ be a particle such that for all arcs $e = uv \in E$ the derivatives of x_e, ℓ_u, $T_e \circ \ell_u$ and τ_e exist and $x'_e(\phi) = f_e^+(\ell_u(\phi)) \cdot \ell'_u(\phi) = f_e^-(\ell_v(\phi)) \cdot \ell'_v(\phi)$ as well as $1 + \tau'_e(\ell_u(\phi)) = \gamma_e(\ell_u(\phi))$. This is given for almost all ϕ.

By (4) we have $\int_0^{\ell_s(\phi)} r(\xi)\, \mathrm{d}\xi = \phi$ and taking the derivative by applying the chain rule, yields $r(\ell_s(\phi)) \cdot \ell'_s(\phi) = 1$, which shows (TF1).

Taking the derivative of (4) at time $\ell_u(\phi)$ by using the differentiation rule for a minimum yields $\ell'_v(\phi) = \min_{e=uv \in E'} T'_e(\ell_u(\phi)) \cdot \ell'_u(\phi)$. By using Lemma 2 (vii) we obtain

$$T'_e(\ell_u(\phi)) \cdot \ell'_u(\phi) = \begin{cases} \frac{f_e^+(\ell_u(\phi))}{\nu_e(T_e(\ell_u(\phi)))} \cdot \ell'_u(\phi) & \text{if } q_e(\ell_u(\phi)) > 0, \\ \max\left\{ \gamma_e(\ell_u(\phi)), \frac{f_e^+(\ell_u(\phi))}{\nu_e(T_e(\ell_u(\phi)))} \right\} \cdot \ell'_u(\phi) & \text{else,} \end{cases}$$
$$= \rho_e(\ell'_u(\phi), x'_e(\phi)),$$

which shows (TF2).

Finally, in the case of $f_e^-(\ell_v(\phi)) \cdot \ell'_v(\phi) = x'_e(\phi) > 0$ we have by (3) that

$$\ell'_v(\phi) = \frac{x'_e(\phi)}{f_e^-(\ell_v(\phi))} = \begin{cases} \dfrac{x'_e(\phi)}{\min\left\{ \frac{f_e^+(\ell_u(\phi))}{\gamma_e(\ell_u(\phi))}, \nu_e(\ell_v(\phi)) \right\}} & \text{if } q_e(\ell_u(\phi)) = 0, \\ \dfrac{x'_e(\phi)}{\nu_e(\ell_v(\phi))} & \text{else,} \end{cases}$$
$$= \begin{cases} \max\left\{ \gamma_e(\ell_u(\phi)) \cdot \ell'_u(\phi), \frac{x'_e(\phi)}{\nu_e(\ell_v(\phi))} \right\} & \text{if } e \in E'_\phi \setminus E^*_\phi, \\ \frac{x'_e(\phi)}{\nu_e(\ell_v(\phi))} & \text{if } e \in E^*_\phi, \end{cases}$$
$$= \rho_e(\ell'_u(\phi), x'_e(\phi)).$$

This shows (TF3) and finishes the proof. □

In order to construct Nash flows over time in time-varying networks, we first have to show that there always exists a thin flow with resetting.

Theorem 2. *Consider an acyclic graph* $G' = (V, E')$ *with source* s*, sink* t*, capacities* $\nu_e > 0$*, speed ratios* $\gamma_e > 0$ *and a subset of arcs* $E^* \subseteq E'$*, as well as a network inflow* $r > 0$*. Furthermore, suppose that every node is reachable from* s*. Then there exists a thin flow* $((x'_e)_{e \in E}, (\ell'_v)_{v \in V})$ *with resetting on* E^*.

This proof works exactly as the proof for the existence of thin flows in the base model presented by Cominetti et al. [4, Theorem 3].

5 Constructing Nash Flows over Time

In the remaining part of this paper we assume that for all $e \in E$ the functions ν_e and λ_e as well as the network inflow rate function r are right-constant. In order to show the existence of Nash flows over time in time-varying networks we use the same α-extension approach as used by Koch and Skutella in [15] for the base model. The key idea is to start with the empty flow over time and expand it step by step by using a thin flow with resetting.

Given a *restricted Nash flow over time* f on $[0, \phi]$, i.e., a Nash flow over time where only the particles in $[0, \phi]$ are considered, we obtain well-defined earliest arrival times $(\ell_v(\phi))_{v \in V}$ for particle ϕ. Hence, by Lemma 4 we can determine the current shortest paths network $G'_\phi = (V, E'_\phi)$ with the resetting arcs E^*_ϕ, the capacities $\nu_e(\ell_v(\phi))$ and speed ratios $\gamma_e(\ell_u(\phi))$ for all arcs $e = uv \in E'$ as well as the network inflow rate $r(\ell_s(\phi))$. By Theorem 2 there exists a thin flow $((x'_e)_{e \in E'}, (\ell'_v)_{v \in V})$ on G'_ϕ with resetting on E^*_ϕ. For $e \notin E'_\phi$ we set $x'_e := 0$. We extend the ℓ- and x-functions for some $\alpha > 0$ by

$$\ell_v(\phi + \xi) := \ell_v(\phi) + \xi \cdot \ell'_v \quad \text{and} \quad x_e(\phi) := x_e(\phi) + \xi \cdot x'_e \qquad \text{for all } \xi \in [0, \alpha)$$

and the in- and outflow rate functions by

$$f_e^+(\theta) := \frac{x'_e}{\ell'_u} \text{ for } \theta \in [\ell_u(\phi), \ell_u(\phi+\alpha)); \qquad f_e^-(\theta) := \frac{x'_e}{\ell'_v} \text{ for } \theta \in [\ell_v(\phi), \ell_v(\phi+\alpha)).$$

We call this extended flow over time *α-extension*. Note that $\ell'_u = 0$ means that $[\ell_u(\phi), \ell_u(\phi + \alpha))$ is empty, and the same holds for ℓ'_v.

An α-extension is a restricted Nash flow over time, which we will prove later on, as long as the α stays within reasonable bounds. Similar to the base model we have to ensure that resetting arcs stay resetting and non-active arcs stay non-active for all particles in $[\phi, \phi + \alpha)$. Since the transit times may now vary over time, we have the following conditions for all $\xi \in [0, \alpha)$:

$$\ell_v(\phi) + \xi \cdot \ell'_v - \ell_u(\phi) - \xi \cdot \ell'_u > \tau_e(\ell_u(\phi) + \xi \cdot \ell'_u)) \quad \text{for every } e \in E^*_\phi, \tag{5}$$
$$\ell_v(\phi) + \xi \cdot \ell'_v - \ell_u(\phi) - \xi \cdot \ell'_u < \tau_e(\ell_u(\phi) + \xi \cdot \ell'_u)) \quad \text{for every } e \in E \setminus E'_\phi. \tag{6}$$

Furthermore, we need to ensure that the capacities of all active arcs and the network inflow rate do not change within the phase:

$$\nu_e(\ell_v(\phi)) = \nu_e(\ell_v(\phi) + \xi \cdot \ell'_v) \quad \text{for every } e \in E'_\phi \text{ and all } \xi \in [0, \alpha). \tag{7}$$
$$r(\ell_s(\phi)) = r(\ell_s(\phi) + \xi \cdot \ell'_s) \quad \text{for all } \xi \in [0, \alpha). \tag{8}$$

Finally, the speed ratios need to stay constant for all active arcs, i.e.,

$$\gamma_e(\ell_u(\phi)) = \gamma_e(\ell_u(\phi) + \xi \cdot \ell'_u) \quad \text{for every } e \in E'_\phi \text{ and all } \xi \in [0, \alpha). \tag{9}$$

We call an $\alpha > 0$ *feasible* if it satisfies (5) to (9).

Lemma 5. *Given a restricted Nash flow over time f on $[0, \phi]$ then for right-constant capacities and speed limits there always exists a feasible $\alpha > 0$.*

Proof. By Lemma 4 we have that $\ell_v(\phi) - \ell_u(\phi) > \tau_e(\phi)$ for all $e \in E^*_\phi$ and $\ell_v(\phi) - \ell_u(\phi) < \tau_e(\phi)$ for all $e \in E \setminus E'_\phi$. Since τ_e is continuous there is an $\alpha_1 > 0$ such that (5) and (6) are satisfied for all $\xi \in [0, \alpha_1)$. Since ν_e, r and λ_e are right-constant so is γ_e, and hence, there is an $\alpha_2 > 0$ such that (7), (8) and (9) are fulfilled for all $\xi \in [0, \alpha_2)$. Clearly, $\alpha := \min\{\alpha_1, \alpha_2\} > 0$ is feasible. □

For the maximal feasible α we call the interval $[\phi, \phi + \alpha)$ a *thin flow phase.*

Lemma 6. *An α-extension is a feasible flow over time and the extended ℓ-labels coincide with the earliest arrival times, i.e., they satisfy Eq. (4) for all $\varphi \in [\phi, \phi + \alpha)$.*

The final step is to show that an α-extension is a restricted Nash flow over time on $[0, \phi + \alpha)$ and that we can continue this process up to ∞.

Theorem 3. *Given a restricted Nash flow over time $f = (f_e^+, f_e^-)_{e \in E}$ on $[0, \phi)$ in a time-varying network and a feasible $\alpha > 0$ then the α-extension is a restricted Nash flow over time on $[0, \phi + \alpha)$.*

Proof. Lemma 3 yields $F_e^+(\ell_u(\varphi)) = F_e^-(\ell_v(\varphi))$ for all $\varphi \in [0, \phi)$, so for $\xi \in [0, \alpha)$ it holds that

$$F_e^+(\ell_u(\phi+\xi)) = F_e^+(\ell_u(\phi)) + \frac{x'_e}{\ell'_u} \cdot \xi \cdot \ell'_u = F_e^-(\ell_v(\phi)) + \frac{x'_e}{\ell'_v} \cdot \xi \cdot \ell'_v = F_e^-(\ell_v(\phi+\xi)).$$

It follows again by Lemma 3 together with Lemma 6 that the α-extension is a restricted Nash flow over time on $[0, \phi + \alpha)$. □

Finally, we obtain our main result:

Theorem 4. *There exists a Nash flow over time in every time-varying network with right-constant speed limits, capacities and network inflow rates.*

Proof. The process starts with the empty flow over time, i.e., a restricted Nash flow over time for $[0, 0)$. We apply Theorem 3 with a maximal feasible α. If one of the α is unbounded we are done. Otherwise, we obtain a sequence $(f_i)_{i \in \mathbb{N}}$, where f_i is a restricted Nash flow over time for $[0, \phi_i)$, with a strictly increasing sequence $(\phi_i)_{i \in \mathbb{N}}$. In the case that this sequence has a finite limit, say $\phi_\infty < \infty$, we define a restricted Nash flow over time f^∞ for $[0, \phi_\infty)$ by using the point-wise limit of the x- and ℓ-labels, which exists due to monotonicity and boundedness of these functions. Note that there are only finitely many different thin flows, and therefore, the derivatives x' and ℓ' are bounded. Then the process can be restarted from this limit point. This so called *transfinite induction* argument works as follows: Let $\mathcal{P}_G$ be the set of all particles $\phi \in \mathbb{R}_{\geq 0}$ for which there exists a restricted Nash flow over time on $[0, \phi)$ constructed as described above. The set $\mathcal{P}_G$ cannot have a maximal element because the corresponding Nash flow over time could be extended by using Theorem 3. But $\mathcal{P}_G$ cannot have an upper

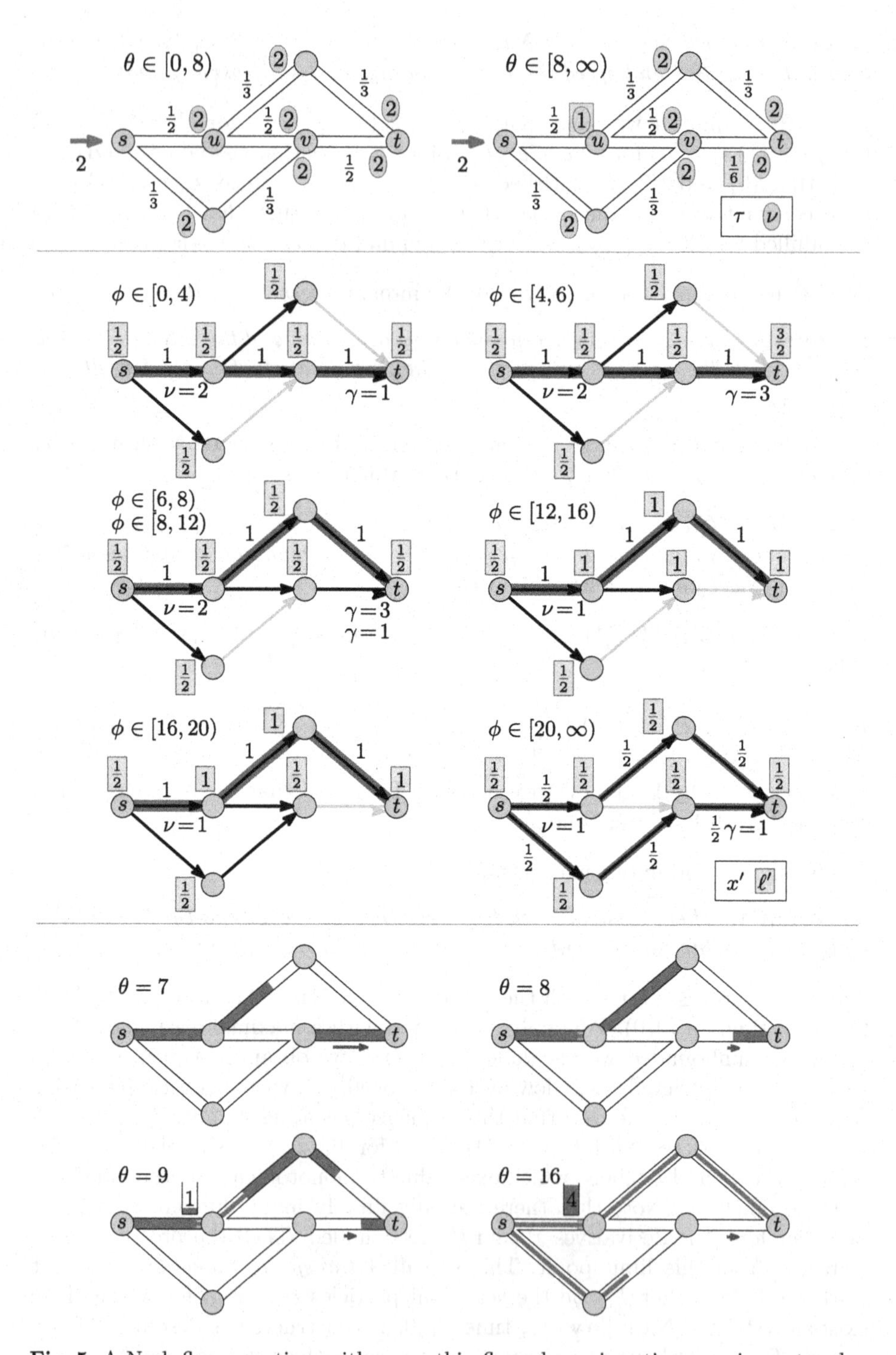

Fig. 5. A Nash flow over time with seven thin flow phases in a time-varying network.

bound either since the limit of any convergent sequence would be contained in this set. Therefore, there exists an unbounded increasing sequence $(\phi_i)_{i=1}^{\infty} \in \mathcal{P}_G$. As a restricted Nash flow over time on $[0, \phi_{i+1}]$ contains a restricted Nash flow over time on $[0, \phi_i]$ we can assume that there exists a sequence of *nested* restricted Nash flow over time. Hence, we can construct a Nash flow over time f on $[0, \infty)$ by taking the point-wise limit of the x- and ℓ-labels, completing the proof. □

Example. An example of a Nash flow over time in a time-varying network together with the corresponding thin flows is shown in Fig. 5 on the next page. *On the top:* The network properties before time 8 (*left side*) and after time 8 (*right side*). *In the middle:* There are seven thin flow phases. Note that the third and forth phase (both depicted in the same network) are almost identical and only the speed ratio of arc vt changes, which does not influence the thin flow at all. *At the bottom:* Some key snapshots in time of the resulting Nash flow over time. The current speed limit λ_{vt} is visualized by the length of the green arrow and, for $\theta \geq 8$, the reduced capacity $\nu_{su}(\theta)$ is displayed by a red bottle-neck.

As displayed at the top the capacity of arc su drops from 2 to 1 at time 8 and, at the same time, the speed limit of arc vt decreases from $\frac{1}{2}$ to $\frac{1}{6}$. The first event for particle 4 is due to a change of the speed ratio leading to an increase of ℓ_t'. For particle 6, the top path becomes active and is taken by all following flow as particles on arc vt are still slowed down. For particle 8, the speed ratio at arc vt changes back to 1 but, as this arc is inactive, this does not change anything. Particle 12 is the first to experience the reduced capacity on arc su. The corresponding queue of this arc increases until the bottom path becomes active. This happens in two steps: first only the path up to node v becomes active for $\phi = 16$, and finally, the complete path is active from $\phi = 20$ onwards.

6 Conclusion and Open Problems

In this paper, we extended the base model that was introduced by Koch and Skutella, to networks which capacities and speed limits that changes over time. We showed that all central results, namely the existence of dynamic equilibria and their underlying structures in form of thin flow with resetting, can be transfered to this new model. With these new insights it is possible to model more general traffic scenarios in which the network properties are time-dependent. In particular, the flooding evacuation scenario, which was mentioned in the introduction, could not be modeled (not even approximately) in the base model.

There are still a lot of open question concerning time-varying networks. For example, it would be interesting to consider other flows over time in this setting, such as earliest arrival flows or instantaneous dynamic equilibria (see [11]) and show their existence. Can the proof for the bound of the price of anarchy [6] be transfered to this model, or is it possible to construct an example where the price of anarchy is unbounded?

References

1. Bhaskar, U., Fleischer, L., Anshelevich, E.: A stackelberg strategy for routing flow over time. Games Econ. Behav. **92**, 232–247 (2015)
2. Cao, Z., Chen, B., Chen, X., Wang, C.: A network game of dynamic traffic. In Proceedings of the 2017 ACM Conference on Economics and Computation, pp. 695–696 (2017)
3. Cominetti, R., Correa, J.R., Larré, O.: Existence and uniqueness of equilibria for flows over time. In: Aceto, L., Henzinger, M., Sgall, J. (eds.) ICALP 2011. LNCS, vol. 6756, pp. 552–563. Springer, Heidelberg (2011). https://doi.org/10.1007/978-3-642-22012-8_44
4. Cominetti, R., Correa, J., Larré, O.: Dynamic equilibria in fluid queueing networks. Oper. Res. **63**(1), 21–34 (2015)
5. Cominetti, R., Correa, J., Olver, N.: Long term behavior of dynamic equilibria in fluid queuing networks. In: Eisenbrand, F., Koenemann, J. (eds.) IPCO 2017. LNCS, vol. 10328, pp. 161–172. Springer, Cham (2017). https://doi.org/10.1007/978-3-319-59250-3_14
6. Correa, J., Cristi, A., Oosterwijk, T.: On the price of anarchy for flows over time. In Proceedings of the 2019 ACM Conference on Economics and Computation, pp. 559–577 (2019)
7. Fleischer, L., Tardos, É.: Efficient continuous-time dynamic network flow algorithms. Oper. Res. Lett. **23**(3–5), 71–80 (1998)
8. Ford, L.R., Fulkerson, D.R.: Constructing maximal dynamic flows from static flows. Oper. Res. **6**, 419–433 (1958)
9. Ford, L.R., Fulkerson, D.R.: Flows in Networks. Princeton University Press, Princeton (1962)
10. Gale, D.: Transient flows in networks. Michigan Math. J. **6**(1), 59–63 (1959)
11. Graf, L., Harks, T., Sering, L.: Dynamic flows with adaptive route choice. Math. Program. (2020)
12. Harks, T., Peis, B., Schmand, D., Tauer, B., Vargas Koch, L.: Competitive packet routing with priority lists. ACM Trans. Econo. Comp. **6**(1), 4 (2018)
13. Koch, R.: Routing Games over Time. Ph.D. thesis, Technische Universität Berlin (2012). https://doi.org/10.14279/depositonce-3347
14. Koch, R., Skutella, M.: Nash equilibria and the price of anarchy for flows over time. In: Mavronicolas, M., Papadopoulou, V.G. (eds.) SAGT 2009. LNCS, vol. 5814, pp. 323–334. Springer, Heidelberg (2009). https://doi.org/10.1007/978-3-642-04645-2_29
15. Koch, R., Skutella, M.: Nash equilibria and the price of anarchy for flows over time. Theor. Comput. Syst. **49**(1), 71–97 (2011)
16. Macko, M., Larson, K., Steskal, L.: Braess's paradox for flows over time. Theor. Comput. Syst. **53**(1), 86–106 (2013)
17. Minieka, E.: Maximal, lexicographic, and dynamic network flows. Oper. Res. **21**(2), 517–527 (1973)
18. Peis, B., Tauer, B., Timmermans, V., Vargas Koch, L.: Oligopolistic competitive packet routing. In: 18th Workshop on Algorithmic Approaches for Transportation Modelling, Optimization, and Systems (2018)
19. Scarsini, M., Schröder, M., Tomala, T.: Dynamic atomic congestion games with seasonal flows. Oper. Res. **66**(2), 327–339 (2018)
20. Sering, L., Skutella, M.: Multi-source multi-sink Nash flows over time. In: 18th Workshop on Algorithmic Approaches for Transportation Modelling, Optimization, and Systems, vol. 65, pp. 12:1–12:20 (2018)

21. Sering, L., Vargas Koch, L.: Nash flows over time with spillback. In: Proceedings of the Thirtieth Annual ACM-SIAM Symposium on Discrete Algorithms, pp. 935–945. SIAM (2019)
22. Skutella, M.: An introduction to network flows over time. In: Cook, W., Lovász, L., Vygen, J. (eds.) Research trends in combinatorial optimization, pp. 451–482. Springer, Heidelberg (2009). https://doi.org/10.1007/978-3-540-76796-1_21
23. Vickrey, W.S.: Congestion theory and transport investment. Am. Econ. Rev. **59**(2), 251–260 (1969)
24. Wardrop, J.G.: Some theoretical aspects of road traffic research. Proc. Inst. Civil Engineers **1**(5), 767–768 (1952)
25. Wilkinson, W.L.: An algorithm for universal maximal dynamic flows in a network. Oper. Res. **19**(7), 1602–1612 (1971)
26. Yagar, S.: Dynamic traffic assignment by individual path minimization and queuing. Transp. Res. **5**(3), 179–196 (1971)

Price of Anarchy in Congestion Games with Altruistic/Spiteful Players

Marc Schröder(✉)

RWTH Aachen University, Aachen, Germany
marc.schroeder@oms.rwth-aachen.de

Abstract. We consider an extension of atomic congestion games with altruistic or spiteful players. Restricting attention to games with affine costs, we study a special class of perception-parameterized congestion games as introduced by Kleer and Schäfer [19]. We provide an upper bound on the price of anarchy for games with players that are sufficiently spiteful, answering an open question posed in [19]. This completes the characterization of the price of anarchy as a function of the level of altruism/spite. We also provide an upper bound on the price of stability when players are sufficiently altruistic, which almost completes the picture of the price of stability as a function of the level of altruism/spite.

Keywords: Price of anarchy · Price of stability · Altruism · Spite

1 Introduction

Congestion games provide a natural model for resource allocation in large-scale decentralized systems such as traffic on roads or data packets in computer networks. A congestion game is given by a set of users that each want to utilize a subset of resources, while the cost of these resources increase as more users make use of them. These games were first introduced by Rosenthal [24], who showed that a pure Nash equilibrium always exists. In fact, Rosenthal proved that congestion games are potential games, which implies that better-response dynamics converge to a Nash equilibrium. This means that not only is there a steady state, selfish users are able to reach such a steady state with natural dynamics. However, these steady states might be inefficient from a societal perspective. The inefficiency of Nash equilibria is quantified by the *price of anarchy* [20] and the *price of stability* [3,26].

The assumption that players are selfish and only care about their own costs but not about others is restrictive and is at odds with altruistic or spiteful behavior that is observed in real-life experiments (see for example, [21] or [22]). Caragiannis et al. [8] proposed a simple extension of congestion games that allows players to be altruistic. They proved, contrary to what one would expect, that the price of anarchy in general increases with the level of altruism. Chen et al. [9] tested this result for several other classes of games and concluded that

T. Harks and M. Klimm (Eds.): SAGT 2020, LNCS 12283, pp. 146–159, 2020.
https://doi.org/10.1007/978-3-030-57980-7_10

the increase in the price of anarchy with altruism is not a universal phenomenon. In fact, if one takes the price of stability as a measure of inefficiency of equilibria, then the price of stability even decreases with the level of altruism.

Kleer and Schäfer [19] generalized the previous two models and unified several extensions of congestion games by means of the class of perception-parameterized congestion games. This class allows to model congestion games in which players are spiteful instead of altruistic. Kleer and Schäfer [19] leaves it as an open question to find tight bounds on the price of anarchy when players are sufficiently spiteful. This paper answers that open question.

1.1 Our Contribution

We study atomic congestion games with affine cost functions, and altruistic or spiteful players. We model this by means of a special class of perception-parameterized congestion games, introduced by Kleer and Schäfer [19], in which the perceived cost of a player choosing a resource e with load x is equal to $a_e \cdot (1 + p \cdot (x-1)) + b_e$ for some $p > 0$. If $p \in (0, 1)$, the model considers a game with players that are spiteful, whereas if $p > 1$, the players are altruistic similar to the models introduced by Caragiannis et al. [8] and Chen et al. [9].

Our main result is the completion of the characterization of the $\mathsf{PoA}(p)$ for all $p > 0$. Caragiannis et al. [8] proved a tight bound of $\frac{1+4p}{1+p}$ if $1 \leq p \leq 2$ and of $1 + p$ if $p \geq 2$. Caragiannis et al. [7] gave a lower bound of $1 + \frac{2}{\sqrt{3}}$ for all $p > 0$ and showed it is tight for $p = \frac{2\sqrt{3}}{9-2\sqrt{3}}$. Kleer and Schäfer [19] extended the result of Caragiannis et al. [8] by giving a tight bound of $\frac{1+4p}{1+p}$ for $\frac{2\sqrt{3}}{9-2\sqrt{3}} \leq p \leq 1$. They posed a tight bound for $0 < p < \frac{2\sqrt{3}}{9-2\sqrt{3}}$ as open question.

We show that the lower bound of Caragiannis et al. [7] of $1 + \frac{2}{\sqrt{3}}$ is tight for $4 - 2\sqrt{3} \leq p \leq \frac{2\sqrt{3}}{9-2\sqrt{3}}$, and that the lower bound of Kleer and Schäfer [19] of $\frac{4}{p(4-p)}$ is tight for $0 < p \leq 4 - 2\sqrt{3}$. Figure 1 plots the price of anarchy as a function of p. Similar to [19], our proof makes use of (λ, μ)-smoothness defined by [25]. So all upper bounds extend to mixed, correlated and coarse-correlated equilibria.

Figure 1 also depicts the price of stability as a function of p. Chen et al. [9] derived an upper bound on the price of stability of $\frac{1+\sqrt{3}}{p-1+\sqrt{3}}$ for $1 \leq p \leq 2$. Kleer and Schäfer [19] obtained a tight bound for the price of stability of $\frac{1+\sqrt{3}}{p-1+\sqrt{3}}$ for $4 - 2\sqrt{3} \leq p \leq 2$ and gave a lower bound for $0 < p \leq 4 - 2\sqrt{3}$. We complete this picture by showing that the lower bound of $\frac{4}{p(4-p)}$ is tight for $0 < p \leq 4 - 2\sqrt{3}$, and we derive an upper bound on the price of stability of $\frac{4-2\sqrt{3}+(\sqrt{3}-1)p}{2}$ for $p \geq 2$. We leave the question whether this upper bound for $p \geq 2$ is tight as open question. Caragiannis et al. [8] derived a lower bound of $\frac{2+4p}{4}$ on the price of stability for $p \geq 2$ using two players and two resources.

1.2 Related Work

The price of anarchy of atomic congestion games was examined in Christodoulou and Koutsoupias [13] for unweighted players, and in Awerbuch et al. [4] for unweighted and weighted players. Aland et al. [1] provided exact bounds for the price of anarchy when costs are polynomial functions. As an extension of results by Harks and Végh [17] and Aland et al. [1], Roughgarden [25] introduced the concept of (λ, μ)-smoothness which provides a simple yet powerful tool to bound the price of anarchy. The bounds obtained by this technique not only hold for pure equilibria, but also for mixed, correlated, and coarse-correlated equilibria. Our proofs make use of (λ, μ)-smoothness.

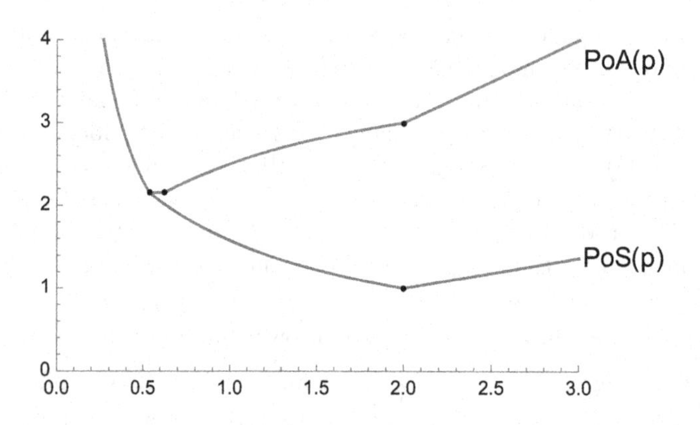

Fig. 1. Upper bounds on the price of anarchy (red) and the price of stability (blue). Almost all bounds are tight, except for the upper bound on the price of stability if $p \geq 2$. (Color figure online)

The price of stability of atomic congestion games was studied in Christodoulou and Koutsoupias [12] and this bound was proven to be tight in Caragiannis et al. [6]. Christodoulou and Gairing [10] provided exact bounds for the price of stability when costs are polynomial functions. Lower and upper bounds for weighted players were given by Christodoulou et al. [11].

Multiple other papers than [8] and [9] studied altruism in congestion games. Hoefer and Skopalik [18] consider an atomic congestion game in which players are allowed to have different levels of altruism. Their focus is on the existence and computation of pure Nash equilibria. Anagnostopoulos et al. [2] generalized the model of [8] by assuming that there is a matrix of coefficients measuring the importance of the other players to a player. Bilo [5] extended and generalized some of the existence and inefficiency results in this model.

Our model is a special class of perception-parameterized congestion games, introduced by Kleer and Schäfer [19]. As already mentioned, this class extends the models of Caragiannis et al. [8] and Chen et al. [9]. Other special classes of perception-parameterized congestion games are the following. Caragiannis et al. [7] studied the impact of universal tax functions on the price of anarchy.

Piliouras et al. [23] considered an atomic congestion game where players using a given resource are randomly ordered, and their costs depend on their position in this order. Cominetti et al. [15] investigated Bernoulli congestion games. These are congestion games in which each player independently participates in the game with a fixed probability. Recently, Cominetti et al. [14] studied the convergence of Bernoulli congestion games to Poisson games as the probability of participation goes to zero.

2 Congestion Games with Spiteful Players

2.1 Atomic Congestion Games

Consider a finite set of players $N = \{1, \dots, n\}$ and a finite set of resources E. Each player $i \in N$ has a set of feasible strategies $S_i \subseteq 2^E$. Given a profile $s \in S := \times_{i \in N} S_i$, the cost for player i is

$$c_i(s) = \sum_{e \in s_i} c_e(n_e(s)) \,. \tag{2.1}$$

Here, $n_e(s)$ is the *load* of resource $e \in E$, defined as the number of players using the resource, that is,

$$n_e(s) = |\{j \in N | e \in s_j\}|, \tag{2.2}$$

and $c_e : \mathbb{N} \to \mathbb{R}_+$ is the cost of the resource e, with $c_e(k)$ being the cost experienced by each player using that resource when the load is k.

The tuple $\Gamma = (N, E, S, (c_e)_{e \in E})$ defines an atomic congestion game. We define the *social cost* to be the sum of all players' costs:

$$c(s) = \sum_{i \in N} c_i(s) = \sum_{e \in E} n_e(s)\, c_e(n_e(s)) \,, \tag{2.3}$$

and we call a *social optimum* any strategy profile s^* that minimizes this social cost.

2.2 Altruistic/Spiteful Players

Given an atomic congestion game Γ, we consider an extension that allows players to be altruistic or spiteful. A player is *altruistic* if his costs increase when the costs of the other players increase. A player is *spiteful* if his costs decrease when the costs of the other players increase. We restrict attention to congestion games with nondecreasing and nonnegative affine costs, that is, we restrict the attention to the class $\mathcal{C}_0$ of costs of the form $c(x) = a\,x + b$ with $a, b \geq 0$.

The model we consider is a special class of the class of perception-parameterized congestion games defined by [19]. Given $p > 0$, the *perceived cost* for player $i \in N$ is given by

$$c_i^p(s) = \sum_{e \in s_i} a_e \cdot (1 + p(n_e(s) - 1)) + b_e. \tag{2.4}$$

We denote the resulting congestion game by Γ^p.

A *pure Nash equilibrium* is a strategy profile $s \in S$ such that no player $i \in N$ can benefit by unilaterally deviating to $s_i \in S_i$, that is, for every player $i \in N$ and every $s_i' \in S_i$, we have

$$c_i^p(s) \leq c_i^p(s_i', s_{-i}). \tag{2.5}$$

The set of pure Nash equilibria of this game is denoted by $\mathsf{NE}(\Gamma^p)$. Rosenthal [24] showed that atomic congestion games are potential games. Given that our extension is still an atomic congestion game but with adjusted cost functions, a potential exists. In particular, the set $\mathsf{NE}(\Gamma^p)$ is nonempty. Rosenthal's potential for our game is defined by

$$\Phi^p(s) = \sum_{e \in E} a_e \cdot \left(n_e(s) + p \cdot \frac{n_e(s)(n_e(s) - 1)}{2} \right) + b_e \cdot n_e(s).$$

It is not immediately clear that the costs in Eq. (2.4) model altruism or spite. Caragiannis et al. [8] defined altruism as follows. The cost of a ξ-altruistic player, where $\xi \in [0, 1]$, equals

$$c_i^\xi(s) = (1 - \xi) \cdot c_i(s) + \xi(c(s) - c_i(s))$$

Later Chen et al. [9] defined a slightly different version of altruism. The cost of an α-altruistic player, where $\alpha \in [0, 1]$, equals

$$c_i^\alpha(s) = (1 - \alpha) \cdot c_i(s) + \alpha \cdot c(s).$$

The next two results show that perception-parameterized congestion games generalize the model of [8] and [9]. This equivalence has already been observed by [19]. The proofs can be found in the appendix.

Lemma 1. *For all $\xi \in [0, 1)$, s is a Nash equilibrium with respect to costs $c_i^\xi(\cdot)$ if and only if s is a Nash equilibrium with respect to costs $c_i^p(\cdot)$, where $p = \frac{1}{1-\xi}$.*

Lemma 2. *For all $\alpha \in [0, 1]$, s is a Nash equilibrium with respect to costs $c_i^\alpha(\cdot)$ if and only if s is a Nash equilibrium with respect to costs $c_i^p(\cdot)$, where $p = 1 + \alpha$.*

Summarizing, the players are spiteful if $p \in (0, 1)$ and the players are altruistic if $p > 1$. To be more precise, if $p \geq 1$, this model is equivalent to the model of [8], and if $p \in [1, 2]$, this model is equivalent to the model of [9].

2.3 Price of Anarchy

The price of anarchy and price of stability are defined as

$$\mathsf{PoA}(\Gamma^p) = \max_{s \in \mathsf{NE}(\Gamma^p)} \frac{c(s)}{c(s^*)} \quad \text{and} \quad \mathsf{PoS}(\Gamma^p) = \min_{s \in \mathsf{NE}(\Gamma^p)} \frac{c(s)}{c(s^*)}, \tag{2.6}$$

where s^* is a social optimum and we recall that

$$c(s) = \sum_{e \in E} n_e(s)\, c_e(n_e(s)). \tag{2.7}$$

For each $p > 0$, we define

$$\mathsf{PoA}(p) = \sup_{\Gamma^p \in \mathcal{G}(\mathcal{C}_0)} \mathsf{PoA}(\Gamma^p),$$

where $\mathcal{G}(\mathcal{C}_0)$ denotes the class of altruistic/spiteful extensions of atomic congestion games with affine costs.

3 Main Results

3.1 Price of Anarchy

Our main result is the completion of the characterization of the $\mathsf{PoA}(p)$ for all $p > 0$. We show that the lower bound of [7] of $1 + \frac{2}{\sqrt{3}}$ is tight for $4 - 2\sqrt{3} \leq p \leq \frac{2\sqrt{3}}{9-2\sqrt{3}}$, and that the lower bound of [19] of $\frac{4}{p(4-p)}$ is tight for $0 < p \leq 4 - 2\sqrt{3}$. In particular, the price of anarchy is smallest when players are spiteful with $4 - 2\sqrt{3} \leq p \leq \frac{2\sqrt{3}}{9-2\sqrt{3}}$. See Fig. 2 for an illustration.

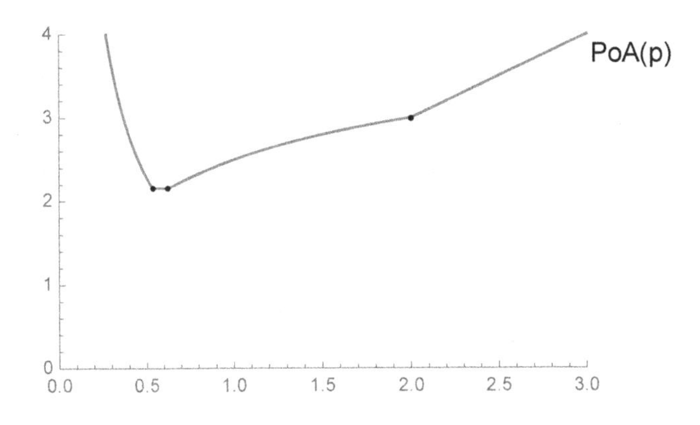

Fig. 2. Tight bounds for the price of anarchy in the four regions.

Theorem 1. *Let* $\bar{p}_0 = 4 - 2\sqrt{3} \approx 0.536$ *and let* $\bar{p}_1 = \frac{2\sqrt{3}}{9-2\sqrt{3}} \approx 0.626$. *Then*

$$PoA(p) = \begin{cases} \frac{4}{p(4-p)} & \text{if } 0 \leq p \leq \bar{p}_0, \\ 1 + \frac{2}{\sqrt{3}} & \text{if } \bar{p}_0 \leq p \leq \bar{p}_1, \\ \frac{1+4p}{1+p} & \text{if } \bar{p}_1 \leq p \leq 2, \\ 1 + p & \text{if } p \geq 2. \end{cases}$$

Proof. Caragiannis et al. [8] proved a tight bound of $\frac{1+4p}{1+p}$ if $1 \leq p \leq 2$ and of $1 + p$ if $p \geq 2$. Caragiannis et al. [7] gave a lower bound of $1 + \frac{2}{\sqrt{3}}$ for all $p > 0$

and showed it is tight for $p = \frac{2\sqrt{3}}{9-2\sqrt{3}}$. Kleer and Schäfer [19] extended the result of Caragiannis et al. [8] by giving a tight bound of $\frac{1+4p}{1+p}$ for $\frac{2\sqrt{3}}{9-2\sqrt{3}} \leq p \leq 1$. So what remains to show is an upper bound of $\frac{4}{p(4-p)}$ for $0 < p \leq 4 - 2\sqrt{3}$, and an upper bound of $1 + \frac{2}{\sqrt{3}}$ for $4 - 2\sqrt{3} \leq p \leq \frac{2\sqrt{3}}{9-2\sqrt{3}}$. We need the following technical lemma.

Lemma 3. *Let* $p \leq \frac{2\sqrt{3}}{9-2\sqrt{3}}$. *If there exists a* $\lambda \geq \frac{1}{p}$ *and* $\mu > 0$ *such that*

$$k\left(m + \frac{1}{p}\right) + \frac{p-1}{p} m \leq \lambda k^2 + \mu m^2 \qquad \forall k, m \in \mathbb{N}. \tag{3.1}$$

then $\mathsf{PoA}(p) \leq \frac{\lambda}{1-\mu}$.

Proof. Let s be a Nash equilibrium and s^* be a social optimum. Observe that

$$\begin{aligned}
\sum_{i \in N} c_i(s) &= \sum_{i \in N} \sum_{e \in s_i} (a_e \cdot (p \cdot n_e(s) + 1 - p) + b_e) \\
&= \sum_{e \in E} (a_e \cdot n_e(s) \cdot (p \cdot n_e(s) + 1 - p) + b_e \cdot n_e(s)) \\
&= p \cdot \sum_{e \in E} (a_e \cdot n_e(s)^2 + b_e \cdot n_e(s)) + \sum_{e \in E} ((a_e + b_e) \cdot n_e(s) \cdot (1-p)) \\
&= p \cdot c(s) + \sum_{e \in E} ((a_e + b_e) \cdot n_e(s) \cdot (1-p)),
\end{aligned}$$

and

$$\begin{aligned}
\sum_{i \in N} c_i(s_i^*, s_{-i}) &= \sum_{i \in N} \sum_{e \in s_i^*} (a_e \cdot (p \cdot n_e(s_i^*, s_{-i}) + 1 - p) + b_e) \\
&\leq \sum_{e \in E} (a_e \cdot n_e(s^*) \cdot (p \cdot n_e(s) + 1) + b_e \cdot n_e(s^*)).
\end{aligned}$$

Hence we have that

$$\begin{aligned}
&c(s) \\
&\leq \sum_{e \in E} \left(a_e \cdot \left(n_e(s^*) \cdot \left(n_e(s) + \frac{1}{p} \right) + n_e(s) \cdot \frac{p-1}{p} \right) + b_e \cdot \left(\frac{n_e(s^*)}{p} + n_e(s) \cdot \frac{p-1}{p} \right) \right) \\
&\leq \sum_{e \in E} (a_e \cdot (\lambda \cdot n_e(s^*)^2 + \mu \cdot n_e(s)^2) + \lambda \cdot b_e \cdot n_e(s^*)) \\
&\leq \lambda \cdot c(s)^* + \mu \cdot c(s).
\end{aligned}$$

where the first inequality follows from $\sum_{i \in N} c_i(s) \leq \sum_{i \in N} c_i(s_i^*, s_{-i})$ from the Nash equilibrium condition and the above two equalities, and the second inequality follows from Eq. (3.1) and $p \leq 1$. Rewriting the last inequality yields the desired result. □

The last two results give precise values for λ and μ satisfying Eq. (3.1) for the two regions that have to be considered.

Lemma 4. *If* $0 < p \leq 2(2-\sqrt{3})$, *then* $\lambda = \frac{1}{p}$ *and* $\mu = \frac{p}{4}$ *satisfies Eq.* (3.1).

Proof. We show that $\lambda = \frac{1}{p}$ and $\mu = \frac{p}{4}$ satisfies Eq. (3.1), or equivalently

$$4(m(1-p)-k) + (2k - mp)^2 \geq 0 \quad \forall k, m \in \mathbb{N}. \tag{3.2}$$

If $m(1-p) \geq k$, the result follows trivially. So we can assume that $m(1-p) < k$ and in particular, $k > 0$. If $k = 1$, Eq. (3.2) is equivalent to $m \cdot (4 - 8p + mp^2) \geq 0$, which is satisfied for $m = 0, 1, 2$ as $0 < p \leq 2(2-\sqrt{3})$.

If $k = 2$, Eq. (3.2) is equivalent to $8 + 4m - 12mp + m^2p^2 \geq 0$. Observe that the left-hand side of Eq. (3.2) is a quadratic in m and is strictly larger than 0 for $m = 0$. The global minimum of the quadratic expression is obtained at $m = \frac{2(3p-1)}{p^2}$, and Eq. (3.2) is thus satisfied for $p \leq 1/3$. For $1/3 < p \leq 2(2-\sqrt{3})$, calculating the global minimum shows that Eq. (3.2) is satisfied.

For $k \geq 2$, the partial derivative of the left-hand side of Eq. (3.2) with respect to k equals $-4+8k-4mp$. Since $k \geq 2$ and $m(1-p) < k$, we have $-4+8k-4mp > 8k - 4 - 4k\frac{p}{1-p} \geq 3k - 4 > 0$, where the second inequality follows because $0 < p \leq 2(2-\sqrt{3})$. Since the left-hand side of Eq. (3.2) is at least 0 for $k = 2$ and the partial derivative with respect to k is larger than 0 for $k \geq 2$, the result follows. □

Lemma 5. *If* $2(2-\sqrt{3}) \leq p \leq \frac{2\sqrt{3}}{9-2\sqrt{3}}$, *then* $\lambda = \frac{2+\sqrt{3}}{2}$ *and* $\mu = \frac{2-\sqrt{3}}{2}$ *satisfies Eq.* (3.1).

Proof. We show that $\lambda = \frac{2+\sqrt{3}}{2}$ and $\mu = \frac{2-\sqrt{3}}{2}$ satisfies Eq. (3.1), or equivalently

$$2(m(1-p)-k) + p\left(\sqrt{2+\sqrt{3}}k - \sqrt{2-\sqrt{3}}m\right)^2 \geq 0 \quad \forall k, m \in \mathbb{N}. \tag{3.3}$$

First observe that $\lambda \geq \frac{1}{p}$ for all $2(2-\sqrt{3}) \leq p \leq \frac{2\sqrt{3}}{9-2\sqrt{3}}$. If $m(1-p) \geq k$, the result follows trivially. So we can assume that $m(1-p) < k$ and in particular, $k > 0$. If $k = 1$ and $m = 0$, Eq. (3.3) is equivalent to $-2 + 2p + \sqrt{3}p \geq 0$, which is satisfied since $p \geq 2(2-\sqrt{3})$. If $k = 1$ and $m = 1$, the left-hand side of Eq. (3.3) is equal to 0. If $k = 1$ and $m \geq 2$, the partial derivative of the left-hand side of Eq. (3.3) with respect to m equals $2 - 2(2 + (-2+\sqrt{3})m)p$. Since $m \geq 2$ and $p \leq \frac{2\sqrt{3}}{9-2\sqrt{3}}$, this partial derivative is larger than 0. Since the left-hand side of Eq. (3.3) is 0 for $m = 2$ and the partial derivative with respect to m is larger than 0 for $m \geq 2$, Eq. (3.3) is satisfied.

For $k \geq 2$, observe that the left-hand side of Eq. (3.3) is a quadratic in m. The global minimum of the quadratic expression is obtained at $m = \frac{1-(1+k)p}{-2+\sqrt{3}p} > 0$. Calculating the global minimum shows that Eq. (3.3) is satisfied when $k \geq 2$ and $2(2-\sqrt{3}) \leq p \leq \frac{2\sqrt{3}}{9-2\sqrt{3}}$. □

Combining Lemma 3 with Lemmas 4 and 5 yields the desired result. □

3.2 Price of Stability

Theorem 1 derives an upper bound on the price of anarchy and hence price of stability of $\frac{4}{p(4-p)}$ if $p \leq 4-2\sqrt{3}$. We complete the picture of the price of stability by deriving an upper bound on the price of stability for $p \geq 2$. We leave it as an open question whether this upper bound is tight. Caragiannis et al. [8] derived a lower bound of $\frac{2+4p}{4}$ on the price of stability for $p \geq 2$. In particular, the price of stability is smallest when players are altruistic with $p = 2$. See Fig. 3 for an illustration.

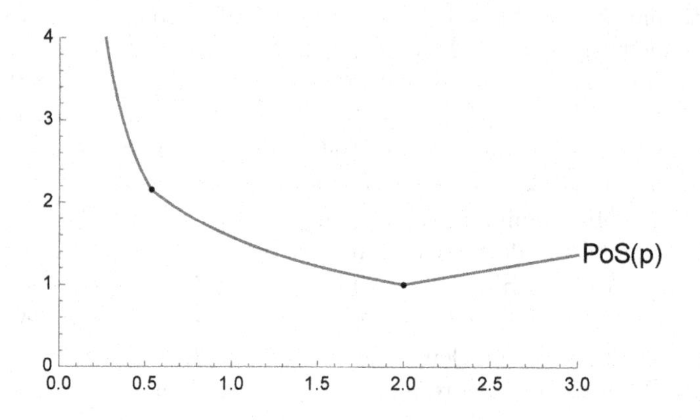

Fig. 3. Upper bounds on the price of stability in the three regions.

Theorem 2. *Let* $\bar{p}_0 = 4 - 2\sqrt{3} \approx 0.536$. *Then*

$$PoS(p) \leq \begin{cases} \frac{4}{p(4-p)} & \text{if } 0 \leq p \leq \bar{p}_0, \\ \frac{1+\sqrt{3}}{p-1+\sqrt{3}} & \text{if } \bar{p}_0 \leq p \leq 2, \\ \frac{4-2\sqrt{3}+(\sqrt{3}-1)p}{2} & \text{if } p \geq 2. \end{cases}$$

Proof. Chen et al. [9] derived an upper bound on the price of stability of $\frac{1+\sqrt{3}}{p-1+\sqrt{3}}$ for $1 \leq p \leq 2$. Kleer and Schäfer [19] obtained a tight bound for the price of stability of $\frac{1+\sqrt{3}}{p-1+\sqrt{3}}$ for $4 - 2\sqrt{3} \leq p \leq 2$ and gave a lower bound for $0 < p \leq 4 - 2\sqrt{3}$, which Theorem 1 shows to be tight. So what remains to show is an upper bound of $\frac{4-2\sqrt{3}+(\sqrt{3}-1)p}{2}$ for $p \geq 2$.

Without loss of generality, we may assume that $a_e = 1$ and $b_e = 0$ for all $e \in E$ (see, e.g., [9]). We need the following technical lemma.

Lemma 6. *If there exists a* $\gamma \geq 0$, *a* $\delta \geq 0$ *and an* $S \geq 1$ *such that*

$$\gamma \cdot [p \cdot (k^2 - m) + (2-p) \cdot (k-m)] + \delta \cdot [(1 + p \cdot m)k - m(1 + p(m-1))] \\ \leq S \cdot k^2 - m^2 \quad \forall k, m \in \mathbb{N}, \tag{3.4}$$

then $\mathsf{PoS}(p) \leq S$.

Proof. Recall that the Rosenthal's potential is given by

$$\Phi^p(s) = \sum_{e \in E} a_e \cdot \left(n_e(s) + p \cdot \frac{n_e(s)(n_e(s)-1)}{2} \right) + b_e \cdot n_e(s).$$

Let s be a global minimizer of the Rosenthal's potential, and let s^* be a social optimum. Since s is a global minimizer of the potential, we have that

$$\sum_{e \in E} p \cdot n_e(s)^2 + (2-p) \cdot n_e(s) \leq \sum_{e \in E} p \cdot n_e(s^*)^2 + (2-p) \cdot n_e(s^*).$$

Since s is a Nash equilibrium, we have that $\sum_{i \in N} c_i(s) \leq \sum_{i \in N} c_i(s_i^*, s_{-i})$ and hence

$$\sum_{e \in E} n_e(s) \cdot (1 + p(n_e(s)-1)) \leq n_e(s^*) \cdot (1 + p \cdot n_e(s)).$$

Assume that $\gamma, \delta \geq 0$ and $S \geq 1$ satisfy Eq. (3.4). Then for all $e \in E$, we have

$$\begin{aligned} 0 &\leq \gamma \cdot [p \cdot (n_e(s^*)^2 - n_e(s)) + (2-p) \cdot (n_e(s^*) - n_e(s))] \\ &\quad + \delta \cdot [n_e(s^*) \cdot (1 + p \cdot n_e(s)) - n_e(s) \cdot (1 + p(n_e(s)-1))] \\ &\leq S \cdot n_e(s^*)^2 - n_e(s)^2, \end{aligned}$$

and thus by summing over $e \in E$, we get $PoS(p) \leq C(s)/C(s^*) \leq S$. □

The last results gives precise values for γ, δ and S satisfying Eq. (3.4).

Lemma 7. *If $p \geq 2$, then $\delta = \frac{(\sqrt{3}-1)(p-2)}{2p}$, $\gamma = \frac{2\sqrt{3}-2+(3-\sqrt{3})p}{4p}$ and $S = \frac{4-2\sqrt{3}+(\sqrt{3}-1)p}{2}$ satisfies Eq. (3.4).*

Proof. Let $p \geq 2$. We show that $\delta = \frac{(\sqrt{3}-1)(p-2)}{2p}$, $\gamma = \frac{2\sqrt{3}-2+(3-\sqrt{3})p}{4p}$ and $S = \frac{4-2\sqrt{3}+(\sqrt{3}-1)p}{2}$ satisfies Eq. (3.4), or equivalently $\forall k, m \in \mathbb{N}$,

$$\frac{1}{4} \cdot (p-2) \cdot \left((3-\sqrt{3})k - (1+\sqrt{3})m + \left(\sqrt{3\sqrt{3}-5k} - \sqrt{1+\sqrt{3}m} \right)^2 \right) \geq 0. \tag{3.5}$$

Observe that $p-2 \geq 0$. If $(3-\sqrt{3})k \geq (1+\sqrt{3})m$, the result follows trivially. So we can assume that $(3-\sqrt{3})k < (1+\sqrt{3})m$ and in particular, $m > 0$. If $m = 1$, Eq. (3.5) is equivalent to $\frac{1}{4} \cdot (p-2) \cdot k \cdot (5 - 3\sqrt{3} + (-5+3\sqrt{3})k) \geq 0$, which is satisfied. If $m \geq 2$, the partial derivative of the left-hand side of Eq. (3.5) with respect to m equals $\frac{1}{4} \cdot (p-2) \cdot \left(-1 - \sqrt{3} - 2\sqrt{1+\sqrt{3}} \left(\sqrt{-5+3\sqrt{3}k} - \sqrt{1+\sqrt{3}m} \right) \right) \geq$ $\frac{1}{4} \cdot (p-2) \cdot \left(-1 - \sqrt{3} - 2\sqrt{1+\sqrt{3}} \left(-\sqrt{1+\sqrt{3}m} + \frac{(1+\sqrt{3})\sqrt{-5+3\sqrt{3}m}}{3-\sqrt{3}} \right) \right) \geq 0$, where the first inequality follows as $(3-\sqrt{3})k < (1+\sqrt{3})m$ and the second inequality as $m \geq 2$.

Since the left-hand side of Eq. (3.5) is at least 0 for $m = 1$ and the partial derivative with respect to m is larger than 0 for $m \geq 2$, Eq. (3.5) is satisfied. □

Combining Lemma 6 with Lemma 7 yields the desired result. □

4 Conclusion

We have studied the inefficiency of Nash equilibria in atomic congestion games which spiteful or altruistic players. We have completed the characterization of the price of anarchy as a function of the parameter that describes the level of spite or altruism of players. In particular, we obtained upper bounds on the price of anarchy when players are sufficiently spiteful, which was left as an open question by Kleer and Schäfer [19]. These bounds immediately imply upper bounds on the price of stability and additionally, we obtained an upper bound on the price of stability when players are sufficiently altruistic. What is interesting is that the best price of anarchy is obtained for players that are moderately spiteful, whereas the best price of stability is obtained when players are moderately altruistic. So there seems to be tradeoff between having a good price of anarchy and a good price of stability. A question that has been recently addressed by Filos-Ratsikas et al. [16] in mechanism design.

Appendix

Proof of Lemma 1. Observe that for all $i \in N$ and all $s \in S$,

$$c_i^{\xi}(s) = (1-2\xi) \cdot \sum_{e \in s_i} (a_e \cdot n_e(s) + b_e) + \xi \cdot \sum_{e \in E} (a_e \cdot n_e(s)^2 + b_e \cdot n_e(s)).$$

The Nash equilibrium condition with respect to costs $c_i^{\xi}(\cdot)$, for all $i \in N$ and all $s_i' \in S_i$,

$$c_i^{\xi}(s) \leq c_i^{\xi}(s_i', s_{-i}),$$

is equivalent to

$$\begin{aligned}
&(1-2\xi) \cdot \sum_{e \in s_i \setminus s_i'} (a_e \cdot n_e(s) + b_e) + \xi \cdot \sum_{e \in s_i \setminus s_i'} (a_e \cdot n_e(s)^2 + b_e \cdot n_e(s)) \\
&\quad + \xi \cdot \sum_{e \in s_i' \setminus s_i} (a_e \cdot n_e(s)^2 + b_e \cdot n_e(s) \\
&\leq (1-2\xi) \cdot \sum_{e \in s_i' \setminus s_i} (a_e \cdot n_e(s) + b_e) + \xi \cdot \sum_{e \in s_i \setminus s_i'} (a_e \cdot (n_e(s)-1)^2 \\
&\quad + b_e \cdot (n_e(s)-1)) + \xi \cdot \sum_{e \in s_i' \setminus s_i} (a_e \cdot (n_e(s)+1)^2 + b_e \cdot (n_e(s)+1)),
\end{aligned}$$

which is equivalent to

$$\sum_{e \in s_i \setminus s_i'} (a_e \cdot (n_e(s) - \xi) + b_e \cdot (1-\xi)) \leq \sum_{e \in s_i' \setminus s_i} (a_e \cdot (n_e(s_i', s_{-i}) - \xi) + b_e \cdot (1-\xi)),$$

or equivalently

$$\sum_{e \in s_i} (a_e \cdot (n_e(s) - \xi) + b_e \cdot (1-\xi)) \leq \sum_{e \in s_i'} (a_e \cdot (n_e(s_i', s_{-i}) - \xi) + b_e \cdot (1-\xi)). \tag{4.1}$$

The Nash equilibrium condition with respect to costs $c_i^p(\cdot)$, for all $i \in N$ and all $s_i' \in S_i$,

$$c_i^p(s) \leq c_i^p(s_i', s_{-i}),$$

is equivalent to

$$\sum_{e\in s_i}(a_e \cdot (p \cdot n_e(s) + 1 - p) + b_e) \leq \sum_{e\in s_i'}(a_e \cdot (p \cdot n_e(s_i', s_{-i}) + 1 - p) + b_e),$$

which for $p = \frac{1}{1-\xi}$ is equivalent to

$$\sum_{e\in s_i}\left(a_e \cdot \left(\frac{n_e(s)}{1-\xi} - \frac{\xi}{1-\xi}\right) + b_e\right) \leq \sum_{e\in s_i'}\left(a_e \cdot \left(\frac{n_e(s_i', s_{-i})}{1-\xi} - \frac{\xi}{1-\xi}\right) + b_e\right). \tag{4.2}$$

Since Eq. (4.1) and Eq. (4.2) are equivalent, the result follows. □

Proof of Lemma 2. Observe that for all $i \in N$ and all $s \in S$,

$$c_i^\alpha(s) = (1-\alpha) \cdot \sum_{e\in s_i}(a_e \cdot n_e(s) + b_e) + \alpha \cdot \sum_{e\in E}(a_e \cdot n_e(s)^2 + b_e \cdot n_e(s)).$$

The Nash equilibrium condition with respect to costs $c_i^\alpha(\cdot)$, for all $i \in N$ and all $s_i' \in S_i$,

$$c_i^\alpha(s) \leq c_i^\alpha(s_i', s_{-i}),$$

is equivalent to

$$\begin{aligned}(1-\alpha) &\cdot \sum_{e\in s_i\setminus s_i'}(a_e \cdot n_e(s) + b_e) + \alpha \cdot \sum_{e\in s_i\setminus s_i'}(a_e \cdot n_e(s)^2 + b_e \cdot n_e(s)) \\ &+ \xi \cdot \sum_{e\in s_i'\setminus s_i}(a_e \cdot n_e(s)^2 + b_e \cdot n_e(s) \\ &\leq (1-\alpha) \cdot \sum_{e\in s_i'\setminus s_i}(a_e \cdot n_e(s) + b_e) + \xi \cdot \sum_{e\in s_i\setminus s_i'}(a_e \cdot (n_e(s)-1)^2 \\ &+ b_e \cdot (n_e(s)-1)) + \xi \cdot \sum_{e\in s_i'\setminus s_i}(a_e \cdot (n_e(s)+1)^2 + b_e \cdot (n_e(s)+1)),\end{aligned}$$

which is equivalent to

$$\sum_{e\in s_i\setminus s_i'}(a_e \cdot ((1+\alpha) \cdot n_e(s) - \alpha) + b_e) \leq \sum_{e\in s_i'\setminus s_i}(a_e \cdot ((1+\alpha) \cdot n_e(s_i', s_{-i}) - \alpha) + b_e),$$

or equivalently

$$\sum_{e\in s_i}(a_e \cdot ((1+\alpha) \cdot n_e(s) - \alpha) + b_e) \leq \sum_{e\in s_i'}(a_e \cdot ((1+\alpha) \cdot n_e(s_i', s_{-i}) - \alpha) + b_e). \tag{4.3}$$

The Nash equilibrium condition with respect to costs $c_i^p(\cdot)$, for all $i \in N$ and all $s_i' \in S_i$

$$c_i^p(s) \leq c_i^p(s_i', s_{-i})$$

is equivalent to

$$\sum_{e \in s_i} (a_e \cdot (p \cdot n_e(s) + 1 - p) + b_e) \leq \sum_{e \in s_i'} (a_e \cdot (p \cdot n_e(s_i', s_{-i}) + 1 - p) + b_e), \quad (4.4)$$

which for $p = 1 + \alpha$ is equivalent to Eq. (4.3). □

References

1. Aland, S., Dumrauf, D., Gairing, M., Monien, B., Schoppmann, F.: Exact price of anarchy for polynomial congestion games. SIAM J. Comput. **40**(5), 1211–1233 (2011). https://doi.org/10.1137/090748986
2. Anagnostopoulos, A., Becchetti, L., De Keijzer, B., Schäfer, G.: Inefficiency of games with social context. Theory Comput. Syst. **57**(3), 782–804 (2015)
3. Anshelevich, E., Dasgupta, A., Kleinberg, J., Tardos, E., Wexler, T., Roughgarden, T.: The price of stability for network design with fair cost allocation. SIAM J. Comput. **38**(4), 1602–1623 (2008)
4. Awerbuch, B., Azar, Y., Epstein, A.: The price of routing unsplittable flow. SIAM J. Comput. **42**(1), 160–177 (2013). https://doi.org/10.1137/070702370
5. Bilò, V.: On linear congestion games with altruistic social context. In: Cai, Z., Zelikovsky, A., Bourgeois, A. (eds.) COCOON 2014. LNCS, vol. 8591, pp. 547–558. Springer, Cham (2014). https://doi.org/10.1007/978-3-319-08783-2_47
6. Caragiannis, I., Flammini, M., Kaklamanis, C., Kanellopoulos, P., Moscardelli, L.: Tight bounds for selfish and greedy load balancing. Algorithmica **61**(3), 606–637 (2011)
7. Caragiannis, I., Kaklamanis, C., Kanellopoulos, P.: Taxes for linear atomic congestion games. ACM Trans. Algorithms (TALG) **7**(1), 1–31 (2010)
8. Caragiannis, I., Kaklamanis, C., Kanellopoulos, P., Kyropoulou, M., Papaioannou, E.: The impact of altruism on the efficiency of atomic congestion games. In: Wirsing, M., Hofmann, M., Rauschmayer, A. (eds.) TGC 2010. LNCS, vol. 6084, pp. 172–188. Springer, Heidelberg (2010). https://doi.org/10.1007/978-3-642-15640-3_12
9. Chen, P.A., Keijzer, B.D., Kempe, D., Schäfer, G.: Altruism and its impact on the price of anarchy. ACM Trans. Econ. Comput. (TEAC) **2**(4), 1–45 (2014)
10. Christodoulou, G., Gairing, M.: Price of stability in polynomial congestion games. ACM Trans. Econ. Comput. (TEAC) **4**(2), 1–17 (2015)
11. Christodoulou, G., Gairing, M., Giannakopoulos, Y., Spirakis, P.G.: The price of stability of weighted congestion games. SIAM J. Comput. **48**(5), 1544–1582 (2019)
12. Christodoulou, G., Koutsoupias, E.: On the price of anarchy and stability of correlated equilibria of linear congestion games. In: Brodal, G.S., Leonardi, S. (eds.) ESA 2005. LNCS, vol. 3669, pp. 59–70. Springer, Heidelberg (2005). https://doi.org/10.1007/11561071_8
13. Christodoulou, G., Koutsoupias, E.: The price of anarchy of finite congestion games. In: STOC'05: Proceedings of the 37th Annual ACM Symposium on Theory of Computing, pp. 67–73. ACM, New York (2005). https://doi.org/10.1145/1060590.1060600

14. Cominetti, R., Scarsini, M., Schröder, M., Stier-Moses, N.: Convergence of large atomic congestion games. arXiv preprint arXiv:2001.02797 (2020)
15. Cominetti, R., Scarsini, M., Schröder, M., Stier-Moses, N.E.: Price of anarchy in stochastic atomic congestion games with affine costs. In: Proceedings of the 2019 ACM Conference on Economics and Computation, EC 2019, New York, NY, USA, pp. 579–580. Association for Computing Machinery (2019). https://doi.org/10.1145/3328526.3329579
16. Filos-Ratsikas, A., Giannakopoulos, Y., Lazos, P.: The pareto frontier of inefficiency in mechanism design. In: Caragiannis, I., Mirrokni, V., Nikolova, E. (eds.) WINE 2019. LNCS, vol. 11920, pp. 186–199. Springer, Cham (2019). https://doi.org/10.1007/978-3-030-35389-6_14
17. Harks, T., Végh, L.A.: Nonadaptive selfish routing with online demands. In: Janssen, J., Prałat, P. (eds.) CAAN 2007. LNCS, vol. 4852, pp. 27–45. Springer, Heidelberg (2007). https://doi.org/10.1007/978-3-540-77294-1_5. http://dl.acm.org/citation.cfm?id=1778487.1778494
18. Hoefer, M., Skopalik, A.: Altruism in atomic congestion games. ACM Trans. Econ. Comput. (TEAC) **1**(4), 1–21 (2013)
19. Kleer, P., Schäfer, G.: Tight inefficiency bounds forperception-parameterized affine congestion games. Theoret. Comput. Sci.**754**, 65–87 (2019).https://doi.org/10.1016/j.tcs.2018.04.025. http://www.sciencedirect.com/science/article/pii/S0304397518302597
20. Koutsoupias, E., Papadimitrou, C.: Worst-case equilibria. In: Annual Symposium on Theoretical Aspects of Computer Science, pp. 404–413 (1999)
21. Ledyard, J.O.: Public goods: a survey of experimental research (1994)
22. Levine, D.K.: Modeling altruism and spitefulness in experiments. Rev. Econ. Dyn. **1**(3), 593–622 (1998)
23. Piliouras, G., Nikolova, E., Shamma, J.S.: Risk sensitivity of price of anarchy under uncertainty. ACM Trans. Econ. Comput. **5**(1), 5:1–5:27 (2016). https://doi.org/10.1145/2930956. http://doi.acm.org/10.1145/2930956
24. Rosenthal, R.W.: A class of games possessing pure-strategy Nash equilibria. Internat. J. Game Theory **2**, 65–67 (1973). https://doi.org/10.1007/BF01737559
25. Roughgarden, T.: Intrinsic robustness of the price of anarchy. J. ACM **62**(5), 1–42 (2015). https://doi.org/10.1145/2806883. Article no. 32
26. Schulz, A.S., Moses, N.S.: On the performance of user equilibria in traffic networks. In: Proceedings of the Fourteenth Annual ACM-SIAM Symposium on Discrete Algorithms, pp. 86–87. Society for Industrial and Applied Mathematics (2003)

Markets and Matchings

Bribery and Control in Stable Marriage

Niclas Boehmer(✉), Robert Bredereck, Klaus Heeger, and Rolf Niedermeier

Algorithmics and Computational Complexity, TU Berlin, Berlin, Germany
{niclas.boehmer,robert.bredereck,heeger,
rolf.niedermeier}@tu-berlin.de

Abstract. We initiate the study of external manipulations in Stable Marriage by considering several manipulative actions as well as several "desirable" manipulation goals. For instance, one goal is to make sure that a given pair of agents is matched in a stable solution, and this may be achieved by the manipulative action of reordering some agents' preference lists. We present a comprehensive study of the computational complexity of all problems arising in this way. We find several polynomial-time solvable cases as well as NP-hard ones. For the NP-hard cases, focusing on the natural parameter "budget" (that is, the number of manipulative actions), we also perform a parameterized complexity analysis and encounter parameterized hardness results.

Keywords: Stable matching · Matching markets · Manipulation · Strategic behavior · Polynomial-time algorithms · Parameterized hardness

1 Introduction

In the Stable Marriage problem, we have two sets of agents, each agent has preferences over all agents from the other set, and the goal is to find a matching between agents of the one set and agents of the other set such that no two agents prefer each other to their assigned partners.

Looking at applications of *stable* marriages and corresponding generalizations in the context of matching markets, we evidence external manipulations in modern applications. For instance, surveys reported that in college admission systems in China, Bulgaria, Moldova, and Serbia, bribes have been performed in order to gain desirable admissions [12,15]. Focusing on the most basic scenario with the same number of agents on both sides and a one-to-one assignment, that is, Stable Marriage, we initiate a thorough study of manipulative actions (bribery and control) from a computational complexity perspective. Notably, bribery scenarios have also been used as motivation in other papers around Stable Marriage, e.g., when finding robust stable matchings [3] or when studying strongly stable matchings in the Hospitals/Residents problem with ties [13].

N. Boehmer—Supported by the DFG project MaMu (NI369/19).
K. Heeger—Supported by DFG Research Training Group 2434 "Facets of Complexity".

T. Harks and M. Klimm (Eds.): SAGT 2020, LNCS 12283, pp. 163–177, 2020.
https://doi.org/10.1007/978-3-030-57980-7_11

External manipulation may have many faces such as deleting agents, adding agents, or changing agents' preference lists. We consider three different manipulation goals and five different manipulative actions.

We introduce the manipulation goals *Constructive-Exists*, *Exact-Exists*, and *Exact-Unique*, where *Constructive-Exists* is the least restrictive goal and asks for modifications such that a desired agent pair is contained in some stable matching. More restrictively, *Exact-Exists* asks for modifications such that a desired matching is stable. Most restrictive, *Exact-Unique* requires that a desired matching becomes the only stable matching.

As manipulative actions, we investigate *Swap*, *Reorder*, *DeleteAcceptability*, *Delete*, and *Add*. The actions *Swap* and *Reorder* model bribery through an external agent. While a single *Reorder* action allows to completely change the preferences of an agent (modeling a briber who can "buy an agent"), a *Swap* action is more fine-granular and only allows to swap two neighboring agents in some agent's preference list (modeling a briber who has to slightly convince agents with increasing costs/effort). For both actions, the external agent might actually change the true preferences of the influenced agent, for example, by advertising some possible partner. However, in settings where the agents' preferences serve as an input for a centralized mechanism computing a stable matching which is subsequently implemented and cannot be changed, it is enough to bribe the agents to cast untruthful preferences. *Delete* and *Add* model control of the instance. They are useful to model an external agent, i.e. the organizer of some matching system, with the power to accept or reject agents to participate or to change the participation rules. While a *Delete* (resp. *Add*) action allows to delete (resp. add) an agent to the instance, a *DeleteAcceptability* action forbids for a specific pair of agents the possibility to be matched to each other and to be blocking. The latter can be seen as a hybrid between bribery and control because it can model an external agent that changes acceptability rules (for example introducing a maximum age gap) as well as it can model a briber who convinces an agent about inacceptability of some other agent at some cost. Note that we do not consider the actions *Delete* and *Add* in the *Exact-Exists* and *Exact-Unique* setting, as these two actions cannot be applied to the natural definitions of these two goals.

Related Work. Since its introduction [10], STABLE MARRIAGE has been intensely studied by researchers from different disciplines and in many contexts [11,14,19].

One topic related to manipulation in stable marriages is the study of *strategic behavior*, which focuses on the question whether agents can misreport their preferences to fool a given matching algorithm to match them to a better partner. Numerous papers have addressed the question of strategic behavior for different matching algorithms, types of agents' preferences and restrictions on the agents that are allowed to misreport their preferences (e.g., [10,20,21,23]; see [19, Chapter 2.9] for a survey). This setting is related to ours in the sense that the preferences of agents are modified to achieve a desired outcome, while it is fundamentally different with respect to the allowed modifications and their

Table 1. Overview of our results, where ℓ denotes the given budget. All stated W[1]- and W[2]-hardness results also imply NP-hardness.

Action/Goal	*Constructive-Exists*	*Exact-Exists*	*Exact-Unique*
Swap	W[1]-h. wrt. ℓ (Theorem 2)	P	NP-c. (Proposition 2)
Reorder	W[1]-h. wrt. ℓ (Theorem 2) 2-approx in P (Proposition 1)	P	W[2]-h. wrt. ℓ (Theorem 4)
Delete Accept.	W[1]-h. wrt. ℓ (Theorem 2)	P	P (Theorem 5)
Delete	P (Theorem 3)	–	–
Add	W[1]-h. wrt. ℓ (Theorem 1) NP-h. even if $\ell = \infty$ (Theorem 1)	–	–

goal: In the context of strategic behavior, an agent is only willing to change *her* preferences if she directly benefits from it.

While we are interested in finding ways to influence a profile to change the set of stable matchings, finding *robust* stable matchings [3,16,17] corresponds to finding stable matchings such that a briber cannot easily make the matching unstable. For instance, Chen et al. [3] introduced the concept of *d-robustness*: A matching is d-robust if it is stable in the given instance and remains stable even if d arbitrary swaps in preference lists are performed.

Conceptually, our work is closely related to the study of bribery and control in elections (see [7] for a survey). In election control problems [1], the goal is to change the structure of a given election, e.g., by modifying the candidate or voter set, such that a designated candidate becomes the winner/looser of the resulting election. In bribery problems [6], the briber is allowed to modify the votes in the election to achieve the goal. Most of the manipulative actions we consider are inspired by either some control operation or bribery action already studied in the context of voting.

Our manipulation goals are also related to problems previously studied in the stable matching literature: For example, the *Constructive-Exists* problem with given budget zero reduces to the STABLE MARRIAGE WITH FORCED PAIRS problem, which aims at finding a stable matching in a given STABLE MARRIAGE instance that includes some given agent pairs. While the problem is polynomial-time solvable for STABLE MARRIAGE instances without ties [19], deciding the existence of a "weakly stable" matching is NP-hard if ties are allowed even if only one pair is forced [18]. This directly implies hardness of the *Constructive-Exists* problem if ties are allowed.

Our Contributions. Providing a complete P vs. NP dichotomy, we settle the computational complexity of all computational problems emanating from our manipulation scenarios. We also conduct a parameterized complexity analysis of these problems using the parameter budget ℓ, that is, the number of elementary manipulative actions that we are allowed to perform. Table 1 gives an overview of our results. Additionally, we prove that CONSTRUCTIVE-EXISTS-SWAP does not admit an $\mathcal{O}(n^{1-\epsilon})$-approximation in $f(\ell)n^{\mathcal{O}(1)}$ time for any $\epsilon > 0$ unless FPT = W[1]. Furthermore, we observe XP-algorithms with respect to the parameter ℓ for all problems. The CONSTRUCTIVE-EXISTS-REORDER and EXACT-UNIQUE-REORDER problem require non-trivial XP-algorithms, which can be found in the full version [2].

Comparing the results for the different combinations of manipulation goals and manipulative actions, we observe a quite diverse complexity landscape: While for all other manipulative actions the corresponding problems are computationally hard, CONSTRUCTIVE-EXISTS-DELETE and EXACT-UNIQUE-DELETEACCEPTABILITY are polynomial-time solvable. Relating the different manipulation goals to each other, we show that specifying a full matching that should be made stable instead of just one agent pair that should be part of some stable matching makes the problem of finding a successful manipulation significantly easier. In contrast to this, providing even more information about the resulting instance by requiring that the given matching is the unique stable matching instead of just one of the stable matchings makes the problem of finding a successful manipulation again harder.

Due to space constraints, we defer the proof of several results (marked by $\star$) and some details to the full version [2].

2 Preliminaries

Parameterized Complexity. A *parameterized problem* consists of a problem instance $\mathcal{I}$ and a parameter value k (in our case the budget ℓ). It is called *fixed-parameter tractable* with respect to k if there exists an *FPT-algorithm*, i.e., an algorithm running in time $f(k)|\mathcal{I}|^{O(1)}$ for a computable function f. Moreover, it lies in XP with respect to k if it can be solved in time $|\mathcal{I}|^{f(k)}$ for some computable function f. There is also a theory of hardness of parameterized problems that includes the notion of W[t]-hardness with W[t] $\subseteq$ W[t'] for $t \leq t'$. If a problem is W[t]-hard for a given parameter for any $t \geq 1$, then it is widely believed not to be fixed-parameter tractable for this parameter.

Stable Marriage. An instance $\mathcal{I}$ of the STABLE MARRIAGE (SM) problem consists of a set $U = \{m_1, \dots m_n\}$ of men and a set $W = \{w_1, \dots, w_n\}$ of women, together with a strict *preference list* $\mathcal{P}_a$ for each $a \in U \cup W$. We call the elements from $U \cup W$ *agents* and denote as $A = U \cup W$ the set of agents. The preference list $\mathcal{P}_a$ of an agent a is a strict order over the agents of the opposite gender. We denote the preference list of an agent $a \in A$ by $a : a_1 \succ a_2 \succ a_3 \succ \dots$, where a_1 is a's most preferred agent, a_2 is a's second most preferred agent, and so on. For the sake of readability, we sometimes only specify parts of the agents' preference relation and end the preferences with "$\succ \overset{\text{(rest)}}{\dots}$". In this case, it is possible to complete the given profile by adding the remaining agents in an arbitrary order. We say that a *prefers* a' to a'' if a ranks a' above a'' in her preference list, i.e., $a' \succ a''$. For two agents $a, a' \in A$ of opposite gender, let $\text{rank}(a, a')$ denote the rank of a' in the preference relation of a, i.e., one plus the number of agents a prefers to a'.

A *matching* M is a subset of $U \times W$ such that each agent is contained in at most one pair. An agent is *unassigned* in a matching if no pair contains this agent. For a matching M and an assigned agent $a \in A$, we denote by $M(a)$ the agent a is matched to in M, i.e., $M(a) = w$ if $(a, w) \in M$ and $M(a) = m$ if $(m, a) \in M$. We slightly abuse notation and write $a \in M$ for an agent a if there exists some agent a' such that $(a, a') \in M$. A matching is called *complete* if no agent is unassigned. For a matching M, a pair $(m, w) \in U \times W$ is *blocking* if both m is unassigned or prefers w to $M(m)$, and w is unassigned or prefers m to $M(w)$. A matching is *stable* if it does not admit a blocking pair. We denote as $\mathcal{M}_\mathcal{I}$ the set of stable matchings in an SM instance $\mathcal{I}$.

The Stable Marriage with Incomplete Lists (SMI) problem is a generalization of the Stable Marriage problem where each agent a is allowed to specify incomplete preferences of agents of the opposite gender and a pair of agents $(m, w) \in U \times W$ can only be part of a stable matching M if they both appear in each others preference list. Let $\text{ma}(M)$ denote the set of agents matched in a matching M. Moreover, for an SMI instance $\mathcal{I}$, let $\text{ma}(\mathcal{I})$ denote the set of agents that are matched in a stable matching in $\mathcal{I}$. Note that by the Rural Hospitals Theorem [22] it holds for all stable matchings $M, M' \in \mathcal{M}_\mathcal{I}$ that: $\text{ma}(\mathcal{I}) = \text{ma}(M) = \text{ma}(M')$.

Manipulative Actions. We introduce five manipulative actions and necessary notation. We denote by $\mathcal{X} \in \{\textit{Swap}, \textit{Reorder}, \textit{DeleteAcceptability}, \textit{Delete}, \textit{Add}\}$ the type of a manipulative action.

Swap. A *Swap* operation changes the order of two neighboring agents in the preference list of an agent.

Reorder. A *Reorder* operation of an agent's preference list reorders her preferences arbitrarily, i.e., one performs an arbitrary permutation.

Delete Acceptability. A *DeleteAcceptability* operation is understood as deleting the mutual acceptability of a man and a woman. This enforces that such a deleted pair cannot be part of any stable matching and cannot be a blocking pair for any stable matching. Thus, after applying a *DeleteAcceptability* action, the given SM instance is transformed into an SMI instance.

Delete. A *Delete* operation deletes an agent from the instance and this agent from the preferences of all remaining agents. Slightly abusing notation, for an SM instance $\mathcal{I} = (U, W, \mathcal{P})$ and a subset of agents $A' \subseteq A$, we write $\mathcal{I} \setminus A'$ to denote the instance $\mathcal{I}'$ that results from deleting the agents A' in $\mathcal{I}$.

Add. An *Add* operation adds an agent from a predefined set of agents to the instance. Formally, the input for a computational problem considering the manipulative action *Add* consists of an SM instance $(U, W, \mathcal{P})$ together with two subsets $U_{\text{add}} \subseteq U$ and $W_{\text{add}} \subseteq W$. U_{add} and W_{add} contain agents that can be added to the instance. All other men $U_{\text{orig}} := U \setminus U_{\text{add}}$ and women $W_{\text{orig}} := W \setminus W_{\text{add}}$ are part of the original instance. Adding a set of agents $X_A = X_U \cup X_W$ with $X_U \subseteq U_{\text{add}}$ and $X_W \subseteq W_{\text{orig}}$ results in the instance

$(U_{\text{orig}} \cup X_U, W_{\text{orig}} \cup X_W, \mathcal{P}')$, where $\mathcal{P}'$ is the restriction of $\mathcal{P}$ to agents from $U_{\text{orig}} \cup X_U \cup W_{\text{orig}} \cup X_W$.

Manipulation Goals. In the *Constructive-Exists* setting, the goal is to modify a given SM instance using manipulative actions of some given type such that a designated man-woman pair is part of some stable matching. For $\mathcal{X} \in$ {*Swap*, *DeleteAcceptability*, *Delete*, *Add*}, the formal definition of the problem is presented below. For the manipulative action *Reorder*, we adapt the definition and forbid to reorder the preferences of m^* and w^*, as otherwise there always exists a trivial solution by reordering the preferences of both m^* and w^*.

CONSTRUCTIVE-EXISTS-$\mathcal{X}$

Input: Given an SM instance $\mathcal{I} = (U, W, \mathcal{P})$, a man-woman pair (m^*, w^*), and a budget $\ell \in \mathbb{N}$.

Question: Is it possible to perform ℓ manipulative actions of type $\mathcal{X}$ such that (m^*, w^*) is part of at least one matching that is stable in the altered instance?

In the *Exact* setting, we are given a complete matching. Within this setting, we consider two different computational problems. First, we consider the EXACT-EXISTS problem where the goal is to modify a given SM instance such that the given matching is stable in the altered instance. Second, we consider the EXACT-UNIQUE problem where the goal is to modify a given SM instance such that the given matching is the *unique* stable matching.

EXACT-EXISTS (UNIQUE)-$\mathcal{X}$

Input: Given an SM instance $\mathcal{I} = (U, W, \mathcal{P})$, a complete matching M^*, and budget $\ell \in \mathbb{N}$.

Question: Is it possible to perform ℓ manipulative actions of type $\mathcal{X}$ such that M^* is a (the unique) stable matching in the altered instance?

There also exist natural optimization variants of all considered decision problems which ask for the minimal number of manipulative actions that are necessary to alter a given SM instance to achieve the specified goal.

3 *Constructive-Exists*

In this section, we analyze the computational complexity of CONSTRUCTIVE-EXISTS-$\mathcal{X}$. We start with showing intractability for $\mathcal{X} \in$ {*Add*, *Swap*, *DeleteAcceptability*, *Reorder*}. Subsequently, we show that CONSTRUCTIVE-EXISTS-DELETE is solvable in $\mathcal{O}(n^2)$ time, and CONSTRUCTIVE-EXISTS-REORDER admits a 2-approximation in the same time.

3.1 A Framework for Computational Hardness

All W[1]-hardness results essentially follow from the same basic idea for a parameterized reduction. We explain the general framework of the reduction, using the manipulative action *Add* as an example. The necessary modifications for the manipulative actions *Swap*, *DeleteAcceptability*, and *Reorder* are described in the full version [2].

We construct a parameterized reduction from the W[1]-hard CLIQUE problem [4], where given an undirected graph G and an integer k, the question is whether G admits a size-k clique, i.e., a set of k vertices that are pairwise adjacent. Fix an instance (G, k) of CLIQUE and denote the set of vertices by $V(G) = \{v_1, \dots, v_n\}$ and the set of edges by $E(G) = \{e_1, \dots, e_m\}$. Let d_v denote the degree of vertex v. Moreover, let $e_1^v, \dots, e_{d_v}^v$ be a list of all edges incident to v.

The high-level approach works as follows. We start by introducing two agents m^* and w^*. The pair (m^*, w^*) shall be contained in a stable matching. Furthermore, we add $q := \binom{k}{2}$ women $w_1^\dagger, \dots, w_q^\dagger$, which we call *penalizing women*. The idea is that m^* prefers all penalizing women to w^*, and thereby, a stable matching containing the pair (m^*, w^*) can only exist if all penalizing women $w_j^\dagger$ are matched to agents they prefer to m^*, as otherwise $(m^*, w_j^\dagger)$ would be a blocking pair for any matching containing (m^*, w^*).

In addition, we introduce one vertex gadget for every vertex and one edge gadget for every edge, which differ for the different manipulative actions. Each vertex gadget includes a *vertex woman* and each edge gadget an *edge man*: A penalizing woman can only be matched to an edge man. However, an edge man can only be matched to a penalizing woman if the gadgets corresponding to the endpoints of the edge and the gadget corresponding to the edge itself are manipulated. Thus, a budget of $\ell = k + \binom{k}{2}$ suffices if and only if G contains a clique of size k.

We implement the ideas of the general approach for the manipulative action *Add*. For each vertex $v \in V$, we introduce a vertex gadget consisting of one vertex woman w_v and two men m_v' and m_v. For each edge $e \in E$, we introduce an edge gadget consisting of an edge man m_e and one man m_e' and one woman w_e. Additionally, we introduce a set of k women $\widetilde{w}_1, \dots, \widetilde{w}_k$. The agents that can be added are $U_{\text{add}} := \{m_v : v \in V\} \cup \{m_e' : e \in E\}$ and $W_{\text{add}} =: \emptyset$, while all other agents are part of the original instance. We set $\ell := k + \binom{k}{2}$. However, we also show that the reduction holds even if $\ell = \infty$. The preferences of the agents are as follows:

$$m_v' : w_v \succ \widetilde{w}_1 \succ \dots \qquad m_v : w_v \succ \overset{\text{(rest)}}{\dots}, \succ \widetilde{w}_k \succ w^* \succ \overset{\text{(rest)}}{\dots},$$

$$w_v : m_v \succ m_{e_1^v} \succ \dots \succ m_{e_{d_v}^v} \succ m_v' \succ \overset{\text{(rest)}}{\dots}, \qquad w_e : m_e' \succ m_e \succ \overset{\text{(rest)}}{\dots},$$

$$m_e : w_e \succ w_u \succ w_v \succ w_1^\dagger \succ \dots \succ w_q^\dagger \succ \overset{\text{(rest)}}{\dots}, \qquad m_e' : w_e \succ \overset{\text{(rest)}}{\dots},$$

$$w_i^\dagger : m_{e_1} \succ \dots \succ m_{e_m} \succ m^* \succ \overset{\text{(rest)}}{\dots}, \qquad \widetilde{w}_i : m_{v_1}' \succ \dots \succ m_{v_n}' \succ \overset{\text{(rest)}}{\dots},$$

$$m^* : w_1^\dagger \succ \dots \succ w_q^\dagger \succ w^* \succ \overset{\text{(rest)}}{\dots}, \qquad w^* : m_{v_1}' \succ \dots \succ m_{v_n}' \succ m^* \succ \overset{\text{(rest)}}{\dots}.$$

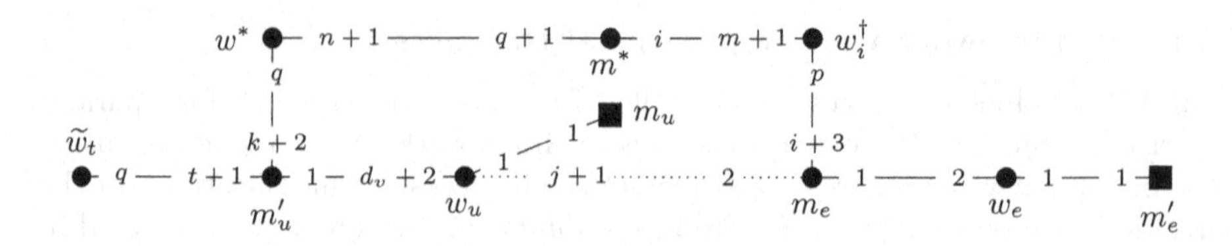

Fig. 1. A vertex gadget and an edge gadget for the hardness reduction for ADD, where $e = e_j^u = e_p$ and $u = v_q$. The squared vertices are the vertices from U_{add} that can be added to the instance. For each $i \in [q]$ (resp. $t \in [k]$), we only exemplarily show one $w_i^\dagger$ (resp. $\widetilde{w}_t$). For an edge $\{x, y\}$, the number on this edge closer to x indicates the rank of y in x's preferences.

Note that in addition to all penalizing women also all men m'_v need to be matched to agents which they prefer to w^*, as otherwise every matching containing (m^*, w^*) is blocked by (m'_v, w^*). This ensures that at most k men m_v can be added to the instance, as there exist only k women $\widetilde{w}_i$ that can be matched to some m'_v from a manipulated vertex gadget. A visualization of parts of the construction is depicted in Fig. 1.

Lemma 1. *If there exists a set X_A of agents (no matter of which size) such that after their addition there exists a stable matching containing (m^*, w^*), then G contains a clique of size k.*

Proof. Let M be a stable matching containing (m^*, w^*). Since the edges $(m^*, w_i^\dagger)$ are not blocking, all penalizing women are matched to an edge man m_e for some $e \in E$. This requires that $m'_e \in X_A$, as otherwise (m_e, w_e) is a blocking pair. Moreover, for each such edge $e = \{u, v\}$, the vertex women w_u and w_v have to be either matched to other edge men or to the men m_u or m_v. Note that in both cases, the corresponding agents m'_u and m'_v are matched to one of the women $\widetilde{w}_i$, as otherwise (m'_u, w^*) or (m'_v, w^*) is a blocking pair. Thus, there exist at most k vertices $v \in V$ where w_v is matched to an edge men or to m_v.

Since there are $\binom{k}{2}$ penalizing women, and each of them is matched to an edge man, it follows that those k vertices form a clique in G, and these edge men correspond to the edges in the clique. □

The reverse direction can be found in the full paper [2]. We conclude that there exists a parameterized reduction from CLIQUE parameterized by k to CONSTRUCTIVE-EXISTS-ADD parameterized by ℓ:

Theorem 1. *Parameterized by budget ℓ, it is W[1]-hard to decide whether* CONSTRUCTIVE-EXISTS-ADD *has a solution with at most ℓ additions or has no solution with an arbitrary number of additions, even if we are only allowed to add agents of one gender.*

It is also possible to implement the general approach described above for the manipulative actions *Reorder*, *DeleteAcceptability*, and *Swap* to show hardness for the respective problems.

Theorem 2 ($\star$). Constructive-Exists-Reorder *and* Constructive-Exists-DeleteAcceptability *are W[1]-hard parameterized by budget ℓ. Unless FPT = W[1],* Constructive-Exists-Swap *does not admit an $\mathcal{O}(n^{1-\epsilon})$-approximation in $f(\ell)n^{\mathcal{O}(1)}$ time for any $\epsilon > 0$.*

3.2 Polynomial-Time Algorithms

In sharp contrast to the hardness results for all other considered manipulative actions, there is a simple algorithm for instances $\mathcal{I}_{\text{del}}$ of Constructive-Exists-Delete consisting of an SM instance $\mathcal{I}$ together with a man-woman pair (m^*, w^*) in linear time in the size of the input. The algorithm is based on the following observation. Let W^* be the set of women preferred by m^* to w^*, and U^* the set of men preferred by w^* to m^*. In all stable matchings M including (m^*, w^*), every woman in W^* needs to be matched to a man which she prefers to m^*, or needs to be deleted. Analogously, every man in U^* needs to be matched to a woman which he prefers to w^*, or needs to be deleted. Consequently, all pairs consisting of an agent $a \in U^* \cup W^*$ and an agent a' which a does not prefer over w^* or m^* cannot be part of any stable matching. This observation motivates a transformation of the given SM instance $\mathcal{I}$ into a SMI instance $\mathcal{I}'$ through the deletion of all such pairs. We also delete w^* and m^* from $\mathcal{I}'$ and compute a stable matching M in the resulting instance.

Let A' be the set of agents from $U^* \cup W^*$ which are unassigned in M. We claim that deleting A' is indeed a minimum number of agents to delete such that (m^*, w^*) is part of a stable matching in $\mathcal{I}$. To show this, we need the following lemma.

Lemma 2 ($\star$). *Let $\mathcal{I}'$ be an* SMI *instance and $a \in A$ some agent. Then, there exists at most one agent $a' \in A$ who was unassigned in $\mathcal{I}'$, i.e., $a' \notin \text{ma}(\mathcal{I}')$, and is matched in $\mathcal{I}' \setminus \{a\}$, i.e., $a' \in \text{ma}(\mathcal{I}' \setminus \{a\})$.*

Using Lemma 2, we can now show the correctness of the algorithm.

Theorem 3. Constructive-Exists-Delete *is solvable in $\mathcal{O}(n^2)$ time.*

Proof. Since a stable matching in an SMI instance can be computed in $\mathcal{O}(n^2)$ time [10], the set A' clearly can be computed in $\mathcal{O}(n^2)$. We claim that $\mathcal{I}_{\text{del}}$ is a YES-instance if and only if $|A'| \leq \ell$.

First assume $|A'| \leq \ell$. Let M' be a stable matching in $\mathcal{I}'$. We add (m^*, w^*) to M', and claim that this is a stable matching in $\mathcal{I} \setminus A'$, showing that $\mathcal{I}$ is a YES-instance. For the sake of a contradiction, assume that there exists a blocking pair (m, w). Since this is not a blocking pair in $\mathcal{I}' \setminus A'$, it contains an agent a from $(U^* \cup W^*) \setminus A'$, and a prefers w^* over w if $a = m$ or m^* over m if $a = w$. Without loss of generality, let $a = m$. However, as m is matched in M', he prefers $M'(m) = M(m)$ to w^*. Thus, m prefers $M(m)$ over w, a contradiction to (m, w) being blocking for M.

Now assume that $|A'| > \ell$. For the sake of contradiction, assume that there exists a set of agents $B' = \{b_1, \ldots, b_k\}$ with $k \leq \ell$ such that $\mathcal{I} \setminus B'$ admits a stable matching M containing (m^*, w^*). For each $i \in [k]$, let M_i be a stable matching in $\mathcal{I}' \setminus \{b_1, \ldots, b_i\}$. By the definition of A', all agents from A' are unassigned in M_0. Note that each agent $a \in A' \subseteq A^*$ is either part of B' or prefers $M(a)$ over m^* or w^* due to the stability of M; in particular, a is matched in M and thereby also in M_k. Since $k \leq \ell < |A'|$, there exists an i such that there exist two agents $a, a' \in A'$ which are unassigned in M_{i-1} and not contained in $\{b_1, \ldots, b_{i-1}\}$ but matched in M_i or contained in $\{b_1, \ldots b_i\}$.

From Lemma 2 it follows that it is not possible that both a and a' are unassigned in M_{i-1} but matched in M_i. Consequently, without loss of generality it needs to hold that $a = b_i$, and $a' \in \text{ma}(\mathcal{I} \setminus \{b_1, \ldots b_i\}) \setminus \text{ma}(\mathcal{I} \setminus \{b_1, \ldots b_{i-1}\})$. However, by deleting an agent that was previously unassigned, the set of matched agents does not change, i.e., $\text{ma}(\mathcal{I} \setminus \{b_1, \ldots b_{i-1}\}) = \text{ma}(\mathcal{I} \setminus \{b_1, \ldots b_i\})$, since a matching that is stable in $\mathcal{I} \setminus \{b_1, \ldots b_{i-1}\}$ is also stable in $\text{ma}(\mathcal{I} \setminus \{b_1, \ldots b_i\})$, as by deleting unassigned agents it is not possible to create new blocking pairs. This contradicts $a' \in \text{ma}(\mathcal{I} \setminus \{b_1, \ldots b_i\}) \setminus \text{ma}(\mathcal{I} \setminus \{b_1, \ldots b_{i-1}\})$. □

Using the same ideas (but observing that a *Reorder* operation can make two previously unassigned agents assigned), we get a 2-approximation for *Reorder*.

Proposition 1 ($\star$). *One can compute a factor-2 approximation of the optimization version of* CONSTRUCTIVE-EXISTS-REORDER *in* $\mathcal{O}(n^2)$ *time.*

4 *Exact-Exists*

In this section, we deal with the problem of making a given matching in an SM instance stable by performing *Swap*, *Reorder*, or *DeleteAcceptability* actions. In fact, it turns out that specifying the full matching instead of one pair makes the problem easier, as for all manipulative actions for which we showed hardness in the previous section, EXACT-EXISTS-$\mathcal{X}$ becomes polynomial-time solvable. The intuitive reason for this difference is that the problem of making a given matching M^* stable reduces to "resolving" all blocking pairs for M^*. We only briefly describe the main ideas here and refer to the full version for details [2].

For *DeleteAcceptability*, this task is straightforward, as it is always optimal to delete the acceptability of all blocking pairs. For *Reorder*, it is possible to delete all blocking pairs involving some agent a at cost one, as it possible to reorder the preferences of a such that $M^*(a)$ becomes her top-choice. Therefore, to find an optimal solution, it is necessary to find a minimal subset of agents that covers all blocking pairs. This reduces to finding a vertex cover in a bipartite graph where we introduce for each agent a vertex and connect two vertices if the corresponding agents form a blocking pair.

For *Swap*, the cost of resolving a blocking pair (m, w) by modifying the preferences of m, and symmetrically for w, is the number of swaps needed to swap $M^*(m)$ with w. However, by resolving some blocking pair involving an agent, also other blocking pairs involving this agent may be resolved. Thereby, the "true" costs of resolving a pair are difficult to compute, and a more involved approach is needed to determine which of the two agents involved in a blocking pair should be manipulated to resolve it. However, it turns out that this problem can be reduced to an instance of the MINIMUM CUT problem, proving that EXACT-EXISTS-SWAP can be solved in $\mathcal{O}(n^3)$ time.

5 *Exact-Unique*

Now, we turn from the task of making a given matching stable to the task of making it the unique stable matching. We show that this change makes the considered computational problems significantly more demanding in the sense that the *Exact-Unique* question is W[2]-hard with respect to ℓ for *Reorder* and NP-complete for *Swap*. In contrast, the problem for *DeleteAcceptability* is still solvable in polynomial time.

The W[2]-hardness result for the manipulative action *Reorder* and the NP-completeness of *Swap* both follow from the same parameterized reduction from the W[2]-complete HITTING SET problem [4] with small modifications. In an instance of HITTING SET, we are given a universe Z, a family $\mathcal{F} = \{F_1, \ldots, F_p\}$ of subsets of Z, and an integer k, and the task is to decide whether there exists a hitting set of size k, i.e., a set $X \subseteq Z$ with $|X| \leq k$ and $X \cap F \neq \emptyset$ for all $F \in \mathcal{F}$. The general idea of the construction is to add, for each set $F \in \mathcal{F}$, a set gadget consisting of two men and two women, and, for each element $z \in Z$, an element gadget consisting of a man-woman pair. We connect all set gadgets to the element gadgets corresponding to the elements in the set. The preferences are constructed in a way such that in each set gadget where none of the connected element gadgets are manipulated, the two women can switch their partners and the resulting matching is still stable given that M^* is stable. In contrast, when a connected element gadget is manipulated, then this switch is blocked and M^* is the unique stable matching in this gadget. Thereby, the manipulated element-gadgets form a hitting set.

Theorem 4. EXACT-UNIQUE-REORDER *parameterized by ℓ is W[2]-hard, even if the given matching M^* is already stable in the original instance and we are only allowed to modify the preferences of agents of one gender.*

Proof sketch. We give the construction of a parameterized reduction from HITTING SET, which is known to be W[2]-complete parameterized by the solution size k [4]. For each element $z \in Z$, we add a man m_z and a woman w_z, which are the first choices of each other. For each set $F = \{z_1, \ldots, z_q\} \in \mathcal{F}$, we add two men m_F^1 and m_F^2 and two women w_F^1 and w_F^2 with the following preferences

$$m_F^1 : w_F^1 \succ w_{z_1} \succ w_{z_2} \succ \cdots \succ w_{z_q} \succ w_F^2 \succ \overset{\text{(rest)}}{\ldots}, \qquad m_F^2 : w_F^2 \succ w_F^1 \succ \overset{\text{(rest)}}{\ldots},$$

$$w_F^1 : m_F^2 \succ m_F^1 \succ \overset{\text{(rest)}}{\ldots}, \qquad w_F^2 : m_F^1 \succ m_F^2 \succ \overset{\text{(rest)}}{\ldots}.$$

We set $M^* := \{(m_z, w_z) : z \in Z\} \cup \{(m_F^1, w_F^1), (m_F^2, w_F^2) : F \in \mathcal{F}\}$, and $\ell := k$. □

It is possible to adapt the reduction from the previous theorem to prove hardness for the manipulative action *Swap*. Here, we utilize the fact that *Reorder* operations can be modeled by (up to $n^2 + n$) *Swap* operations. To do so, we adapt the reduction such that it is only possible to modify the preferences of women w_z and add an "activation cost" to modifying the preferences of w_z such that only the preferences of a fixed number of women can be modified but for these we can modify them arbitrarily.

Proposition 2 (⋆). Exact-Unique-Swap *is NP-complete, even if the given matching M^* is already stable in the original instance and we are only allowed to modify the preferences of agents of one gender.*

In contrast to the hardness results for the other two manipulative actions, Exact-Unique-DeleteAcceptability is solvable in polynomial time. On an intuitive level, one reason for this is that it is not possible to manipulate whether an agent a ranks some other agent a' above or below $M^*(a)$.

The polynomial-time algorithm for Exact-Unique-DeleteAcceptability uses the theory of rotations, which we briefly recap. In a stable matching M, for a man $m \in U$, let $s_M(m)$ denote the first woman w succeeding $M(m)$ in m's preference list that prefers m to $M(w)$. If no such woman exists, then we set $s_M(m) := \emptyset$. A *rotation exposed in a stable matching* M is a sequence $\rho = (m_{i_0}, w_{j_0}), \ldots, (m_{i_{r-1}}, w_{j_{r-1}})$ such that for each $k \in [0, r-1]$ it holds that $(m_{i_k}, w_{j_k}) \in M$ and $w_{j_{k+1}} = s_M(m_{i_k})$, where additions are taken modulo r. We call such a rotation a *man-rotation* and $s_M(m)$ the *rotation successor* of m. We define $s_W(w)$ for $w \in W$ analogously and call a rotation where the roles of men and women are switched *woman-rotation*. As a matching is unique if and only if it exposes neither a man-rotation nor a woman-rotation [11], it is possible to reformulate the goal of Exact-Unique-DeleteAcceptability: Modify the given SM instance by deleting the acceptability of at most ℓ pairs such that neither a man-rotation nor a woman-rotation is exposed in M^*.

First of all, note that it is possible to solve the problem of removing all man-rotations exposed in M^* and the problem of removing all woman-rotations separately. To remove man-rotations, we only delete the acceptability of pairs (m, w) where m prefers $M^*(m)$ to w and w prefers m to $M^*(w)$. For woman-rotations, the situation is symmetric. We solve both problems by reducing them to the Minimum Weight Spanning Anti-Arborescence problem, which can be solved in $\mathcal{O}(m + n \log n)$ time [5,9]. In an instance of the Minimum Weight Spanning Anti-Arborescence problem, we are given a directed graph G with arc costs and a budget $k \in \mathbb{N}$. The question is whether there exists a spanning anti-arborescence, i.e., an acyclic subgraph of G such that all vertices of G but one have out-degree exactly one, of cost at most k.

The basic idea of the algorithm is the following for man-rotations (and symmetrically for woman-rotations). For a set of deleted acceptabilities F, let $s_{M^*}^F(m)$ denote the rotation successor of m after the deletion of F. We need to find a set of deleted acceptabilities F such that $M^* \cup \{(m, s^F_{M^*}(m)) : m \in U \text{ with } s^F_{M^*}(m) \neq \emptyset\}$ is acyclic. Note that we can change $s_{M^*}(m)$ only by deleting the pair $(m, s_{M^*}(m))$. In this case, the new rotation successor becomes the first woman w' succeeding $s_{M^*}(m)$ in m's preference list that prefers m over $M^*(w')$. Thus, we know the costs of making a woman w' the rotation successor of m. The problem of making $M^* \cup \{(m, s^F_{M^*}(m)) : m \in U \text{ with } s^F_{M^*}(m) \neq \emptyset\}$ acyclic can thus be translated to finding a minimum weight spanning anti-arborescence where we add a "sink" t to represent the case that $s^F_{M^*}(m) = \emptyset$.

Theorem 5. Exact-Unique-DeleteAcceptability *can be solved in $\mathcal{O}(n^2)$ time.*

Proof. Clearly, any solution needs to delete all blocking pairs. Thus, we assume without loss of generality that M^* is a stable matching.

We reduce the problem to two instances of the Minimum Weight Spanning Anti-Arborescence problem. The first instance of this problem is constructed as follows. We contract all pairs (m, w) of M^* to a vertex $\{m, w\}$ and add a sink t. We add an arc $(\{m, w\}, \{m', w'\})$ if w' prefers m to m' and m prefers w to w'. The weight of this arc is the number of women w^* such that m prefers w^* to w' and w to w^*, and w^* prefers m to $M^*(w^*)$. We call this graph H_U. Similarly, we construct a graph H_W (where the roles of men and women are exchanged).

We claim that M^* can be made the unique stable matching after the deletion of ℓ arcs if and only if the minimum weight anti-arborescences in H_U and H_W together have weight at most ℓ.

($\Rightarrow$) Let $F \subseteq U \times W$ be the set of pairs whose deletion makes M^* the unique stable matching. Let $F_U := \{(m, w) \in F : w \succ_m M^*(m)\}$ and $F_W := \{(m, w) \in F : m \succ_w M^*(w)\}$. For any man m, let $e_m := \{w', M^*(w')\}$, where w' is the woman best-ranked by m succeeding $M^*(m)$ such that w prefers m to $M^*(w)$ after the manipulation, i.e,. w' is the rotation successor of m after the manipulation. If no such woman exists, then we set $e_m := t$. We construct an anti-arborescence in H_U of cost at most $|F_W|$ by adding for each pair (m, w) the arc $(\{m, w\}, e_m)$ to the anti-arborescence. We claim that H_U is an anti-arborescence. Every vertex but t has exactly one outgoing arc, so it is enough to show that there does not exist a cycle. As we have inserted for each man an arc from the node including him to the node including his rotation successor, there cannot exist any cycle in the anti-arborescence, as such a cycle would induce a man-rotation in the modified SM instance. Such a man-rotation cannot exist, as we have assumed that M^* is the unique stable matching after the modifications.

In the same way one can construct an anti-arborescence of cost $|F_U|$ in H_W. The constructed anti-arborescences together have weight at most $|F_W| + |F_U| \leq |F|$, as any arc in $F_W \cap F_U$ would be a blocking pair for M^*.

($\Leftarrow$) Let $\mathcal{A}_U$ be an anti-arborescence in H_U, and $\mathcal{A}_W$ be an anti-arborescence in H_W. For every arc $(\{m, w\}, \{\widetilde{m}, \widetilde{w}\}) \in \mathcal{A}_U$, we delete the acceptability of all pairs $\{m, w'\}$ with m preferring w' to $\widetilde{w}$, and w to w', and w' preferring m to $M^*(w')$. After these deletions, $\widetilde{w}$ is the rotation successor of m. Let F_U denote the set of pairs deleted. We proceed with $\mathcal{A}_W$ analogously, and denote as F_W the set of deleted pairs. Clearly, $\mathcal{A}_U$ has cost $|F_U|$, and $\mathcal{A}_W$ has cost $|F_W|$.

Assume that M^* is not the unique stable matching after deleting the pairs from $F_U \cup F_W$. Then, without loss of generality, a man-rotation is exposed in M^*: $(m_{i_0}, w_{j_0}), \dots, (m_{i_{r-1}}, w_{j_{r-1}})$. As we already observed, the anti-aborescence $\mathcal{A}_U$ contains all arcs $(\{m, w\}, \{\widetilde{m}, \widetilde{w}\})$ where $\widetilde{w}$ is m's rotation-successor (after the deletion of F_U). Thus, $\mathcal{A}_U$ contains the arcs $(\{m_{i_k}, w_{j_k}\}, \{m_{i_{k+1}}, w_{j_{k+1}}\})$ for all $k \in [0, r-1]$ (all indices are taken modulo r). This implies that $\mathcal{A}_U$ contains a cycle, a contradiction to $\mathcal{A}_U$ being an anti-arborescence. □

6 Conclusion

We provided a first comprehensive study of the computational complexity of several manipulative actions and goals in the context of the Stable Marriage problem. Our rich and diverse set of theoretical results is surveyed in Table 1.

Several challenges for future research remain. In contrast to the setting considered here, there is also a destructive view on manipulation, where the goal is to prevent a certain constellation—our algorithmic results and some of our hardness results for the constructive case seem to carry over. Moreover, for the *Constructive-Unique* scenario not presented here, most of our hardness results still hold. A very specific open question is whether the Exact-Unique-Swap problem is fixed-parameter tractable when parameterized by the budget. Additionally, there is clearly a lot of room for investigating more manipulative actions or to extend the study of external manipulation to stable matching problems beyond Stable Marriage. Also weighted matchings might be of special interest.

On the practical side, we performed some preliminary experimental work with some of the algorithms derived in this paper. Experimenting with several forms of synthetic data and one set of real-world data collected in the context of the analysis of speed dating [8], we draw the following main conclusions for two of our settings: First, in the *Constructive-Exists* setting, we observed that for more than half of all possible agent pairs it was sufficient to delete around 15% or less of the agents to ensure that this agent pair is part of a stable matching. This suggests that the *Delete* operation is pretty powerful in this setting. Second, in the *Exact-Exists* setting, the given instance needs to be significantly changed to make a randomly drawn complete matching stable. Surprisingly, for the powerful action *Reorder* on average close to half of the agents had to be modified. In this regard, observe that there always exists a trivial solution where the preferences of all agents from one gender are reordered.

References

1. Bartholdi III, J.J., Tovey, C.A., Trick, M.A.: How hard is it to control an election? Math. Comput. Model. **16**(8–9), 27–40 (1992)
2. Boehmer, N., Bredereck, R., Heeger, K., Niedermeier, R.: Bribery and control in stable marriage. arXiv preprint arXiv:2007.04948 [cs.GT] (2020)

3. Chen, J., Skowron, P., Sorge, M.: Matchings under preferences: strength of stability and trade-offs. In: Proceedings of the 2019 ACM Conference on Economics and Computation (EC 2019), pp. 41–59 (2019)
4. Downey, R.G., Fellows, M.R.: Fundamentals of Parameterized Complexity. TCS. Springer, London (2013). https://doi.org/10.1007/978-1-4471-5559-1
5. Edmonds, J.: Optimum branchings. J. Res. Natl. Bur. Stand. B **71**(4), 233–240 (1967)
6. Faliszewski, P., Hemaspaandra, E., Hemaspaandra, L.A.: How hard is bribery in elections? J. Artif. Intell. Res. **35**, 485–532 (2009)
7. Faliszewski, P., Rothe, J.: Control and bribery in voting. In: Handbook of Computational Social Choice, pp. 146–168. Cambridge University Press (2016)
8. Fisman, R., Iyengar, S.S., Kamenica, E., Simonson, I.: Gender differences in mate selection: evidence from a speed dating experiment. Q. J. Econ. **121**(2), 673–697 (2006)
9. Gabow, H.N., Galil, Z., Spencer, T.H., Tarjan, R.E.: Efficient algorithms for finding minimum spanning trees in undirected and directed graphs. Combinatorica **6**(2), 109–122 (1986)
10. Gale, D., Shapley, L.S.: College admissions and the stability of marriage. Am. Math. Mon. **120**(5), 386–391 (2013)
11. Gusfield, D., Irving, R.W.: The Stable Marriage Problem - Structure and Algorithms. Foundations of Computing Series. MIT Press, Cambridge (1989)
12. Heyneman, S., Anderson, K., Nuraliyeva, N.: The cost of corruption in higher education. Comp. Educ. Rev. **52**(1), 1–25 (2008)
13. Irving, R.W., Manlove, D.F., Scott, S.: Strong stability in the hospitals/residents problem. In: Alt, H., Habib, M. (eds.) STACS 2003. LNCS, vol. 2607, pp. 439–450. Springer, Heidelberg (2003). https://doi.org/10.1007/3-540-36494-3_39
14. Knuth, D.E.: Mariages stables et leurs relations avec d'autres problèmes combinatoires. Les Presses de l'Université de Montréal, Montreal, Que. (1976). Introduction à l'analyse mathématique des algorithmes, Collection de la Chaire Aisenstadt
15. Liu, Q., Peng, Y.: Corruption in college admissions examinations in China. Int. J. Educ. Dev. **41**, 104–111 (2015)
16. Mai, T., Vazirani, V.V.: Finding stable matchings that are robust to errors in the input. In: Proceedings of the 26th Annual European Symposium on Algorithms (ESA 2018), pp. 60:1–60:11 (2018)
17. Mai, T., Vazirani, V.V.: Stable matchings, robust solutions, and finite distributive lattices. arXiv preprint arXiv:1804.05537 [cs.DM] (2018)
18. Manlove, D., Irving, R.W., Iwama, K., Miyazaki, S., Morita, Y.: Hard variants of stable marriage. Theor. Comput. Sci. **276**(1–2), 261–279 (2002)
19. Manlove, D.F.: Algorithmics of Matching Under Preferences. Series on Theoretical Computer Science, vol. 2. WorldScientific, Singapore (2013)
20. Pini, M.S., Rossi, F., Venable, K.B., Walsh, T.: Manipulation complexity and gender neutrality in stable marriage procedures. Auton. Agents Multi-Agent Syst. **22**(1), 183–199 (2011)
21. Roth, A.E.: The economics of matching: stability and incentives. Math. Oper. Res. **7**(4), 617–628 (1982)
22. Roth, A.E.: On the allocation of residents to rural hospitals: a general property of two-sided matching markets. Econometrica J. Econ. Soc. **54**(2), 425–427 (1986)
23. Teo, C., Sethuraman, J., Tan, W.: Gale-shapley stable marriage problem revisited: strategic issues and applications. Manag. Sci. **47**(9), 1252–1267 (2001)

Approximating Stable Matchings with Ties of Bounded Size

Jochen Koenemann[1], Kanstantsin Pashkovich[2], and Natig Tofigzade[1](✉)

[1] University of Waterloo, Waterloo, ON N2L 3G1, Canada
{jochen,natig.tofigzade}@uwaterloo.ca
[2] University of Ottawa, Ottawa, ON K1N 6N5, Canada
kpashkov@uottawa.ca

Abstract. Finding a stable matching is one of the central problems in algorithmic game theory. If participants are allowed to have ties and incomplete lists, computing a stable matching of maximum cardinality is known to be NP-hard. In this paper we present a $(3L-2)/(2L-1)$-approximation algorithm for the stable matching problem with ties of size at most L and incomplete preferences. Our result matches the known lower bound on the integrality gap for the associated LP formulation.

Keywords: Stable matching · Approximation algorithms · Combinatorial optimization

1 Introduction

In an instance of the classical stable matching problem we are given a (complete) bipartite graph $G = (A \cup B, E)$ where, following standard terminology, the nodes in A will be referred to as *men*, and the nodes in B represent *women*. Each man $a \in A$ possesses a (strict, and complete) preference order over women in B, and similarly, all women in B have a preference order over men in A. A matching M in G is called *stable* if there are no *blocking pairs* (a, b); i.e. there do not exist $(a, b) \notin M$ where both a and b prefer each other over their current partners in M (if there are any). In their celebrated work [4], Gale and Shapley proposed an efficient algorithm for finding a stable matching, providing a constructive proof that stable matchings *always* exist.

Stable matchings have wide-spread applications (e.g., see Manlove [3]), and many of these are large-scale. Therefore, as McDermid [15] points out, assuming that preferences are complete and strict is not realistic. Thus, in this paper, we will focus on stable matchings in the setting where preference lists are allowed to be incomplete and contain ties. Here, a woman is allowed to be indifferent between various men, and similarly, a man may be indifferent between several women. In this setting we consider the *maximum-cardinality stable matching* problem where the goal is to find a stable matching of maximum cardinality.

It is well-known that, in the settings where G is either complete or preferences do not contain ties, all stable matchings have the same cardinality [5].

T. Harks and M. Klimm (Eds.): SAGT 2020, LNCS 12283, pp. 178–192, 2020.
https://doi.org/10.1007/978-3-030-57980-7_12

Moreover, a straightforward extension of the algorithm in [4] solves our problem in these cases. When ties and incomplete preferences are permitted simultaneously, on the other hand, the problem of finding a maximum-cardinality stable matching is well-known to be NP-hard [14]. Furthermore, Yanagisawa [17] showed that it is NP-hard to find a $(33/29 - \varepsilon)$-approximate, maximum-cardinality stable matching. The same author also showed that assuming the *unique games* conjecture (UGC) it is hard to achieve performance guarantee of $4/3 - \varepsilon$.

On the positive side, maximum-cardinality stable matchings with ties and incomplete preferences have attracted significant attention [1,6–12,15,16]. The best-known approximation algorithms for the problem achieve an approximation ratio of 3/2 [9,15,16].

How does the hardness of maximum-cardinality stable matching depend on the maximum allowed size of ties in the given instance? Huang and Kavitha [6] recently considered the case where the *size of any tie* is bounded by $L = 2$. The authors proposed an algorithm and showed that its performance guarantee is at most 10/7. Chiang and Pashkovich [2] later provided an improved analysis for the same algorithm, showing that its real performance ratio is at most 4/3, and this result is tight under the UGC [17]. Lam and Plaxton [13] very recently designed a $1 + (1 - 1/L)^L$-approximation algorithm for the so-called *one-sided* special case of our problem, where only preferences of men are allowed to have ties.

1.1 Our Contribution

Our main result is captured in the following theorem. Note that the integrality gap of the natural LP relaxation for the problem is at least $(3L-2)/(2L-1)$ [8]. Hence, the performance ratio of our algorithm matches the known lower bound on the integrality gap.

Theorem 1. *Given an instance of the maximum-cardinality stable matching problem with incomplete preferences, and ties of size at most L; the polynomial-time algorithm described in Sect. 2 finds a stable matching M with*

$$|M| \geq \frac{2L-1}{3L-2}|\mathit{OPT}|,$$

where OPT is an optimal stable matching.

Our algorithm is an extension of that by Huang and Kavitha [6] for ties of size two: every man has L proposals where each proposal goes to the acceptable women. Women can *accept* or *reject* these proposals under the condition that no woman holds more than L proposals at any point during the algorithm. Similar to the algorithm in [6], we use the concept of *promotion* introduced by Király [9] to grant men repeat chances in proposing to women. In comparison to [6], the larger number of proposals in our algorithm leads to subtle changes to the forward and rejection mechanisms of women, and to further modifications to the way we obtain the output matching.

Our analysis is inspired by the analyses of both, Chiang and Pashkovich [2], and Huang and Kavitha [6], but requires several new ideas to extend it to the setting with larger ties. In both [6] and [2], the analyses are based on *charging schemes*: some objects are first assigned some values, called charges, and then charges are redistributed to nodes by a cost function. After a charging scheme is determined, relations between the generated total charges, and the sizes of output and optimal matchings are established, respectively, that lead to an approximation ratio. The analysis in [6] employs a complex charging scheme that acts *globally*, possibly distributing charges over the entire graph. In contrast, the charging scheme in [2] is *local* in nature, and exploits only the local structure of the output and optimal matchings, respectively.

We do not know of a direct way to extend the local *cost*-based analysis of [2] to obtain an approximation algorithm whose performance beats the best known 3/2-approximation for the general case. Indeed we believe that any such improvement must involve a non-trivial change in the charging scheme employed. As a result, we propose a new analysis that combines local and global aspects from [2,6]. The central technical novelty in the analysis is captured by Lemma 4 that provides an improved lower bound on the *cost* of components. As we will see later, our new charging scheme allows for a more fine-grained accounting of augmenting paths for the output matching of our algorithm.

2 Algorithm for Two-Sided Ties of Size Up to L

We introduce some notational conventions. Let $a', a'' \in A$ be on the preference list of $b \in B$. We write $a' \simeq_b a''$ if b is *indifferent* between a' and a'', and we write $a' >_b a''$, or $a' \geq_b a''$ if b *strongly*, or *weakly* prefers a' over a'', respectively. The preferences of men over women are defined analogously. For $c \in A \cup B$, we let $N(c)$ denote the set of nodes adjacent to c in G.

2.1 How Men Propose

Each man $a \in A$ has L proposals $p_a^1, p_a^2, \ldots, p_a^L$. A man starts out as *basic*, and later becomes *1-promoted* before he is eventually elevated to *2-promoted* status. Each man $a \in A$ has a *rejection history* $R(a)$ which records the women who rejected a proposal from a during his current promotion status. Initially, we let $R(a) = \varnothing$, for all $a \in A$.

Each proposal p_a^i for $a \in A$ and $i = 1, 2, \ldots, L$ goes to a woman in $N(a) \setminus R(a)$ most preferred by a, and ties are broken arbitrarily. If a proposal p_a^i for $a \in A$ and $i = 1, 2, \ldots, L$ is rejected by a woman $b \in B$, b is added to the rejection history of a, and subsequently, p_a^i is sent to a most preferred remaining woman in $N(a) \setminus R(a)$.

Suppose now that $R(a)$ becomes equal to $N(a)$ for some man $a \in A$. If a is either basic or 1-promoted then a's rejection history is cleared, and a is promoted. Otherwise, if a is already 2-promoted, a stops making proposals.

2.2 How Women Decide

Each woman $b \in B$ can hold up to L proposals, and among these more than one can come from the same man. Whenever she holds less than L proposals, newly received proposals are automatically accepted. Otherwise, b first tries to *bounce* one of her proposals, and if that fails, she will try to *forward* one of her proposals. If b can neither bounce nor forward a proposal, then b rejects a proposal.

We continue describing the details. In the following, we let $P(b)$ and $A(b)$ denote the set of proposals held by $b \in B$ at the current point, and the set of men corresponding to these, respectively. Suppose that $|P(b)| = L$, and that b receives a new proposal p_a^i for some $a \in A$ and $i = 1, \ldots, L$.

Bounce Step. If there is a man $\alpha \in A(b) \cup \{a\}$ and a woman $\beta \in B \setminus \{b\}$ such that $\beta \simeq_\alpha b$, and β currently holds less than L proposals, then we move one of α's proposals from b to β, and we call the bounce step *successful.*

Forward Step. If there is a man $\alpha \in A(b) \cup \{a\}$ and a woman $\beta \in B \setminus \{b\}$ such that $\beta \simeq_\alpha b$, at least two proposals from α are present in $P(b)$, no proposal from α is present in $P(\beta)$ and β is not in $R(\alpha)$, then b *forwards* a proposal $p_\alpha^j \in P(b) \cup \{p_a^i\}$ for some $j = 1, \ldots, L$ to β and the forward step is called *successful.* As a consequence of a successful forward step, α makes the proposal p_α^j to β.

We point out that bounce and forward steps do not lead to an update to the rejection history of an involved man. To describe the rejection step, we introduce the following notions. For a woman $b \in B$, a proposal $p_{a'}^{i'}$ is called *more desirable* than $p_{a''}^{i''}$ for $a', a'' \in A$ and $i', i'' = 1, \ldots, L$ if b strongly prefers a' to a'', or if b is indifferent between a' and a'' and a' has higher promotion status than a''. A proposal $p_{a'}^{i'} \in P(b)$ is *least desirable* in $P(b)$ if $p_{a'}^{i'}$ is not more desirable than any proposal in $P(b)$. Whenever $b \in B$ receives a proposal p_a^i, $|P(b)| = L$, and neither *bounce* nor *forward* steps are successful, we execute a rejection step.

Rejection Step. If there is unique *least desirable proposal* in $P(b) \cup \{p_a^i\}$, then b rejects that proposal. Otherwise, if there are more than one least desirable proposal in $P(b)$, b rejects a proposal from a man with the largest number of least desirable proposals in $P(b) \cup \{p_a^i\}$. If there are several such men, then we break ties arbitrarily. Subsequently, b is added to the rejection history of the man whose proposal is rejected.

2.3 The Algorithm

An approximate maximum-cardinality stable matching for a given instance $G = (A \cup B, E)$ is computed in two stages.

Stage 1. Please see Algorithm 1 for the pseudo code for Stage 1.

Men propose in an arbitrary order and women bounce, forward or reject proposals as described above. The first stage finishes, when for each man $a \in A$, one of the following two conditions is satisfied: all proposals of a are accepted; $R(a)$ becomes equal to $N(a)$ for the third time.

We represent the outcome of the first stage as a bipartite graph $G' = (A \cup B, E')$ with the node set $A \cup B$ and the edge set E', where each edge $(a, b) \in E'$ denotes a proposal from a held by b at the end of the first stage. Note that G' may be a multigraph in which an edge of the form (a, b) appears with multiplicity equal to the number of proposals that b holds from a. Clearly, each node u in G' has degree at most L, denoted by $\deg_{G'}(u) \leq L$, since every man has at most L proposals that may be accepted and every woman can hold at most L proposals at any point in the first stage.

Stage 2. We compute a maximum-cardinality matching M in G' such that all nodes of degree L in G' are matched. The existence of such matching is guaranteed by Lemma 1. The result of the second stage is such a matching M, that is the output of the algorithm.

Lemma 1. *There exists a matching in the graph G' such that all nodes of degree L in G' are matched. Moreover, there is such a matching M, where all nodes of degree L in G' are matched and we have*

$$|M| \geq |E'|/L\,.$$

Proof. Consider the graph $G' = (A \cup B, E')$ and the following linear program

$$\begin{aligned} \max \quad & \sum_{e \in E'} x_e \\ \text{s.t.} \quad & \sum_{e \in \delta(u)} x_e \leq 1 \quad (u \in A \cup B) \\ & \sum_{e \in \delta(u)} x_e = 1 \quad (u \in A \cup B, \deg_{G'}(u) = L) \\ & x \geq 0. \end{aligned}$$

It is well-known that the feasible region of the above LP is an integral polyhedron. Moreover, the above LP is feasible as is easily seen by considering the point that assigns $1/L$ to each edge in E'. Hence there exists an integral point optimal for this linear program. Notice, that every integral point feasible for this linear program is a characteristic vector of a matching in G', which matches all nodes of degree L in G'. To finish the proof, notice that the value of the objective function calculated at $x^\star$ equals $|E'|/L$. Thus the value of this linear program is at least $|E'|/L$, finishing the proof. □

2.4 Stability of Output Matching

Let the above algorithm terminate with a matching M. We first argue that it is stable.

Algorithm 1. Pseudo code for Stage 1 of the algorithm

1: let $G = (A \cup B, E)$ be an instance graph, and $N(c)$ denote the set of nodes adjacent to $c \in A \cup B$ in G
2: let $G' = (A \cup B, E')$ be a multigraph with E' initialized to the empty multiset of edges
3: let $\deg_{G'}(u)$ denote the degree of node u in G', and $A(b)$ denote the set of nodes adjacent to $b \in B$ in G'
4: **for all** $a \in A$ **do**
5: $\quad R(a) := \varnothing$ ▷ $R(a)$ is the rejection history of man a
6: $\quad stat_a := 0$ ▷ $stat_a$ is the promotion status of man a
7: **end for**
8: **while** $\exists a \in A$ s.t. $\deg_{G'}(a) < L$ and $R(a) \neq N(a)$ **do**
9: $\quad$ let $b \in N(a) \setminus R(a)$ be a woman s.t. $b \geq_a b'$ for all $b' \in N(a) \setminus R(a)$
10: $\quad$ PROPOSE(a, b)
11: **end while**
12: **return** E'
{The following subroutine describes how b accepts the proposal from a, or bounces, forwards, or rejects a proposal}
13: **procedure** PROPOSE(a, b)
14: $\quad$ **if** $\deg_{G'}(b) < L$ **then**
15: $\quad\quad E' := E' \cup \{(a, b)\}$
16: $\quad$ **else if** $\exists \alpha \in A(b) \cup \{a\}$ and $\exists \beta \in N(\alpha)$ s.t. $\beta \simeq_\alpha b$ and $\deg_{G'} \beta < L$ **then**
17: $\quad\quad E' := E' \cup \{(a, b), (\alpha, \beta)\} \setminus \{(\alpha, b)\}$ ▷ bounce
18: $\quad$ **else if** $\exists \alpha \in A(b) \cup \{a\}$ and $\exists \beta \in N(\alpha) \setminus R(\alpha)$ s.t. $\beta \simeq_\alpha b$, $|(E' \cup \{(a, b)\}) \cap \{(\alpha, b)\}| \geq 2$ and $\alpha \notin A(\beta)$ **then**
19: $\quad\quad E' := E' \cup \{(a, b)\} \setminus \{(\alpha, b)\}$
20: $\quad\quad$ PROPOSE(α, β) ▷ forward
21: $\quad$ **else**
22: $\quad\quad$ let $\mathcal{A}$ denote $\{\alpha \in A(b) \cup \{a\}$: for all $a' \in A(b) \cup \{a\}$, $\alpha \leq_b a'$ and if $\alpha \simeq_b a'$, then $stat_\alpha \leq stat_{a'}\}$
23: $\quad\quad$ let α_0 be a man in $\arg\max_{\alpha \in \mathcal{A}} |(E' \cup \{(a, b)\}) \cap \{(\alpha, b)\}|$
24: $\quad\quad E' := E' \cup \{(a, b)\} \setminus \{(\alpha_0, b)\}$ ▷ reject
25: $\quad\quad R(\alpha_0) := R(\alpha_0) \cup \{b\}$
26: $\quad\quad$ **if** $R(\alpha_0) = N(\alpha_0)$ **then**
27: $\quad\quad\quad$ **if** $stat_{\alpha_0} < 2$ **then**
28: $\quad\quad\quad\quad stat_{\alpha_0} := stat_{\alpha_0} + 1$
29: $\quad\quad\quad\quad R(\alpha_0) := \varnothing$
30: $\quad\quad\quad$ **end if**
31: $\quad\quad$ **end if**
32: $\quad$ **end if**
33: **end procedure**

Lemma 2. *The output matching M is stable in $G = (A \cup B, E)$.*

Proof. Suppose for contradiction that M is not stable, i.e. suppose that there exists an edge $(a, b) \in E$ that blocks M. If b rejected a proposal from a during the algorithm, then b holds L proposals when the algorithm terminates and all these proposals are from men, who are weakly preferred by b over a. Thus the degree of b in G' is L implying that b is matched in M with a man, who is not less preferred by b than a. We get a contradiction to the statement that (a, b) blocks M.

Conversely, if b did not reject any proposal from a during the algorithm, then the algorithm terminates with all L proposals of a being accepted, particularly, by women, who are weakly preferred by a over b. Therefore the degree of a in G' is L implying that a is matched in M with a woman, who is not less preferred by a than b. Again, we get a contradiction to the statement that (a, b) is a blocking pair for M. □

2.5 Running Time

We now show that each stage of the algorithm has polynomial execution time. For the first stage, we illustrate that only a polynomial number of proposals are bounced, forwarded, or rejected during this stage. For the second stage, the proof of Lemma 2 implies that it is sufficient to find an optimal extreme solution for a linear program of polynomial size.

First, we show that proposals are bounced only polynomially many times. For every $b \in B$, at most L proposals may be bounced to b. Indeed, with each proposal bounced to b, the number of proposals held by b increases; also, the number of proposals held by b never decreases or exceeds L during the algorithm. Hence at most $L|B|$ proposals are bounced during the first stage.

Second, we illustrate that proposals are forwarded only polynomially many times. For each $a \in A$, promotion status of a, and $b \in B$ such that $(a, b) \in E$, at most one proposal of a may be forwarded to b. To see this, let b' be a woman forwarding a proposal of a to b. Notice that b cannot bounce the proposal after b receives it because, otherwise, b' could bounce it by the transitivity of indifference. Observe also that b may forward a proposal from a only if she holds another proposal from him. Then it follows from the forward step that no woman can forward a proposal of a to b as long as b holds a proposal from him. If b rejects the proposal, then she is added to the rejection history of a, and so b does not receive any proposal from a unless the promotion status of a changes. Hence at most $3|A||B|$ proposals are forwarded during the first stage.

Finally, for each $a \in A$, promotion status of a, and $b \in B$ such that $(a, b) \in E$, b may reject at most L proposals from a. Indeed, b holds at most L proposals at any point in time, and since b is added to the rejection history of a after she rejected him, b does not receive any proposal from a unless the promotion status of a changes. Hence at most $3L|A||B|$ proposals are rejected during the first stage.

3 Tight Analysis

Recall that OPT is a maximum-cardinality stable matching in G, and let M be the output matching defined above. If $a \in A$ is matched with $b \in B$ in OPT, we write $\mathsf{OPT}(a) := b$ and $\mathsf{OPT}(b) := a$. Similarly, we use the notations $M(a) := b$

and $M(b) := a$ when $a \in A$ is matched with $b \in B$ in M. Note that our analysis is based on graph G' and therefore all graph-related objects will assume G'.

Definition 1. *A man $a \in A$ is called* successful *if the algorithm terminates with all of his L proposals being accepted. Likewise, a woman b is called* successful *if she holds L proposals when the algorithm stops. In other words, a person $c \in A \cup B$ is* successful *if the degree of c in G' is L, and* unsuccessful *otherwise.*

Definition 2. *A woman is called* popular *if she rejected a proposal during the algorithm, and* unpopular *otherwise.*

Remarks 1 and 2 below directly follow from the algorithm and are consequences of the bouncing step, and the rejection step, respectively.

Remark 1. Let $a \in A$ and $b, b' \in B$ be such that b holds a proposal from a when the algorithm finishes, b' is unsuccessful, and $b' \simeq_a b$. Then b is unpopular.

Proof. Suppose for contradiction that b is popular. Then at some point she could not bounce or forward any one of her proposals and so she was to reject a proposal. This implies that after b became popular, whenever she received a new proposal that could be bounced, that proposal would immediately be bounced. But then, when the algorithm terminates, b holds a proposal from a, that could successfully be bounced to b', a contradiction. □

Remark 2. Let $a, a' \in A$ and $b \in B$ be such that b holds at least two proposals from a when the algorithm finishes, b rejected a proposal from a' at some point, a is basic, and $a' \simeq_b a$. Then there is an edge (a', b) in G'.

Proof. Suppose for a contradiction that $(a', b) \notin G'$ holds. Let t be the most recent point in time when b rejects a proposal from a'. Then it follows from the algorithm that, at t, $a'' \geq_b a'$ holds for all $a'' \in A(b)$. The rejection step also implies that, at t, there is no $a'' \in A$ such that $a' \simeq_b a''$, a'' is basic, and b holds more than one proposal from a''. Moreover, the algorithm implies that, after t, whenever she receives a new proposal from a man a'' such that $a'' <_b a'$, she will immediately reject it unless she successfully bounces or forwards it. Now, consider a point in time after t when there is a man a'' such that $a' \simeq_b a''$, b already holds a proposal from a'', and receives another proposal from a''. Then the rejection step implies that she will reject one of the proposals from a'' unless she successfully bounces or forwards it. But then, when the algorithm terminates, b holds at least two proposals from a, a contradiction.

3.1 Analytical Techniques

In the following, we define *inputs*, *outputs*, and *costs* – notions that are central in the analysis of our charging scheme. Before we take a closer look at these notions and define them formally, let us discuss phenomena captured by them.

We use two different objects, inputs and outputs, to differentiate between two different viewpoints on proposals accepted when the algorithm ends. In particular, inputs are associated with the viewpoint of women on the proposals whereas outputs are associated with the viewpoint of men. The choice of terms "inputs" and "outputs" is due to the analysis in [6] where the edges of G' are directed from men to women, and so each proposal becomes an "input" for the woman, and analogously becomes an "output" for the corresponding man.

Now we describe the ideas that motivated our definitions concerning outputs and inputs. Let $M + \mathsf{OPT}$ denote the multiset that contains the edges in M and the edges in OPT. To establish the approximation guarantee of our algorithm, we analyze each connected component in $M + \mathsf{OPT}$. In order to show that M-augmenting paths in $M + \mathsf{OPT}$ do not lead to a large approximation guarantee, we introduce the notions of *bad* and *good inputs* as well as *bad* and *good outputs*. For example, a certain number of bad inputs and bad outputs are generated by the edges incident to the endpoints of an M-augmenting path in $M + \mathsf{OPT}$. Indeed, as we will see later, if $a_0 - b_0 - a_1 - \ldots - a_k - b_k$ is an M-augmenting path in $M + \mathsf{OPT}$ of length $2k + 1$, $k \geq 2$ where $a_0 \in A$, then b_0 has at least $L - 2$ bad inputs and a_k has at least $L - 2$ bad outputs. Then to show the approximation guarantee of $(3L - 2)/(2L - 1)$, we provide a way to obtain a lower bound on the number of bad inputs and bad outputs of men and women in each M-augmenting path; and later we provide an upper bound on the total number of bad inputs and bad outputs of all men and women.

To implement the above ideas, we use a charging scheme. Our charging scheme associates a cost with each man and each woman. These costs keep track of bad inputs and bad outputs: bad inputs lead to an increase of the corresponding woman's cost and bad outputs lead to an increase of the corresponding man's cost. We show that the total cost of all men and women is bounded above by $2L|M|$. On the other side, we provide a lower bound on the total cost by giving a lower bound on the cost of each connected component in $M + \mathsf{OPT}$. These upper and lower bounds lead to the desired approximation guarantee of $(3L - 2)/(2L - 1)$.

3.2 Inputs and Outputs

In our analysis inputs and outputs are fundamental edge-related objects for our charging scheme. Each edge in G' generates a certain number of charges. For example, as we will see in Sect. 3.3, if an edge (a, b) in G' belongs either to M or to OPT, two charges are generated by (a, b) so that one is carried to node a and one is carried to node b by cost function. To define similar charging mechanisms for the remaining types of edges in G', we first distinguish them as in the following definitions.

Definition 3. *Given an edge (a, b) in G', we say that (a, b) is an* output from *$a \in A$ and an* input to *$b \in B$ if (a, b) is not in $M + \mathsf{OPT}$.*

To illustrate how outputs and inputs are determined, for example, let $(a, b) \in M$, $a \in A$, $b \in B$ and $n_{(a,b)}$ be the number of edges of the form (a, b) in the multigraph G', then the edge (a, b) gives rise to the following number $s_{(a,b)}$ of inputs (and to the same number of outputs)

$$s_{(a,b)} := \begin{cases} n_{(a,b)} - 1 & \text{if } (a,b) \notin \mathsf{OPT} \\ 0 & \text{if } n_{(a,b)} = 1 \\ n_{(a,b)} - 2 & \text{otherwise.} \end{cases}$$

Definition 4. *An input* (a, b) *to* $b \in B$ *is called a* bad input *if one of the following is true:*

- *b is popular and* $a >_b \mathsf{OPT}(b)$.
- *b is popular,* $a \simeq_b \mathsf{OPT}(b)$, *but* $\mathsf{OPT}(b)$ *is unsuccessful.*
- *b is popular, a is 1-promoted,* $\mathsf{OPT}(b)$ *is successful and* $M(b) \simeq_b \mathsf{OPT}(b) \simeq_b a$.

An input (a, b) *to* $b \in B$ *is a* good input *if it is not a bad input. In other words, an input* (a, b) *to* $b \in B$ *is a* good input *if one of the following is true:*

- *b is unpopular.*
- *b is popular and* $\mathsf{OPT}(b) >_b a$.
- *b is popular,* $a \simeq_b \mathsf{OPT}(b)$, $\mathsf{OPT}(b)$ *is successful and a is not 1-promoted.*
- *b is popular,* $a \simeq_b \mathsf{OPT}(b)$, $\mathsf{OPT}(b)$ *is successful, but not* $M(b) \simeq_b \mathsf{OPT}(b) \simeq_b a$.

An output (a, b) *from a man a is called a* bad output *if one of the following is true:*

- *b is unpopular.*
- *b is popular,* $b >_a \mathsf{OPT}(a)$, *a is 1-promoted, but not* $M(b) \simeq_b \mathsf{OPT}(b) \simeq_b a$.
- *b is popular,* $b >_a \mathsf{OPT}(a)$ *and a is basic.*

An output from a man a is a good output *if that is not a bad output. In other words, an output* (a, b) *from a man* $a \in A$ *is a* good output *if one of the following is true:*

- *b is popular and* $\mathsf{OPT}(a) \geq_a b$.
- *b is popular,* $b >_a \mathsf{OPT}(a)$ *and a is 2-promoted.*
- *b is popular,* $b >_a \mathsf{OPT}(a)$, *a is 1-promoted and* $M(b) \simeq_b \mathsf{OPT}(b) \simeq_b a$.

Lemma 3. *There is no edge which is both a bad input and a bad output.*

Proof. Assume that an edge (a, b), $a \in A$, $b \in B$ is both a bad input to b and a bad output from a. First, consider the first case from the definition of a bad output. It trivially contradicts all the cases from the definition of a bad input. Second, consider the first case from the definition of a bad input and either the second or the third case from the definition of a bad output. Then the case (1) below is implied. Third, consider the second case from the definition of a bad input and either the second or the third case from the definition of a bad output. Then the case (2) below is implied. Finally, consider the third case from the definition of a bad input. It trivially contradicts both the second and the third case from the definition of a bad output. Thus one of the following cases is true:

1. $a >_b \mathsf{OPT}(b)$; $b >_a \mathsf{OPT}(a)$.
2. $a \simeq_b \mathsf{OPT}(b)$, and $\mathsf{OPT}(b)$ is unsuccessful; a is not 2-promoted.

In case (1), the edge (a, b) is a blocking pair for OPT, contradicting the stability of OPT.

In case (2), since $\mathsf{OPT}(b)$ is unsuccessful, $\mathsf{OPT}(b)$ was rejected by b as a 2-promoted man. On the other hand, $a \simeq_b \mathsf{OPT}(b)$, a is not 2-promoted, and b holds a proposal from a when the algorithm terminates, contradicting the rejection step. □

Corollary 1. *The number of good inputs is at least the number of bad outputs.*

Proof. Assume for a contradiction that the number of good inputs is smaller than the number of bad outputs. Then there is an edge in G' which is a bad output but not a good input. In other words, there is an edge in G' which is both a bad output and a bad input, contradicting Lemma 3. □

3.3 Cost

In our charging scheme, cost is a function that assigns charges, that originate from the edges, to the nodes. More specifically, the cost of a man a is obtained by counting the edges in G' incident to a, where bad outputs contribute 2 and all other edges contribute 1. Similarly, the cost of a woman b is obtained by counting the edges in G' incident to b, to which good inputs contribute 0 and all other edges contribute 1.

In the following, let $\deg(u)$ be the degree of the node u in G'. For $a \in A$, we define his *cost* as follows:

$$\mathsf{cost}(a) := \deg(a) + k\,, \quad \text{where k is the number of bad outputs from a;}$$

for $b \in B$, we define her cost as follows:

$$\mathsf{cost}(b) := \deg(b) - k\,, \quad \text{where k is the number of good inputs to b,}$$

For a node set $S \subseteq A \cup B$, $\mathsf{cost}(S)$ is defined as the sum of costs of all the nodes in S.

The above definitions lead to next three remarks.

Remark 3. Let $b \in B$ be matched in M and have at least k bad inputs. Then $\mathsf{cost}(b) \geq k + 1$.

Proof. Let k' be the number of good inputs to b. Since b is matched in M, the edge $(M(b), b)$ is contained in G' and therefore it is not an input to b. Thus $\deg(b) \geq k + k' + 1$. Hence, by definition of cost, $\mathsf{cost}(b) = \deg(b) - k' \geq k + 1$ holds. □

Remark 4. Let $b \in B$ be matched in OPT, have at least k bad inputs, and $(\mathsf{OPT}(b), b) \in E'$ where E' is the edge set of G'. Then $\mathsf{cost}(b) \geq k + 1$.

Proof. Let k' be the number of good inputs to b. Since the edge $(\mathsf{OPT}(b), b)$ is in G', it is not an input to b. Thus $\deg(b) \geq k + k' + 1$. So, by definition of cost, $\mathsf{cost}(b) = \deg(b) - k' \geq k + 1$ holds. □

Remark 5. Let $b \in B$ be matched in both OPT and M, $\mathsf{OPT}(b) \neq M(b)$, and $(\mathsf{OPT}(b), b) \in E'$ where E' is the edge set of G'. Then $\mathsf{cost}(b) \geq 2$.

Proof. Let k and k' be the numbers of bad inputs and good inputs to b, respectively. Since the edges $(\mathsf{OPT}(b), b)$ and $(M(b), b)$ are contained in G', they are not inputs to b. Thus $\deg(b) \geq k + k' + 2$. So, by definition of cost, $\mathsf{cost}(b) = \deg(b) - k' \geq k + 2 \geq 2$ holds. □

3.4 The Approximation Ratio

Let $\mathcal{C}(M + \mathsf{OPT})$ denote the set of connected components in a graph induced by the edge set $M + \mathsf{OPT}$. Lemma 4 below bounds the cost of $M + \mathsf{OPT}$. For the proof of this lemma, we refer the reader to the full version of the paper on arXiv

Lemma 4. $\sum_{C \in \mathcal{C}(M+OPT)} cost(C) \geq (L+1)|OPT| + (L-2)(|OPT| - |M|)$.

We are ready to prove our main theorem, and restate it here for completeness.

Theorem 1. *Given an instance of the maximum-cardinality stable matching problem with incomplete preferences, and ties of size at most L; the polynomial-time algorithm described in Sect. 2 finds a stable matching M with*

$$|M| \geq \frac{2L-1}{3L-2}|OPT|,$$

where OPT is an optimal stable matching.

Proof. By Lemma 1, we have

$$|M| \geq \frac{|E'|}{L} = \sum_{u \in A \cup B} \frac{\deg(u)}{2L}.$$

By definition of cost and by Corollary 1, we obtain

$$\sum_{u \in A \cup B} \deg(u) \geq \mathsf{cost}(A \cup B).$$

Combining the above inequalities, we get

$$2L|M| \geq \sum_{u \in A \cup B} \deg(u) \geq \mathsf{cost}(A \cup B) = \sum_{C \in \mathcal{C}(M+\mathsf{OPT})} \mathsf{cost}(C),$$

By Lemma 4, we obtain

$$2L|M| \geq \sum_{C \in \mathcal{C}(M+\mathsf{OPT})} \mathsf{cost}(C) \geq (L+1)|\mathsf{OPT}| + (L-2)(|\mathsf{OPT}| - |M|).$$

By rearranging the terms, we obtain

$$2L|M| + (L-2)|M| \geq (L+1)|\mathsf{OPT}| + (L-2)|\mathsf{OPT}|,$$

and so we obtain the desired inequality

$$(3L-2)|M| \geq (2L-1)|\mathsf{OPT}|.$$

□

3.5 Tightness of the Analysis

The following example shows that the bound in Theorem 1 is tight.

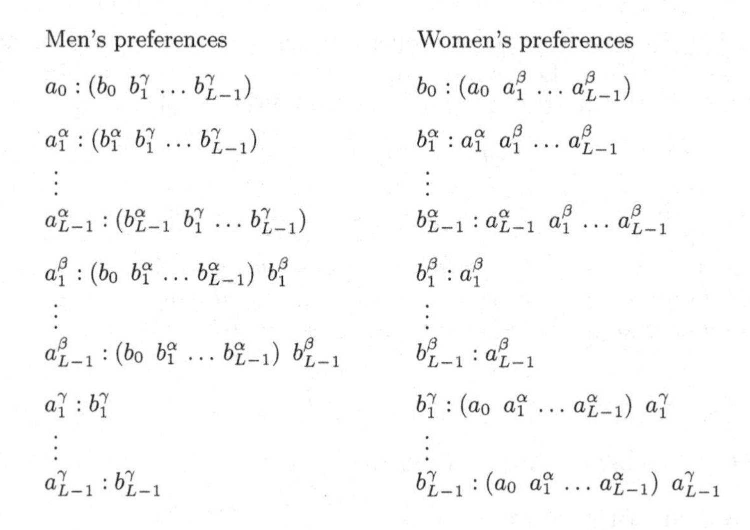

Men's preferences
$a_0 : (b_0\ b_1^\gamma \dots b_{L-1}^\gamma)$
$a_1^\alpha : (b_1^\alpha\ b_1^\gamma \dots b_{L-1}^\gamma)$
$\vdots$
$a_{L-1}^\alpha : (b_{L-1}^\alpha\ b_1^\gamma \dots b_{L-1}^\gamma)$
$a_1^\beta : (b_0\ b_1^\alpha \dots b_{L-1}^\alpha)\ b_1^\beta$
$\vdots$
$a_{L-1}^\beta : (b_0\ b_1^\alpha \dots b_{L-1}^\alpha)\ b_{L-1}^\beta$
$a_1^\gamma : b_1^\gamma$
$\vdots$
$a_{L-1}^\gamma : b_{L-1}^\gamma$
Women's preferences
$b_0 : (a_0\ a_1^\beta \dots a_{L-1}^\beta)$
$b_1^\alpha : a_1^\alpha\ a_1^\beta \dots a_{L-1}^\beta$
$\vdots$
$b_{L-1}^\alpha : a_{L-1}^\alpha\ a_1^\beta \dots a_{L-1}^\beta$
$b_1^\beta : a_1^\beta$
$\vdots$
$b_{L-1}^\beta : a_{L-1}^\beta$
$b_1^\gamma : (a_0\ a_1^\alpha \dots a_{L-1}^\alpha)\ a_1^\gamma$
$\vdots$
$b_{L-1}^\gamma : (a_0\ a_1^\alpha \dots a_{L-1}^\alpha)\ a_{L-1}^\gamma$

Fig. 1. An instance with ties of size at most L, $L \geq 2$ for which the algorithm outputs a stable matching M with $|\mathsf{OPT}|/|M| = (3L-2)/(2L-1)$

Example 1. In Fig. 1, the preference list of each individual is ordered from a most preferred person to a least preferred one, where individuals within parentheses are tied. For example, a_1^β is indifferent between all the women in his preference list except b_1^β, who is less preferred than the others.

It is straightforward to check that there exists a unique maximum-cardinality stable matching, namely $\mathsf{OPT} = \{(a_0, b_0)\} \cup \{(a_i^j, b_i^j) \mid i = 1, \dots, L-1,\ j = \alpha, \beta, \gamma\}$. We show that there exists an execution of the algorithm which outputs the matching $M = \{(a_0, b_0)\} \cup \{(a_i^\alpha, b_i^\gamma) \mid i = 1, \dots, L-1\} \cup \{(a_i^\beta, b_i^\alpha) \mid i = 1, \dots, L-1\}$, leading to the ratio $|\mathsf{OPT}|/|M| = (3L-2)/(2L-1)$.

Proof. The following is an execution of the algorithm which leads either to the matching M or a matching with the size of M.

- a_0 makes one proposal to every woman in his list; the women accept.
- a_i^α for all $i = 1, \ldots, L-1$ makes one proposal to every woman in his list; the women accept.
- a_i^β for all $i = 1, \ldots, L-1$ makes one proposal to every woman except the last one in his list; the women accept.
- a_i^γ starts to propose b_i^γ for all $i = 1, \ldots, L-1$, but each time a_i^γ makes a proposal, the proposal is rejected; a_i^γ gives up.

□

References

1. Bauckholt, F., Pashkovich, K., Sanità, L.: On the approximability of the stable marriage problem with one-sided ties. ArXiv e-prints (2018)
2. Chiang, R., Pashkovich, K.: On the approximability of the stable matching problem with ties of size two. Algorithmica 1–19 (2020). https://doi.org/10.1007/s00453-020-00703-9
3. David, M.: Algorithmics of Matching Under Preferences, vol. 2. World Scientific, Singapore (2013)
4. Gale, D., Shapley, L.: College admissions and the stability of marriage. Am. Math. Monthly **69**(1), 9–15 (1962)
5. Gale, D., Sotomayor, M.: Some remarks on the stable matching problem. Discrete Appl. Math. **11**(3), 223–232 (1985)
6. Huang, C.-C., Kavitha, T.: Improved approximation algorithms for two variants of the stable marriage problem with ties. Math. Program. 353–380 (2015). https://doi.org/10.1007/s10107-015-0923-0
7. Iwama, K., Miyazaki, S., Yamauchi, N.: A 1.875-approximation algorithm for the stable marriage problem. In: 18th Symposium on Discrete Algorithms (SODA), pp. 288–297 (2007)
8. Iwama, K., Miyazaki, S., Yanagisawa, H.: A 25/17-approximation algorithm for the stable marriage problem with one-sided ties. Algorithmica **68**(3), 758–775 (2014). https://doi.org/10.1007/s00453-012-9699-2
9. Király, Z.: Better and simpler approximation algorithms for the stable marriage problem. Algorithmica **60**(1), 3–20 (2011)
10. Király, Z.: Linear time local approximation algorithm for maximum stable marriage. Algorithms **6**(3), 471–484 (2013)
11. Lam, C.K.: Algorithms for stable matching with indifferences. Ph.D. thesis, University of Texas at Austin (2019)
12. Lam, C.K., Plaxton, C.G.: A $(1 + 1/e)$-approximation algorithm for maximum stable matching with one-sided ties and incomplete lists. In: Proceedings of the Thirtieth Annual ACM-SIAM Symposium on Discrete Algorithms, pp. 2823–2840. SIAM, Philadelphia (2019). https://doi.org/10.1137/1.9781611975482.175
13. Lam, C.-K., Plaxton, C.G.: Maximum stable matching with one-sided ties of bounded length. In: Fotakis, D., Markakis, E. (eds.) SAGT 2019. LNCS, vol. 11801, pp. 343–356. Springer, Cham (2019). https://doi.org/10.1007/978-3-030-30473-7_23
14. Manlove, D.F., Irving, R.W., Iwama, K., Miyazaki, S., Morita, Y.: Hard variants of stable marriage. Theoret. Comput. Sci. **276**(1–2), 261–279 (2002). https://doi.org/10.1016/S0304-3975(01)00206-7

15. McDermid, E.: A 3/2-approximation algorithm for general stable marriage. In: Albers, S., Marchetti-Spaccamela, A., Matias, Y., Nikoletseas, S., Thomas, W. (eds.) ICALP 2009, Part I. LNCS, vol. 5555, pp. 689–700. Springer, Heidelberg (2009). https://doi.org/10.1007/978-3-642-02927-1_57
16. Paluch, K.: Faster and simpler approximation of stable matchings. Algorithms **7**(2), 189–202 (2014)
17. Yanagisawa, H.: Approximation algorithms for stable marriage problems. Ph.D. thesis, Kyoto University (2007)

Envy-Freeness and Relaxed Stability: Hardness and Approximation Algorithms

Prem Krishnaa[1], Girija Limaye[1], Meghana Nasre[1], and Prajakta Nimbhorkar[2,3](✉)

[1] Indian Institute of Technology Madras, Chennai, India
premkrishnaa.jaganmohan@cohesity.com, {girija,meghana}@cse.iitm.ac.in
[2] Chennai Mathematical Institute, Chennai, India
prajakta@cmi.ac.in
[3] UMI ReLaX, Chennai, India

Abstract. We consider the problem of matchings under two-sided preferences in the presence of maximum as well as minimum quota requirements for the agents. When there are no minimum quotas, *stability* is the de-facto notion of optimality. In the presence of minimum quotas, ensuring stability and simultaneously satisfying lower quotas is not an attainable goal in many instances.

To address this, a relaxation of stability known as *envy-freeness*, is proposed in literature. In our work, we thoroughly investigate envy-freeness from a computational view point. Our results show that computing envy-free matchings that match maximum number of agents is computationally hard and also hard to approximate up to a constant factor. Additionally, it is known that envy-free matchings satisfying lower-quotas may not exist. To circumvent these drawbacks, we propose a new notion called *relaxed stability*. We show that relaxed stable matchings are guaranteed to exist even in the presence of lower-quotas. Despite the computational intractability of finding a largest matching that is feasible and relaxed stable, we give an efficient algorithm that computes a constant factor approximation to this matching in terms of size.

Keywords: Matchings under preferences · Lower quota · Envy-freeness · Relaxed stability · Approximation

1 Introduction

Matching problems with two-sided preferences have been extensively investigated for matching markets where agents (hospitals/residents or colleges/students) have upper quotas that cannot be exceeded. Stability [6] is a widely accepted notion of optimality in this scenario. An allocation is said to be *stable* if no pair of agents has an incentive to deviate from it. However, the case when the agents

This work was partially supported by the grant CRG/2019/004757.
P. Krishnaa—Part of this work was done when the author was a student at IIT Madras.

T. Harks and M. Klimm (Eds.): SAGT 2020, LNCS 12283, pp. 193–208, 2020.
https://doi.org/10.1007/978-3-030-57980-7_13

have maximum as well as minimum quotas poses new challenges and there is still a want of satisfactory mechanisms that take minimum quotas into account. Practically, lower quotas are important, since it is natural for a hospital to require a minimum number of residents to run the hospital smoothly. Lower quotas are crucial in applications like course-allocation, and assigning teaching assistants (TAs) in academic institutions where a minimum guarantee is essential.

Ensuring stability while satisfying lower quotas may not be attainable always. On one hand, disregarding preferences in the interest of satisfying the lower quotas gives rise to social unfairness (for instance agents envying each other); on the other hand, too much emphasis on fairness can lead to *wastefulness* [5]. Hence, it is necessary to strike a balance between these three mutually conflicting goals – optimality with respect to preferences, feasibility for minimum quotas and minimizing wastefulness. The main contribution of this paper is to propose a mechanism to achieve this balance.

Envy-freeness [3,5,7,11,12] is a widely accepted notion for achieving fairness from a social perspective. Unfortunately, even envy-freeness and feasibility may not be simultaneously achievable. Whether feasible envy-free matchings exist can be answered efficiently by the characterization of Yokoi [20]. Fragiadakis et al. [5] explore strategyproofness and the trade-off between envy-freeness and wastefulness for a restricted setting of agent preferences. In our work, we thoroughly investigate envy-freeness from a computational view point. Our results show that computing a maximum size envy-free matching is computationally hard and such matchings can be wasteful. To circumvent these drawbacks, we propose a new notion called *relaxed stability.* We show that relaxed stable matchings are guaranteed to exist even in the presence of lower-quotas. Despite the computational intractability of finding a largest feasible relaxed stable matching, we give an efficient constant-factor approximation algorithm for it.

We state the problem formally in terms of a setting known as the HRLQ setting in literature. An HRLQ instance consists of a bipartite graph $G = (\mathcal{R} \cup \mathcal{H}, E)$, $\mathcal{R}$ and $\mathcal{H}$ being the sets of residents and hospitals respectively, and an edge $(r, h) \in E$ denotes that r and h are mutually acceptable. Each $h \in \mathcal{H}$ has an upper-quota $q^+(h)$ and a lower-quota $q^-(h)$, respectively denoting the maximum and minimum number of residents that can be assigned to h. Every vertex in $\mathcal{R} \cup \mathcal{H}$ ranks its neighbors in a strict order, referred to as its *preference list.* If a vertex a prefers its neighbor b_1 over b_2, we denote it by $b_1 >_a b_2$.

A matching $M \subseteq E$ in G is an assignment of residents to hospitals such that each resident is matched to at most one hospital, and every hospital h is matched to at most $q^+(h)$-many residents. Let $M(r)$ denote the hospital that r is matched to in M, and $M(h)$ denote the set of residents matched to h in M. We let $M(r) = \perp$ if r is unmatched in M, and $\perp$ is considered as the least preferred choice of each $r \in \mathcal{R}$. We say that a hospital h is *under-subscribed* in M if $|M(h)| < q^+(h)$, is *fully-subscribed* if $|M(h)| = q^+(h)$ and is *deficient* if $|M(h)| < q^-(h)$. A matching is *feasible* for an HRLQ instance if no hospital is deficient in M. The goal in the HRLQ setting is to find a feasible matching M that is *optimal* with respect to the preference lists. The HRLQ problem is a generalization of the well-studied HR problem (introduced by Gale and Shapley [6]) where there are no lower quotas. In the HR problem, *stability* is a de-facto notion of optimality and is defined by the absence of *blocking pairs.*

Definition 1 (Stable matchings). *A pair $(r, h) \in E \setminus M$ is a blocking pair w.r.t. the matching M if $h >_r M(r)$ and h is either under-subscribed in M or there exists at least one resident $r' \in M(h)$ such that $r >_h r'$. A matching M is* stable *if there is no blocking pair w.r.t. M.*

Existence of Stable Feasible Matchings: Given an HRLQ instance, it is natural to ask "does the instance admit a stable feasible matching?" Unlike HR instances, an HRLQ instance may not admit a stable, feasible matching. Figure 1 shows an example. The stable matching $M_s = \{(r_1, h_1)\}$ is not feasible since h_2 is deficient in M_s, and the feasible matchings are not stable. The well-known Rural Hospitals Theorem [18] implies that the *number* of residents matched to a hospital is invariant across all stable matchings of the instance. Hence, for any HRLQ instance, either all stable matchings are feasible or all are infeasible. In light of the fact that stable and feasible matchings may not exist, relaxations of stability, like popularity and envy-freeness have been proposed in the literature [16,17,20]. Envy-freeness is defined by the absence of *envy-pairs.*

$r_1 : h_1, h_2$	[0,1] $h_1 : r_1, r_2$
$r_2 : h_1$	[1,1] $h_2 : r_1$

Fig. 1. An HRLQ instance with no feasible and stable matching. Here $\mathcal{R} = \{r_1, r_2\}, \mathcal{H} = \{h_1, h_2\}$ and quotas are denoted as [lower-quota, upper-quota] pair preceding each hospital.

Definition 2 (Envy-free matchings). *Given a matching M, a resident r has a* justified envy *(here onwards called envy) towards a matched resident r', where $M(r') = h$ and $(r, h) \in E$ if $h >_r M(r)$ and $r >_h r'$. The pair (r, r') is an envy-pair w.r.t. M. A matching is envy-free if there is no envy-pair w.r.t. it.*

Note that an envy-pair implies a blocking pair but the converse is not true and hence envy-freeness is a relaxation of stability. In the example in Fig. 1, the matching $\{(r_1, h_2)\}$ is envy-free and feasible, although not stable. Thus, envy-free matchings provide an alternative to stability in such instances. Envy-freeness is motivated by fairness from a social perspective. Importance of envy-free matchings has been recognized in the context of constrained matchings [3,5,7,11,12], and their structural properties have been investigated in [19].

Size of Envy-Free Matchings: In terms of size, there is a sharp contrast between stable matchings in the HR setting and envy-free matchings in the HRLQ setting. While all the stable matchings in an HR instance have the same size, the envy-free matchings in an HRLQ instance may have significantly different sizes. For example,

$$\forall i \in [n],\ r_i : h_1, h_2 \qquad \begin{array}{l} [0, n]\ h_1 : r_1, \ldots, r_n \\ [1, 1]\ h_2 : r_1, \ldots, r_n \end{array}$$

Fig. 2. An HRLQ instance with two envy-free matchings of different sizes.

in Fig. 2, there are two envy-free matchings, $N_1 = \{(r_1, h_2)\}$ of size one and $N_n = \{(r_1, h_1), (r_2, h_1), \ldots, (r_{n-1}, h_1), (r_n, h_2)\}$ of size n.

Shortcomings of Envy-Free Matchings: It is interesting to note that a feasible, envy-free matching itself may not exist – e.g. in Fig. 1, if both h_1, h_2 have a unit lower-quota, then the unique feasible matching is not envy-free. If a stable matching is not feasible in an HRLQ instance, *wastefulness* may be inevitable for attaining feasibility. A matching is *wasteful* if there exists a resident who prefers a hospital to her current assignment and that hospital has a vacant position [5]. Envy-free matchings can be significantly wasteful (e.g. the matching N_1 in Fig. 2). Therefore, it would be ideal to have a notion of optimality which is guaranteed to exist, is efficiently computable and avoids wastefulness.

Quest for a Better Optimality Criterion: We propose a new notion of *relaxed stability* which always exists for any HRLQ instance. We observe that in the presence of lower quotas, there can be at most $q^-(h)$-many residents that are forced to be matched to h, even though they have higher preferred under-subscribed hospitals in their list. Our relaxation allows these forced residents to participate in blocking-pairs,[1] however, the matching is still stable when restricted to the remaining residents. We now make this formal below.

Definition 3 (Relaxed stable matchings). *A matching M is* relaxed stable *if, for every hospital h, at most $q^-(h)$ residents from $M(h)$ participate in blocking pairs and no unmatched resident participates in a blocking pair.*

In Fig. 1, the matching $\{(r_1, h_2), (r_2, h_1)\}$ (which was not envy-free) is feasible, relaxed stable and non-wasteful. We show that a feasible relaxed stable matching always exists in an HRLQ instance. However, computing a largest relaxed stable matching is NP-hard. We present an efficient algorithm that computes a matching that is at least as large as any stable matching in the instance, thus addressing wastefulness. In fact, a relaxed stable matching may be even larger than the stable matching in the instance. In the instance shown in Fig. 3, $M_s = \{(r_1, h_1), (r_2, h_2)\}$ is an infeasible stable matching. Matchings $M_1' = \{(r_1, h_3), (r_2, h_2)\}$ and $M_2' = \{(r_1, h_1), (r_2, h_3), (r_3, h_2)\}$ both are feasible, relaxed stable and M_2' is larger than M_s. This is in contrast to maximum size envy-free matching which cannot be larger than a stable matching (see Sect. 2.2).

$r_1 : h_1, h_3$ $\qquad$ $[0,1]\ h_1 : r_1$

$r_2 : h_2, h_3$ $\qquad$ $[0,1]\ h_2 : r_2, r_3$

$r_3 : h_2$ $\qquad$ $[1,1]\ h_3 : r_1, r_2$

Fig. 3. An HRLQ instance with two relaxed stable matchings of different sizes, one larger than stable matching

In the spirit of allowing blocking pairs, different notions have been proposed in [2,9]. In [9], the goal is to compute a feasible matching with the least number

[1] Our initial idea was to allow them to participate in envy-pairs. We thank anonymous reviewer for suggesting this modification which is stricter than our earlier notion.

of blocking pairs or blocking residents; however, both these problems are NP-hard even for CL-restriction [9] (i.e., every hospital with positive lower-quota must rank every resident, and hence, every resident must rank every hospital with positive lower-quota) whereas a feasible relaxed stable matching can be efficiently computed. *Popular* matchings [17] allow blocking pairs to address feasibility and guaranteed existence in the HRLQ setting. However, there is no known bound on the number of blocking residents in a largest popular matching, whereas the number of blocking residents in a relaxed stable matching is at most the sum of lower quotas of all the hospitals. Lower quotas with constraints [4,10] and in a model where hospitals can be closed [1] are investigated.

Our Contributions: We denote the problem of computing a maximum size feasible envy-free matching (respectively a maximum size feasible relaxed stable matching) as the MAXEFM (respectively the MAXRSM) problem. Throughout the paper, we assume that our input HRLQ instance admits a feasible matching. In the interest of space, proofs of Theorems and Lemmas marked with ($\star$) are deferred to the full-version [15].

Results on Envy-Freeness: We show that the MAXEFM problem is NP-hard, and is hard to approximate below a constant factor.

Theorem 1 ($\star$). *The* MAXEFM *problem is* NP*-hard and cannot be approximated within a factor of* $\frac{21}{19} - \epsilon$ *for* $\epsilon > 0$ *unless* P = NP *even when every hospital has a quota of at most one.*

In light of the above negative result, we turn our attention to the approximation and tractable special cases. In practice it is common to have incomplete preference lists and in many cases the preference lists of residents may also be of constant size. A matching M is a *maximal envy-free matching* if addition of an edge to M violates either the upper-quota or envy-freeness. Prior to our work, no size guarantee of a maximal envy-free matching was known. Let ℓ_1 and ℓ_2 be the length of the longest preference list of a resident and a hospital respectively.

Theorem 2. *A maximal envy-free matching is*

(I) an ℓ_1*-approximation of* MAXEFM *when hospital quotas are at most one.*
(II) ($\star$) an $(\ell_1 \cdot \ell_2)$*-approximation of* MAXEFM *when quotas are unrestricted.*

Next, we consider the HRLQ instances with the CL-restriction [9]. In contrast to the NP-hardness results in [9], the MAXEFM problem is tractable under the CL-restriction.

Theorem 3. *There is a simple linear-time algorithm for the* MAXEFM *problem for* CL*-restricted* HRLQ *instances.*

Results on Relaxed Stability: We prove that the MAXRSM problem is NP-hard and is also hard to approximate, but has a better approximation behavior than the MAXEFM problem.

Theorem 4. *The* MAXRSM *problem is* NP*-hard and cannot be approximated within a factor of* $\frac{21}{19} - \epsilon$ *for* $\epsilon > 0$ *unless* P = NP *even when every hospital has a quota of at most one.*

We complement the above negative result with the following.

Theorem 5. *Any feasible* HRLQ *instance always admits a relaxed stable matching. Moreover, there is a polynomial-time algorithm that outputs a* $\frac{3}{2}$*-approximation to the maximum size relaxed stable matching.*

We summarize our results in Table 1.

Table 1. Summary of our results

Problem	Inapproximability	Approximation	Restricted settings
MAXEFM	$(\frac{21}{19} - \epsilon)$-inapproximability	$(\ell_1 \cdot \ell_2)$-approximation	P-time for CL-restriction, ℓ_1-approximation for 0/1 quotas
MAXRSM	$(\frac{21}{19} - \epsilon)$-inapproximability	$\frac{3}{2}$-approximation	–

Organization of the Paper: Our algorithmic results for envy-free matchings and relaxed stable matchings are presented in Sect. 2 and in Sect. 3 respectively. The NP-hardness and inapproximability results are presented in Sect. 4.

2 Envy-Freeness: Algorithmic Results

In this section, we first focus on the approximation guarantee of maximal envy-free matchings and then present an efficient algorithm for the MAXEFM problem on the CL-restricted HRLQ instances.

2.1 Approximation to MAXEFM

A maximal envy-free matching can be efficiently computed; Krishnapriya et al. [16] present one such algorithm which extends a given envy-free matching. The results in [16] are empirical and no theoretical guarantee is known about the size of a maximal envy-free matching. Below we prove the guarantee for the instances where hospital quotas are at most one.

Proof (of Theorem 2(I)). Let M and OPT be respectively a maximal and a maximum size envy-free matching. Let R_{OPT} and R_M denote the set of residents matched in OPT and M respectively. Let X_1 be the set of residents matched in both M and OPT. Let X_2 be the set of residents matched in OPT but not matched in M. Thus, $|R_{OPT}| = |X_1| + |X_2|$. Since $X_1 = R_{OPT} \cap R_M \subseteq R_M$, so $|X_1| \leq |R_M|$. Our goal is to show that $|X_2| \leq |R_M| \cdot (\ell_1 - 1)$. Recall that ℓ_1 is the length of the longest preference list of a resident. Once we establish that, it is immediate that a maximal envy-free matching is an ℓ_1-approximation.

We show that for every resident $r \in X_2$ we can associate a unique hospital h_r such that h_r is unmatched in M and there exists a resident r' in the neighbourhood of h_r such that r' is matched in M. Denote the set of such hospitals as Y_2. Note that due to the uniqueness assumption $|X_2| = |Y_2|$. Since each resident has a preference list of length at most ℓ_1, any r' who is matched in M can have at most $\ell_1 - 1$ neighbouring hospitals which are unmatched in M. Thus $|X_2| = |Y_2| \leq |R_M| \cdot (\ell_1 - 1)$ which establishes the approximation guarantee. To finish the proof we show a unique hospital h_r with desired properties that can be associated with each $r \in X_2$. Let $r \in X_2$ such that $h = OPT(r)$. We have following two exhaustive cases.

Case 1: If h is unmatched in M, then due to maximality of M, there must exist a resident r' matched in M such that adding (r, h) causes envy to r'. Thus, h has a neighboring resident r' matched in M, and we let $h_r = h$.

Case 2: If h is matched in M, then since M and OPT are both envy-free, there must exist a path $\langle r, h, r_1, h_1, \ldots, r_i, h_i \rangle$ such that $(r, h) \in OPT$, for each $k = 1, \ldots, i$, we have $(r_k, h_k) \in OPT$, $(r_1, h) \in M$, for each $k = 2, \ldots, i$, we have $(r_k, h_{k-1}) \in M$ and h_i is unmatched in M. Thus, h_i has a neighboring resident r_i matched in M, and we let $h_r = h_i$.

Uniqueness Guarantee: For any $r \in X_2$ such that case 1 applies, the associated h_i is unique since hospital quotas are at most 1. For two distinct $r, r' \in X_2$ such that case 2 applies for both, the paths mentioned above are disjoint since hospital quotas are at most 1, which guarantees uniqueness within case 2. The h_i associated in case 2 cannot be associated in case 1 to $OPT(h_i)$ since $OPT(h_i) = r_i \notin X_2$. This completes the proof of existence of the unique hospital. □

2.2 Polynomial Time Algorithm for the CL-Restricted Instances

In this section, we consider the MAXEFM problem on CL-restricted HRLQ instances with general quotas. Recall that under CL-restriction [9], hospitals with positive lower-quota rank every resident and vice versa. It follows from the characterization of Yokoi [20] that every HRLQ instance with CL-restriction admits a feasible envy-free matching. We now present a simple modification to the standard Gale and Shapley algorithm [6] that computes a maximum size envy-free matching. We start with an empty matching M. Throughout the algorithm, we maintain two parameters:

- d: denotes the deficiency of the matching M, that is, the sum of deficiencies of all hospitals with positive lower-quota.
- k: the number of unmatched residents w.r.t. M.

In every iteration, an unmatched resident r who has not yet exhausted its preference list, proposes to the most preferred hospital h. If h is deficient w.r.t. M, h accepts r's proposal. Otherwise, if h is under-subscribed w.r.t. M, h accepts the r's proposal only if there are enough unmatched residents to satisfy the deficiency of the other hospitals, that is, $k > d$. If h is fully-subscribed, then h rejects the least preferred resident in $M(h) \cup \{r\}$. This process continues as long as some unmatched resident has not exhausted its preference list.

Algorithm 1: MAXEFM in CL-restricted HRLQ instances.

Input: An HRLQ instance $G = (\mathcal{R} \cup \mathcal{H}, E)$ with CL-restriction
Output: Maximum size envy-free matching

```
1  let M = ∅;   d = Σ_{h:q⁻(h)>0} q⁻(h);   k = |R|;
2  while there is an unmatched resident r who has at least one hospital not yet
   proposed to do
3      r proposes to the most preferred hospital h;
4      if |M(h)| < q⁻(h) then
5          M = M ∪ {(r, h)};
6          reduce d and k each by 1;
7      else
8          if |M(h)| == q⁺(h) then
9              let r' be the least preferred resident in M(h) ∪ {r};
10             M(h) = M(h) ∪ {r} \ {r'};
11         if |M(h)| < q⁺(h) and k == d then
12             let r' be the least preferred resident in M(h) ∪ {r};
13             M(h) = M(h) ∪ {r} \ {r'};
14         else
               // we have |M(h)| < q⁺(h) and k > d
15             M = M ∪ {(r, h)};
16             reduce k by 1;
17 return M;
```

Since the input instance is feasible, we start with $k \geq d$ and this inequality is maintained throughout the algorithm. If no resident is rejected due to $k = d$ in line 11, then our algorithm degenerates to the Gale and Shapley algorithm [6] and hence outputs a stable matching. It is straightforward to verify that Algorithm 1 runs in linear time in the size of the instance. Lemma 1 proves the correctness of our algorithm and establishes Theorem 3.

Lemma 1. *The matching M computed by Algorithm 1 is feasible and maximum size envy-free.*

Proof. We first prove that the output is feasible. Suppose not. Then at termination, $d > 0$, that is, there is at least one hospital h that is deficient w.r.t. M. It implies that $k \geq 1$. Thus there is some resident r unmatched w.r.t. M. Note that r could not have been rejected by every hospital with positive lower-quota since h appears in the preference list of r, and h is deficient. This contradicts the termination of our algorithm and proves the feasibility of our matching.

Next, we prove that M is envy-free. Suppose for the sake of contradiction, M contains an envy-pair (r', r) such that $(r, h) \in M$ where $r' >_h r$ and $h >_{r'} M(r')$. This implies that r' must have proposed to h and h rejected r'. If h rejected r' because $|M(h)| = q^+(h)$, h is matched with better preferred residents than r', a contradiction to the fact that $r' >_h r$. If h rejected r' because $k = d$, then there are two cases. Either r was matched to h when r' proposed to h. In this case, in line 11 our algorithm rejected the least preferred resident in $M(h)$. This contradicts that $r' >_h r$. Similarly if r proposed to h later, since $k = d$, the algorithm rejected the least preferred resident again contradicting the presence of any envy-pair.

Finally, we show that M is a maximum size envy-free matching. We have $k \geq d$ at the start of the algorithm. If during the algorithm, $k = d$ at some point, then at the end of the algorithm we have $k = d = 0$, implying that, we have an $\mathcal{R}$-perfect matching and hence the maximum size matching. Otherwise, $k > d$ at the end of the algorithm and then we output a stable matching which is maximum size envy-free by Lemma 2. □

Lemma 2 ($\star$). *A stable matching when feasible, is an optimal solution of* MAXEFM.

Note that Algorithm 1 is similar to the ESDA algorithm presented in [5]. The ESDA algorithm needs a stricter assumption that the underlying graph is complete, whereas we assume the weaker CL-restriction. Moreover, only empirical results without theoretical guarantees on the size of the output matching are presented in [5].

3 Relaxed Stability: Algorithmic Results

In this section, we present Algorithm 2 that computes a relaxed stable matching in an HRLQ instance and prove that it gives a $\frac{3}{2}$-approximation to MAXRSM. Furthermore, we show that the output of Algorithm 2 is at least as large as any stable matching in the instance (disregarding lower-quotas).

We say a feasible matching M_0 is *minimal* if, for any edge $e \in M_0$, $M_0 \setminus \{e\}$ is infeasible. Thus, if M_0 is minimal, then for every hospital h, $|M_0(h)| = q^-(h)$. Algorithm 2 begins by computing a feasible matching M_0 in the instance G disregarding the preferences of the residents and hospitals. Such a feasible matching can be computed by the standard reduction from bipartite matchings to flows with demands on edges [14]. Let $M = M_0$. We now associate levels with the residents – all residents matched in M are set to have level-0; all residents unmatched in M are assigned level-1. We now execute the Gale and Shapley resident proposing algorithm, with the modification that a hospital prefers any level-1 resident over any level-0 resident (irrespective of the preference list of h). Furthermore, if a level-0 resident becomes unmatched during the course of the proposals, then it gets assigned a level-1 and it starts proposing from the beginning of its preference list. Amongst two residents of the same level, the hospital uses its preference list to order them. Our algorithm terminates when every resident is either matched or has exhausted its preference list when proposing hospitals at level-1. The two level idea is somewhat similar to the one used in Király [13] for stable matchings with ties and incomplete lists. It is clear that our algorithm runs in polynomial time since it only computes a feasible matching (using a reduction to flows) and executes a modification of Gale and Shapley algorithm. We prove the correctness of our algorithm below.

Algorithm 2: Algorithm to compute $\frac{3}{2}$-approximation of MAXRSM

Input: Input: HRLQ instance $G = (\mathcal{R} \cup \mathcal{H}, E)$
Output: A relaxed stable matching that is a $\frac{3}{2}$-approximation of MAXRSM

1 M_0 is a minimal feasible matching in G. Let $M = M_0$;
2 For every matched resident r, set level of r to level-0;
3 For every unmatched resident r, set level of r to level-1;
4 **while** *there is an unmatched resident* r *who has not exhausted his preference list* **do**
5   r proposes to the most preferred hospital h to whom he has not yet proposed;
6   **if** h *is under-subscribed* **then**
7         $M = M \cup \{(r, h)\}$;
8   **else**
9   **if** $M(h)$ *has at least one level-*0 *resident* r' **then**
10             $M = M \setminus \{(r', h)\} \cup \{(r, h)\}$;
11   Set level of r' to level-1 and r' starts proposing from the beginning of his list;
12   else
13   h rejects the least preferred resident in $M(h) \cup \{r\}$;
14 Return M;

Remark 1. If r is unmatched in M then r is a level-1 resident and all the hospitals in r's preference list are fully-subscribed with level-1 residents preferred over r.

Lemma 3. *Matching M output by Algorithm 2 is feasible and relaxed stable.*

Proof. We note that M_0 is feasible and since residents propose it is clear that for any hospital h, we have $|M(h)| \geq |M_0(h)| = q^-(h)$. Thus M is feasible.

To show relaxed stability, we claim that when the algorithm terminates, a resident at level-1 does not participate in a blocking pair. Whenever a level-1 resident r proposes to a hospital h, resident r always gets accepted except when h is fully-subscribed and all the residents matched to h are level-1 and are better preferred than r. When a matched level-1 resident r is rejected by a hospital h, h gets a better preferred resident than r. Thus, a level-1 resident does not participate in a blocking pair. By Remark 1, an unmatched resident (being at level-1) does not participate in a blocking pair. Recall that all residents matched in M_0 are level-0 residents and M_0 is minimal. This implies that for every hospital h, at most $q^-(h)$ many residents assigned to h in M_0 participate in a blocking pair. We show that in M, the number of level-0 residents assigned to any hospital does not increase. To see this, if r is matched to h in M, but not matched to h in M_0, it implies that either r was unmatched in M_0 or r was matched to some h' in M_0. In either case r becomes level-1 when it gets assigned to h in M. Thus the number of level-0 residents assigned to any hospital h in M is at most $q^-(h)$, all of which can potentially participate in blocking pairs. This completes the proof that M is relaxed stable. □

Lemma 4. *Matching M output by Algorithm 2 is a $\frac{3}{2}$-approximation to the maximum size relaxed stable matching.*

Proof. Let OPT denote the maximum size relaxed stable matching in G. To prove the lemma we show that in $M \oplus OPT$ there does not exist any one-length as well as any three-length augmenting path. Suppose that (r, h) is a one-length augmenting path w.r.t. M in $M \oplus OPT$ implying that r is unmatched in M. Then by Remark 1, h is fully-subscribed - a contradiction that (r, h) is an augmenting path. Thus, there is no one-length augmenting path in $M \oplus OPT$.

For the three-length augmenting paths, we first convert the matchings M and OPT as one-to-one matchings, by making *clones* of the hospital. In particular we make $q^+(h)$ many copies of the hospital h for every h where the first $q^-(h)$ copies are called *lower-quota copies* and the $q^-(h) + 1$ to $q^+(h)$ copies are called *non lower-quota copies* of h. Let M_1 denote the one-to-one matching corresponding to M. To obtain M_1, we assign every resident $r \in M(h)$ to a unique copy of h as follows: first, all the residents in $M(h)$ who participate in blocking pair w.r.t. M are assigned unique lower-quota copies of h arbitrarily. The remaining residents in $M(h)$ are assigned to the rest of the copies of h, ensuring all lower-quota copies get assigned some resident. We get OPT_1 from OPT in the same manner.

Now, suppose there exists a three-length augmenting path w.r.t. M which starts at an under-subscribed hospital, say h_j and ends at an unmatched resident in M. Since h_j is under-subscribed in M, and there is an augmenting path starting at h_j, it implies that there exists a copy h_j^d such that (i) h_j^d is matched in OPT_1 and unmatched in M_1, say $OPT_1(h_j^d) = r_d$ and (ii) the resident r_d is matched in M_1 (otherwise there is a one-length augmenting path w.r.t. M_1, which does not exist); let $M_1(r_d) = h_i^c$, and (iii) the copy h_i^c is matched in OPT_1 and $OPT_1(h_i^c) = r_c$ is unmatched in M_1 (else the claimed three-length augmenting path does not exist).

We first note that h_i^c and h_j^d are not copies of the same hospital, that is, $i \neq j$, otherwise there is a one-length augmenting path (r_c, h_i) w.r.t. M. Since r_c is unmatched, by Remark 1, r_d is a level-1 resident and $r_d >_{h_i} r_c$. Thus, r_d proposed to hospitals from the beginning of its preference list. Since h_j is under-subscribed, it must be the case that $h_i >_{r_d} h_j$. Thus, (r_d, h_i) is a blocking pair w.r.t. OPT. By the construction of OPT_1 from OPT, we must have assigned r_d

to a lower-quota copy of h_j. However, copy h_j^d is a non lower-quota copy, since it is unassigned in M_1, a contradiction. Thus, the claimed three-length augmenting path does not exist. □

We note that our analysis is tight [15]. We now show that the matching M computed by Algorithm 2 is at least as large as any stable matching.

Lemma 5 (⋆). *A resident matched in a stable matching M_s is also matched in M output by Algorithm 2. Hence $|M| \geq |M_s|$.*

Proof (Sketch). Suppose there exists a resident r matched in M_s to hospital h but unmatched in M. We start constructing a path starting at r using edges from M_s and M alternately. We show that such a path can neither terminate at a resident nor at a hospital and hence cannot exist. Thus, every resident matched in M_s is matched in M and hence $|M| \geq |M_s|$. □

4 Hardness Results

In this section we give an overview of the techniques used in proving the hardness and inapproximability results. Theorem 1 and Theorem 4 are proved using suitable reductions from the Minimum Vertex Cover (MVC) problem. We present the proof for Theorem 4 below.

Proof (of Theorem 4). Given a graph $G = (V, E)$, which is an instance of the MVC problem, we construct an instance G' of the MAXRSM problem. Corresponding to each vertex v_i in G, G' contains a gadget with three residents r_1^i, r_2^i, r_3^i, and three hospitals h_1^i, h_2^i, h_3^i. All hospitals have an upper-quota of 1 and h_3^i has a lower-quota of 1. Assume that the vertex v_i has d neighbors in G, namely $v_{j_1}, \ldots, v_{j_d}$. The preference lists of the residents and hospitals are shown in Fig. 4. We impose an arbitrary but fixed ordering on the vertices which is used as a strict ordering of neighbors in the preference lists of resident r_1^i and hospital h_2^i in G'. Note that G' has $N = 3|V|$ residents and hospitals.

$r_1^i : h_3^i, h_2^{j_1}, h_2^{j_2}, \ldots, h_2^{j_d}, h_1^i$ $\qquad$ $[0,1]\ h_1^i : r_1^i$

$r_2^i : h_2^i, h_3^i$ $\qquad$ $[0,1]\ h_2^i : r_2^i, r_1^{j_1}, r_2^{j_2}, \ldots, r_2^{j_d}, r_3^i$

$r_3^i : h_2^i$ $\qquad$ $[1,1]\ h_3^i : r_2^i, r_1^i$

Fig. 4. Preferences of agents corresponding to a vertex v_i in G.

Lemma 6. *If $VC(G)$ denotes a minimum vertex cover of G and $OPT(G')$ denotes a maximum size relaxed stable matching in G', then $|OPT(G')| = 3|V| - |VC(G)|$.*

Proof. We first prove that $|OPT(G')| \geq 3|V| - |VC(G)|$. Given a minimum vertex cover $VC(G)$ of G we construct a relaxed stable matching M for G' as follows. $M = \{(r_1^i, h_3^i), (r_2^i, h_2^i) \mid v_i \in VC(G)\} \cup \{(r_1^i, h_1^i), (r_2^i, h_3^i), (r_3^i, h_2^i) \mid v_i \notin VC(G)\}$. Thus, $|OPT(G')| \geq |M| = 2|VC(G)| + 3(|V| - |VC(G)|) = 3|V| - |VC(G)|$.

Claim. M is relaxed stable in G'.

Proof. When $v_i \in VC(G)$, residents r_1^i and r_2^i both are matched to their top choice hospitals and hospital h_2^i is matched to its top choice resident r_2^i. Thus, when $v_i \in VC(G)$, no resident from the i-th gadget participates in a blocking pair. When $v_i \notin VC(G)$, hospitals h_1^i and h_3^i are matched to their top choice residents and we ignore blocking pair (r_2^i, h_2^i) because r_2^i is matched to a lower-quota hospital h_3^i, thus there is no blocking pair within the gadget for $v_i \notin VC(G)$. Now suppose that there is a blocking pair (r_1^i, h_2^j) for some j such that $(v_i, v_j) \in E$. Note that either v_i or v_j is in $VC(G)$. If $v_i \in VC(G)$, r_1^i is matched to its top choice hospital h_3^i, thus cannot participate in a blocking pair. If $v_i \notin VC(G)$, it implies that $v_j \in VC(G)$. Then for v_j's gadget, h_2^j is matched to its top choice r_2^j, thus cannot form a blocking pair. □

Now we prove that $OPT(G') \leq 3|V| - |VC(G)|$. Let $M = OPT(G')$ be a maximum size relaxed stable matching in G'. Consider a vertex $v_i \in V$ and the corresponding residents and hospitals in G'. Refer Fig. 5 for the possible patterns caused by v_i. Hospital h_3^i must be matched to either resident r_1^i (Pattern 1) or resident r_2^i (Pattern 2 to Pattern 7). If $(r_1^i, h_3^i) \in M$, then the resident r_2^i must be matched to a higher preferred hospital h_2^i in M. If $(r_2^i, h_3^i) \in M$ then h_2^i may be matched with either r_3^i or r_1^j of some neighbour v_j or may be left unmatched. Similarly, r_1^i can either be matched to h_1^i or h_2^j of some neighbour v_j. This leads to 6 combinations as shown in Fig. 5b to Fig. 5g.

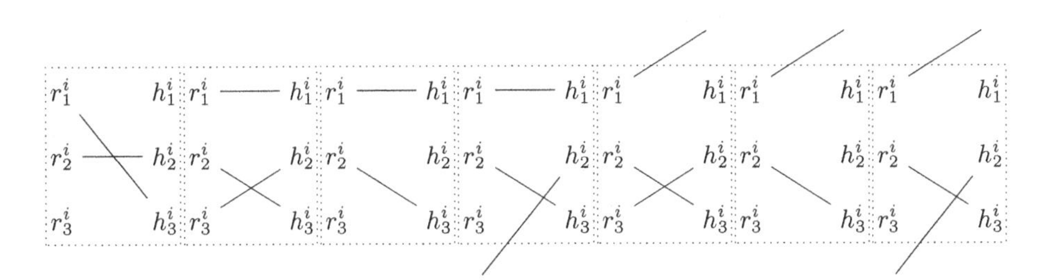

Fig. 5. Seven patterns possibly caused by vertex v_i

Claim. A vertex cannot cause pattern 5.

Proof. Assume for the sake of contradiction that a vertex v_i causes pattern 5. Then, there must exist a vertex v_j adjacent to v_i such that v_j causes either pattern 4 or pattern 7.

Case 1: If vertex v_j causes pattern 4, then (r_1^j, h_2^i) form a blocking pair, a contradiction.

Case 2: If vertex v_j causes pattern 7, then there must exist vertices $v_{j+1}, \ldots, v_t$ such that there are following edges in G: $(v_i, v_j), (v_j, v_{j+1}), (v_{j+1}, v_{j+2}), \ldots, (v_{t-1}, v_t)$ and vertices v_j to v_{t-1} cause pattern 7 and v_t causes pattern 4. See Fig. 6. In the vertex ordering, we must have $v_{j+1} > v_i$ otherwise (r_1^j, h_2^i) form a blocking pair. But, since h_2^j is matched to r_1^i, $v_{j+2} > v_j$. Continuing this way, $v_t > v_{t-2}$ but this causes (r_1^t, h_2^{t-1}) form a blocking pair. Thus, the claimed set of edges cannot exist. □

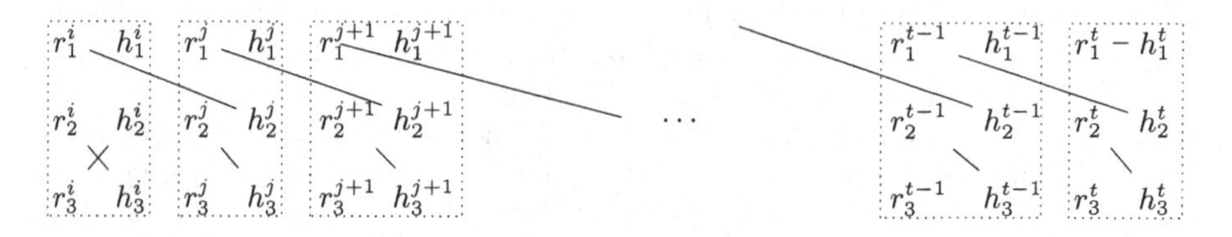

Fig. 6. Pattern combination that is not relaxed stable if v_i causes pattern 5

Claim. A vertex cannot cause pattern 3 or pattern 6 or pattern 4.

Proof. In pattern 3 and 6, r_3^i participates in a blocking pair (r_3^i, h_2^i), contradicting that M is relaxed stable. If a vertex v_i causes pattern 4, then there exists a set of t vertices $v_{i+1}, \ldots, v_{i+t}$ such that for $0 \leq k < t, (v_{i+k}, v_{i+k+1})$ is an edge in G and v_{i+t} causes pattern 6. But, since pattern 6 cannot occur, pattern 4 cannot occur. □

Thus, a vertex can cause either pattern 1 or 2 and thus match all the residents and hospitals within its own gadget or pattern 7 and match r_1 and h_2 outside its own gadget. Accordingly there are following cases.

Case 1: A vertex causing pattern 7 contributes size 1 for (r_2^i, h_3^i) edge and 0.5 each for two edges matched to another vertex causing pattern 7, contributing an average matching size of 2.

Case 2: It is clear that a vertex causing pattern 1 or 2 contributes to matching size of 2 or 3 respectively.

Vertex Cover C of G Corresponding to M: Using M, we now construct the set C of vertices in G which constitute a vertex cover of G. If v_i causes pattern 2, we do not include it in the C; Otherwise, we include it. We prove that C is a vertex cover. Suppose not, then there exists an edge (v_i, v_j) such that both v_i and v_j cause pattern 2. But, this means that (r_1^i, h_2^j) and (r_1^j, h_2^i) form a blocking pair, a contradiction since M is relaxed stable. Now, it is easy to see that $|OPT(G')| = 2|C| + 3(|V| - |C|) = 3|V| - |C|$. Thus, $VC(G) \leq |C| = 3|V| - |OPT(G')|$. This completes the proof of the lemma. □

The rest of our proof is similar to the approach of Halldórsson et al. [8] to prove inapproximability of the stable matchings with ties and incomplete lists; however our gadgets above are entirely different. Lemma 7 is analogous to Theorem 3.2 and Corollary 3.4 from [8] and its proof can be reproduced in a similar way [15].

Lemma 7. *It is* NP*-hard to approximate the* MAXRSM *problem within a factor of* $\frac{21}{19} - \epsilon$*, for any constant* $\epsilon > 0$*, even when the quotas of all hospitals are either 0 or 1.*

Acknowledgement. We thank the anonymous reviewers for their useful comments which has improved the presentation of the paper.

References

1. Biró, P., Fleiner, T., Irving, R.W., Manlove, D.: The college admissions problem with lower and common quotas. Theoret. Comput. Sci. **411**(34–36), 3136–3153 (2010). https://doi.org/10.1016/j.tcs.2010.05.005
2. Biró, P., Manlove, D., Mittal, S.: Size versus stability in the marriage problem. Theoret. Comput. Sci. **411**(16–18), 1828–1841 (2010). https://doi.org/10.1016/j.tcs.2010.02.003
3. Ehlers, L., Hafalir, I.E., Yenmez, M.B., Yildirim, M.A.: School choice with controlled choice constraints: hard bounds versus soft bounds. J. Econ. Theory **153**, 648–683 (2014). https://doi.org/10.1016/j.jet.2014.03.004
4. Fleiner, T., Kamiyama, N.: A matroid approach to stable matchings with lower quotas. Math. Oper. Res. **41**(2), 734–744 (2016). https://doi.org/10.1287/moor.2015.0751
5. Fragiadakis, D., Iwasaki, A., Troyan, P., Ueda, S., Yokoo, M.: Strategyproof matching with minimum quotas. ACM Trans. Econ. Comput. **4**(1), 6:1–6:40 (2015). http://doi.acm.org/10.1145/2841226
6. Gale, D., Shapley, L.S.: College admissions and the stability of marriage. Am. Math. Mon. **69**(1), 9–15 (1962). http://www.jstor.org/stable/2312726
7. Goto, M., Iwasaki, A., Kawasaki, Y., Kurata, R., Yasuda, Y., Yokoo, M.: Strategyproof matching with regional minimum and maximum quotas. Artif. Intell. **235**, 40–57 (2016). https://doi.org/10.1016/j.artint.2016.02.002
8. Halldórsson, M.M., Iwama, K., Miyazaki, S., Yanagisawa, H.: Improved approximation results for the stable marriage problem. ACM Trans. Algorithms **3**(3), 30 (2007). https://doi.org/10.1145/1273340.1273346
9. Hamada, K., Iwama, K., Miyazaki, S.: The hospitals/residents problem with lower quotas. Algorithmica **74**(1), 440–465 (2016). https://doi.org/10.1007/s00453-014-9951-z
10. Huang, C.: Classified stable matching. In: Proceedings of the Twenty-First Annual ACM-SIAM Symposium on Discrete Algorithms, SODA 2010, pp. 1235–1253 (2010). https://doi.org/10.1137/1.9781611973075.99
11. Kamada, Y., Kojima, F.: Efficient matching under distributional constraints: theory and applications. Am. Econo. Rev. **105**(1), 67–99 (2015). https://www.aeaweb.org/articles?id=10.1257/aer.20101552
12. Kamada, Y., Kojima, F.: Stability concepts in matching under distributional constraints. J. Econ. Theory **168**, 107–142 (2017). https://doi.org/10.1016/j.jet.2016.12.006

13. Király, Z.: Linear time local approximation algorithm for maximum stable marriage. Algorithms **6**(3), 471–484 (2013). https://doi.org/10.3390/a6030471
14. Kleinberg, J., Tardos, E.: Algorithm Design. Addison-Wesley Longman Publishing Co. Inc., Boston (2005)
15. Krishnaa, P., Limaye, G., Nasre, M., Nimbhorkar, P.: Envy-freeness and relaxed stability under lower quotas. CoRR abs/1910.07159 (2019). http://arxiv.org/abs/1910.07159
16. Krishnapriya, A.M., Nasre, M., Nimbhorkar, P., Rawat, A.: How good are popular matchings? In: 17th International Symposium on Experimental Algorithms, SEA 2018, pp. 9:1–9:14 (2018). https://doi.org/10.4230/LIPIcs.SEA.2018.9
17. Nasre, M., Nimbhorkar, P.: Popular matchings with lower quotas. In: 37th IARCS Annual Conference on Foundations of Software Technology and Theoretical Computer Science, FSTTCS 2017, pp. 44:1–44:15 (2017). https://doi.org/10.4230/LIPIcs.FSTTCS.2017.44
18. Roth, A.E.: On the allocation of residents to rural hospitals: a general property of two-sided matching markets. Econometrica **54**(2), 425–427 (1986). http://www.jstor.org/stable/1913160
19. Wu, Q., Roth, A.E.: The lattice of envy-free matchings. Games Econ. Behav. **109**, 201–211 (2018). https://doi.org/10.1016/j.geb.2017.12.016
20. Yokoi, Y.: Envy-free matchings with lower quotas. Algorithmica **82**(2), 188–211 (2020). https://doi.org/10.1007/s00453-018-0493-7

Scheduling and Games on Graphs

Targeted Intervention in Random Graphs

William Brown(✉) and Utkarsh Patange

Columbia University, New York, NY 10027, USA
{w.brown,utkarsh.patange}@columbia.edu

Abstract. We consider a setting where individuals interact in a network, each choosing actions which optimize utility as a function of neighbors' actions. A central authority aiming to maximize social welfare at equilibrium can intervene by paying some cost to shift individual incentives, and the optimal intervention can be computed using the spectral decomposition of the graph, yet this is infeasible in practice if the adjacency matrix is unknown. In this paper, we study the question of designing intervention strategies for graphs where the adjacency matrix is unknown and is drawn from some distribution. For several commonly studied random graph models, we show that there is a single intervention, proportional to the first eigenvector of the expected adjacency matrix, which is near-optimal for almost all generated graphs when the budget is sufficiently large. We also provide several efficient sampling-based approaches for approximately recovering the first eigenvector when we do not know the distribution. On the whole, our analysis compares three categories of interventions: those which use no data about the network, those which use some data (such as distributional knowledge or queries to the graph), and those which are fully optimal. We evaluate these intervention strategies on synthetic and real-world network data, and our results suggest that analysis of random graph models can be useful for determining when certain heuristics may perform well in practice.

Keywords: Random graphs · Intervention · Social welfare · Sampling

1 Introduction

Individual decision-making in many domains is driven by personal as well as social factors. If one wants to decide a level of time, money, or effort to exert on some task, the behaviors of one's friends or neighbors can be powerful influencing factors. We can view these settings as games where agents in a network are playing some game, each trying to maximize their individual utility as a function of their "standalone value" for action as well as their neighbors' actions. The actions of agents who are "central" in a network can have large ripple effects. Identifying and understanding the role of central agents is of high importance for tasks ranging from microfinance [6] and vaccinations [7], to tracking the spread of information throughout a community [8]. We view our work as providing theoretical support for heuristic approaches to intervention in these settings.

T. Harks and M. Klimm (Eds.): SAGT 2020, LNCS 12283, pp. 211–225, 2020.
https://doi.org/10.1007/978-3-030-57980-7_14

A model for such a setting is studied in recent work by Galeotti, Golub, and Goyal [20], where they ask the natural question of how a third party should "intervene" in the network to maximally improve social welfare at the equilibrium of the game. Interventions are modeled by assuming that the third party can pay some cost to adjust the standalone value parameter of any agent, and must decide how to allocate a fixed budget. This may be interpreted as suggesting that these targeted agents are subjected to advertizing, monetary incentives, or some other form of encouragement. For their model, they provide a general framework for computing the optimal intervention subject to any budget constraint, which can be expressed in terms of the spectral decomposition of the graph. For large budgets, the optimal intervention is approximately proportional to the first eigenvector of the adjacency matrix of the graph, a common measure of network centrality.

While this method is optimal, and computable in polynomial time if the adjacency matrix is known, it is rare in practice that we can hope to map all connections in a large network. For physical networks, edges representing personal connections may be far harder to map than simply identifying the set of agents, and for large digital networks we may be bound by computational or data access constraints. However, real-world networks are often well-behaved in that their structure can be approximately described by a simple generating process. If we cannot afford to map an entire network, is optimal targeted intervention feasible at all? A natural target would be to implement interventions which are competitive with the optimal intervention, i.e. obtaining almost the same increase in social welfare, without access to the full adjacency matrix. Under what conditions can we use knowledge of the distribution a graph is drawn from to compute a near-optimal intervention without observing the realization of the graph? Without knowledge of the distribution, how much information about the graph is necessary to find such an intervention? Can we ever reach near-optimality with no information about the graph? These are the questions we address.

1.1 Contributions

Our main result shows that for random graphs with independent edges, the first eigenvector of the "expected adjacency matrix", representing the probability of each edge being included in the graph, constitutes a near-optimal intervention simultaneously for almost all generated graphs, when the budget is large enough and the expected matrix satisfies basic spectral conditions. We further explore graphs with given expected degrees, Erdős-Rényi graphs, power law graphs, and stochastic block model graphs as special cases for which our main result holds. In these cases, the first eigenvector of the expected matrix can often be characterized by a simple expression of parameters of the graph distribution.

Yet in general, this approach still assumes a fair amount of knowledge about the distribution, and that we can map agents in the network to their corresponding roles in the distribution. We give several sampling-based methods for approximating the first eigenvector of a graph in each of the aforementioned special cases, which do not assume knowledge of agent identities or distribution parameters, other than group membership in the stochastic block model. These methods assume different query models for accessing information about the realized graph, such as the ability to query the existence of an edge or to observe a random neighbor of an agent. Using the fact that the graph was drawn from *some* distribution, we can reconstruct an approximation of the first eigenvector more efficiently than we could reconstruct the full matrix. The lower-information settings we consider can be viewed as assumptions about qualitative domain-specific knowledge, such as a degree pattern which approximately follows an (unknown) power law distribution, or the existence of many tight-knit disjoint communities.

We evaluate our results experimentally on both synthetic and real-world networks for a range of parameter regimes. We find that our heuristic interventions can perform quite well compared to the optimal intervention, even at modest budget and network sizes. These results further illustrate the comparative efficacies of interventions requiring varying degrees of graph information under different values for distribution parameters, budget sizes, and degrees of network effects.

On the whole, our results suggest that explicit mapping of the connections in a network is unnecessary to implement near-optimal targeted interventions in strategic settings, and that distributional knowledge or limited queries will often suffice.

1.2 Related Work

Recent work by Akbarpour, Malladi, and Saberi [3] has focused on the challenge of overcoming network data barriers in targeted interventions under a diffusion model of social influence. In this setting, for $G(n, p)$ and power law random graphs, they derive bounds on the additional number of "seeds" needed to match optimal targeting when network information is limited. A version of this problem where network information can be purchased is studied in [19]. Another similar model was employed by Candogan, Bimpikis, and Ozdaglar [11] where they study optimal pricing strategies to maximize profit of a monopolist selling service to consumers in a social network where the consumer experiences a positive local network effect, where notions of centrality play a key role. Similar targeting strategies are considered in [18], where the planner tries to maximize aggregate action in a network with complementarities. [23] studies the efficacy of blind interventions in a pricing game for the special case of Erdős-Rényi graphs. In [25], targeted interventions are also studied for "linear-quadratic games", quite similar to those from [20], in the setting of infinite-population graphons, where a concentration result is given for near-optimal interventions.

Our results can be viewed as qualitatively similar findings to the above results in the model of [20]. While they have showed that exact optimal interventions can be constructed on a graph with full information, we propose that local information is enough to construct an approximately optimal intervention for many distributions of random graphs. It is argued in [9] that collecting data of this kind (aggregate relational data) is easier in real networks compared to obtaining full network information. We make use of concentration inequalities for the spectra of random adjacency matrices; there is a great deal of work studying various spectral properties of random graphs (see e.g. [2,13–16]). Particularly relevant to us is [16], which characterizes the asymptotic distributions of various centrality measures for random graphs. There is further relevant literature for studying centrality in graphons, see e.g. [4]. Of relevance to our sampling techniques, a method for estimating eigenvector centrality via sampling is given in [26], and the task of finding a "representative" sample of a graph is discussed in [24].

2 Model and Preliminary Results

Here we introduce the "linear-quadratic" network game setting from [20], also studied in e.g. [25], which captures the dynamics of personal and social motivations for action in which we are interested.

2.1 Setting

Agents are playing a game on an undirected graph with adjacency matrix A. Each agent takes an action $a_i \in \mathbb{R}$ and obtains individual utility given by:

$$u_i(a, A) = b_i a_i - \frac{1}{2} a_i^2 + \beta \sum_j A_{ij} a_i a_j$$

Here, b_i represents agent i's "standalone marginal value" for action. The parameter β controls the effect of strategic dynamics, where a positive sign promotes complementary behavior with neighbors and a negative sign promotes acting in opposition to one's neighbors. In this paper we focus on the case where each value A_{ij} is in $\{0, 1\}$ and $\beta > 0$. The assumption that $\beta > 0$ corresponds to the case where agents' actions are complementary, meaning that an increase in action by an agent will increase their neighbors' propensities for action.[1] We assume that $b_i \geq 0$ for each agent as well.

The matrix $M = (I - \beta A)$ can be used to determine the best response for each agent given their opponents' actions. The best response vector a^*, given current actions a, can be computed as:

$$a^* = b + \beta A a.$$

Upon solving for $a^* = a$, we get that $a^* = (I - \beta A)^{-1} b = M^{-1} b$, giving us the Nash equilibrium for the game as all agents are simultaneously best responding to each other. We show in Appendix D of [10] that when agents begin with null action values, repeated best responses will converge to equilibrium, and further that the new equilibrium is likewise reached after intervention.

[1] When $\beta < 0$, neighbors' actions act as substitutes, and one obtains less utility when neighbors increase levels of action. In that case, the optimal intervention for large budgets is approximated by the last eigenvector of the graph, which measures its "bipartiteness".

Our results will apply to cases where all eigenvalues of M are almost surely positive, ensuring invertibility.[2] The social welfare of the game $W = \sum_i u_i$ can be computed as a function of the equilibrium actions:

$$W = \frac{1}{2}(a^*)^\top a^*$$

Given the above assumptions, equilibrium actions a_i^* will always be non-negative.

2.2 Targeted Intervention

In this game, a central authority has the ability to modify agents' standalone marginal utilities from b_i to $\hat{b}_i$ by paying a cost of $(b_i - \hat{b}_i)^2$, and their goal is to maximize social welfare subject to a budget constraint C:

$$\max \sum_i u_i \quad \text{subject to} \quad \sum_i (b_i - \hat{b}_i)^2 \leq C.$$

Here, an *intervention* is a vector $y = \hat{b} - b$ such that $\|y\|^2 \leq C$. Let $W(y)$ denote the social welfare at equilibrium following an intervention y.[3] It is shown in [20] that the optimal budget-constrained intervention for any C can be computed using the eigenvectors of A, and that in the large-budget limit as C tends to infinity, the optimal intervention approaches $\sqrt{C} \cdot v_1(A)$. Throughout, we assume $v_i(A)$ is the unit ℓ_2-norm eigenvector associated with λ_i, the ith largest eigenvalue of a matrix A. We also define $\alpha_i = \frac{1}{(1-\beta\lambda_i)^2}$, which is the square of the corresponding eigenvalue of M^{-1}. Note that we do not consider eigenvalues to be ordered by absolute value; this is done to preserve the ordering correspondence between eigenvalues of A and M^{-1}. A may have negative eigenvalues, but all eigenvalues of M^{-1} will be positive when $\beta\lambda_1 < 1$, as we will ensure throughout.

The key result we use from [20] states that when β is positive, as the budget increases the cosine similarity between the optimal intervention y^* and the first eigenvector of a graph, which we denote by $\rho(v_1(A), y^*)$,[4] approaches 1 at a rate depending on the (inverted) spectral gap of the adjacency matrix.[5]

Our results will involve quantifying the *competitive ratio* of an intervention y, which we define as $\frac{W(y)}{W(y^*)}$, where $W(\cdot)$ denotes the social welfare at equilibrium after an intervention vector is applied, and where $y^* = \arg\max_{x\,:\,\|x\|=\sqrt{C}} W(x)$. This ratio is at most 1, and maximizing it will be our objective for evaluating interventions.

---

[2] If $\beta > 0$ and M is not invertible, equilibrium actions will be infinite for all agents in some component of the graph.

[3] Unless specified otherwise, $\|\cdot\|$ refers to the ℓ_2 norm. When the argument is a matrix, this denotes the associated spectral norm.

[4] The cosine similarity of two non-zero vectors z and y is $\rho(z, y) = \frac{z \cdot y}{\|z\|\|y\|}$. For unit vectors x, y, by the law of cosines, $\|x - y\|^2 = 2(1 - \rho(x, y))$, and so $1 - \frac{\|x-y\|^2}{2} = \rho(x, y)$. Thus $\|x - y\| < \epsilon$ for $\epsilon > 0$ if and only if $\rho(x, y) > 1 - \epsilon^2/2$.

[5] It will sometimes be convenient for us to work with what we call the *inverted spectral gap* of a matrix A, which is the smallest value κ such that $|\lambda_i(A) \leq \kappa \cdot \lambda_1(A)|$.

2.3 Random Graph Models

We introduce several families of random graph distributions which we consider throughout. All of these models generate graphs which are undirected and have edges which are drawn independently.

Definition 1 (Random Graphs with Independent Edges). *A distribution of random graphs with independent edges is specified by a symmetric matrix* $\overline{A} \in [0,1]^{n \times n}$. *A graph is sampled by including each edge* (i,j) *independently with probability* $\overline{A}_{ij}$.

Graphs with given expected degrees ($G(w)$, or Chung-Lu graphs) and stochastic block model graphs, often used as models of realistic "well-behaved" networks, are notable cases of this model which we will additionally focus on.

Definition 2 ($G(w)$ Graphs). *A* $G(w)$ *graph is an undirected graph with an expected degree sequence given by a vector* w, *whose length (which we denote by* n) *defines the number of vertices in the graph. For each pair of vertices* i *and* j *with respective expected degrees* w_i *and* w_j, *the edge* (i,j) *is included independently with probability* $\overline{A}_{ij} = \frac{w_i w_j}{\sum_{k \in [n]} w_k}$.

Without loss of generality, we impose an ordering on w_i values so that $w_1 \geq w_2 \geq \ldots \geq w_n$. To ensure that each edge probability as described above is in $[0,1]$, we assume throughout that for all vectors w we have that $w_1 \leq \sqrt{\sum_{k \in [n]} w_k}$.

$G(n,p)$ graphs and power law graphs are well-studied examples of graphs which can be generated by the $G(w)$ model.[6] For $G(n,p)$ graphs, w is a uniform vector where $w_i = np$ for each i. Power law graphs are another notable special case where w is a *power law sequence* $\{w_i\}_{i=1}^{n}$ such that $w_i = c\,(i + i_0)^{-\frac{1}{\sigma - 1}}$ for $\sigma > 2$, some constant $c > 0$, and some integer $i_0 \geq 0$. In such a sequence, the number of elements with value x is asymptotically proportional to $\frac{1}{x^\sigma}$.

Definition 3 (Stochastic Block Model Graphs). *A stochastic block model graph with* n *vertices is undirected and has* m *groups for some* $m \leq n$. *Edges are drawn independently according to a matrix* $\overline{A}$, *and the probability of an edge between two agents depends only on their group membership. For any two groups* i *and* j, *there is an edge probability* $p_{ij} \in [0,1]$ *such that* $\overline{A}_{kl} = p_{ij}$ *for any agent* k *in group* i *and agent* l *in group* j.[7]

For each graph model, one can choose to disallow self-loops by setting $\overline{A}_{ii} = 0$ for $1 \leq i \leq n$, as is standard for $G(n,p)$ graphs. Our results will apply to both cases.

---

[6] There are several other well-studied models of graphs with power law degree sequences, such as the BA preferential attachment model, as well as the fixed-degree model involving a random matching of "half-edges". Like the $G(w)$ model, the latter model can support arbitrary degree sequences. We restrict ourselves to the independent edge model described above.

[7] If $m = n$, the stochastic block model can express any distribution of random graphs with independent edges, but will be most interesting when there are few groups.

3 Approximately Optimal Interventions

The main idea behind all of our intervention strategies is to target efforts proportionally to the first eigenvector of the expected adjacency matrix. Here we assume that this eigenvector is known exactly. In Sect. 4, we discuss cases when an approximation of the eigenvector can be computed with zero or minimal information about the graph. Our main theorem for random graphs with independent edges shows conditions under which an intervention proportional to the first eigenvector of the expected matrix $\overline{A}$ is near-optimal.

We define a property for random graphs which we call (ϵ, δ)-*concentration* which will ensure that the expected first eigenvector constitutes a near-optimal intervention. In essence, this is an explicit quantification of the asymptotic properties of "large enough eigenvalues" and "non-vanishing spectral gap" for *sequences* of random graphs from [16]. Intuitively, this captures graphs which are "well-connected" and not too sparse. One can think of the first eigenvalue as a proxy for density, and the (inverse) second eigenvalue as a proxy for regularity or degree of clustering (it is closely related to a graph's mixing time). Both are important in ensuring concentration, and they trade off with each other (via the spectral gap condition) for any fixed values of ϵ and δ.

Definition 4 (**(ϵ, δ)-Concentration**). *A random graph with independent edges specified by $\overline{A}$ satisfies (ϵ, δ)-concentration for $\epsilon, \delta \in (0, 1)$ if:*

1. *The largest expected degree $d_{\max} = \max_i \sum_{j \in [n]} \overline{A}_{ij}$ is at least $\frac{4}{9} \log(2n/\delta)$*
2. *The inverted spectral gap of $\overline{A}$ is at most κ*
3. *The quantity $\lambda_1(\overline{A}) \cdot (1 - \kappa^2)$ is at least $\frac{1024\sqrt{d_{\max} \log(2n/\delta)}}{\epsilon^2}$*

Theorem 1. *If $\overline{A}$ satisfies (ϵ, δ)-concentration, then with probability at least $1 - \delta$, the competitive ratio of $y = \sqrt{C} v_1(\overline{A})$ for a graph drawn from $\overline{A}$ is at least $1 - \epsilon$ for a sufficiently large budget C if the spectral radius of the sampled matrix A is less than $1/\beta$.*

The concentration conditions are used to show that the relevant spectral properties of generated graphs are almost surely close to their expectations, and the constraint on β is necessary to ensure that actions and utilities are finite at equilibrium.[8] The sufficient budget will depend on the size of the *spectral gap* of $\overline{A}$, as well as the standard marginal values. For example, if $\lambda_1 > 2|\lambda_i|$ holds in the realized graph for all $i > 1$, then a budget of $C = 256 \cdot \|b\|^2 / (\epsilon \beta \lambda_1(\overline{A}))^2$ will suffice. Intuitively, a large β would mean more correlation between neighbors' actions at equilibrium. A large $\lambda_1 (\overline{A})$ would mean a denser graph (more connections between agents) in expectation and a large ϵ would mean that the realized graph is more likely to be close to expectation. All of these conditions reduce the required budget because a small intervention gets magnified by agent

[8] The spectral radius condition holds with probability $1 - \delta$ when $1/\beta$ is at least $\lambda_1(\overline{A}) + \sqrt{4d_{\max} \log(2n/\delta)}$ (follows from e.g. [15], see Appendix E.2 of [10] for details).

interaction. Further, the smaller the magnitude of initial b, the easier it is to change its direction.

The proof of Theorem 1 is deferred to the full version of the paper [10], along with proofs for the lemmas stated throughout. At a high level, our results proceed by first showing that the first eigenvector is almost surely close to $v_1(\overline{A})$, then showing that the spectral gap is almost surely large enough such that the first eigenvector is close to the optimal intervention for appropriate budgets. A key lemma for completing the proof shows that interventions which are close to the optimal intervention in cosine similarity have a competitive ratio close to 1.

Lemma 1. *Let b be the vector of standalone values, and assume that $C > \max(\|b\|^2, 1)$. For any y where $\|y\|^2 = C$ and $\rho(y, y^*) > \gamma$ for some γ, the competitive ratio of y is at least $1 - 4\sqrt{2(1-\gamma)}$.*

The main idea behind this lemma is a smoothness argument for the welfare function. When considering interventions as points on the sphere of radius $\sqrt{C}$, small changes to an intervention cannot change the resulting welfare by too much. This additionally implies that when a vector y is close to y^*, the exact utility of y^* for some budget C can be achieved by an intervention proportional to y with a budget C' which is not much larger than C.

In the full paper [10], we give a specialization of Theorem 1 to the case of $G(w)$ graphs. There, the expected first eigenvector is proportional to w when self-loops are not removed. We give more explicit characterizations of the properties for $G(w)$, $G(n, p)$, and power law graphs which ensure the above spectral conditions (i.e. without relying on eigenvalues), as well as a budget threshold for near-optimality. We discuss the steps of the proof in greater detail, and they are largely symmetric to the steps required to prove Theorem 1.

4 Centrality Estimation

The previous sections show that interventions proportional to $v_1(\overline{A})$ are often near-optimal simultaneously for almost all graphs generated by $\overline{A}$. While we often may have domain knowledge about a network which helps characterize its edge distribution, we still may not be able to precisely approximate the first eigenvector of $\overline{A}$ *a priori*. In particular, even if we believe our graph comes from a power law distribution, we may be at a loss in knowing which vertices have which expected degrees.

In this section, we discuss approaches for obtaining near-optimal interventions without initial knowledge of $\overline{A}$. We first observe that "blind" interventions, which treat all vertices equally in expectation, will fail to approach optimality. We then consider statistical estimation techniques for approximating the first eigenvector which leverage the special structure of $G(w)$ and stochastic block model graphs. In each case, we identify a simple *target intervention*, computable directly from the realized graph, which is near-optimal when (ϵ, δ)-concentration is satisfied. We then give efficient sampling methods for approximating these target interventions. Throughout Sect. 4, our focus is to give a broad overview of these techniques rather than to present them as concrete algorithms, and we frequently omit constant-factor terms with asymptotic notation.

4.1 Suboptimality of Blind Interventions

Here we begin by showing that when the spectral gap is large, all interventions which are far from the optimal intervention in cosine similarity will fail to be near-optimal even if the budget is very large.

Lemma 2. *Assume that C is sufficiently large such that the role of standalone values is negligible. For any y where $\|y\|^2 = C$ and $\rho(y, y^*) < \gamma$, the competitive ratio is bounded by*

$$\gamma^2 \left(1 - \frac{\alpha_2}{\alpha_1}\right) + \frac{\alpha_2}{\alpha_1} + 2\sqrt{\frac{\alpha_2}{\alpha_1}},$$

where α_i is the square of the ith largest eigenvalue of M^{-1}.

This tells us that if one were to design an intervention without using any information about the underlying graph, the intervention is unlikely to do well compared to the optimal one for the same budget unless eigenvector centrality is uniform, as in the case of $G(n, p)$ graphs. Thus, there is a need to try to learn graph information to design a close-to-optimal intervention. We discuss methods for this next.

4.2 Degree Estimation in $G(w)$ Graphs

For $G(w)$ graphs, we have seen that expected degrees suffice for near-optimal interventions, and we show that degrees can suffice as well.

Lemma 3. *If a $G(w)$ graph specified by $\overline{A}$ satisfies (ϵ, δ)-concentration, then with probability at least $1 - O(\delta)$,*

$$\|w - w^*\| \leq O(\epsilon \|w\|),$$

where w^* is the empirical degree vector, and the intervention proportional to w^* obtains a competitive ratio of $1 - O(\epsilon)$ when the other conditions for Theorem 1 are satisfied.

Thus, degree estimation is our primary objective in considering statistical approaches. As we can see from the analysis in Theorems 1 and 2 of [10], if we can estimate the unit-normalized degree vector w^* to within ϵ ℓ_2-distance, our competitive ratio for the corresponding proportional intervention will be $1 - O(\epsilon)$. Our approaches focus on different query models, representing the types of questions we are allowed to ask about the graph; these query models are also studied for the problem of estimating the average degree in a graph [17,22]. If we are allowed to query agents' degrees, near-optimality follows directly from the above lemma, so we consider more limited models.

Edge Queries. Suppose we are allowed to query whether an edge exists between two vertices. We can then reduce the task of degree estimation to the problem of estimating the mean of n biased coins, where for each vertex, we "flip" the corresponding coin by picking another vertex uniformly at random to query. By Hoeffding and union bounds, $O\left(\frac{n}{\epsilon^2}\log\left(\frac{n}{\delta}\right)\right)$ total queries suffice to ensure that with probability $1-\delta$, each degree estimate is within ϵn additive error. Particularly in the case of dense graphs, and when ϵ is not too small compared to $1/n$, this will be considerably more efficient than reconstructing the entire adjacency matrix. In particular, if $\|w\|_1 = \Theta(n^2)$, the above error bound on additive error for each degree estimate directly implies that the estimated degree vector $\hat{w}$ is within ℓ_1 (and thus ℓ_2) distance of $O(\epsilon\|w\|_2)$.

Random Neighbor Queries. Suppose instead we are restricted to queries which give us a uniformly random neighbor of a vertex. We give an approach wherein queries are used to conduct a random walk in the graph. The stationary distribution is equivalent to the the first eigenvector of the *diffusion matrix* $P = AD^{-1}$, where D is the diagonal matrix of degree counts.[9] We can then learn estimates of degree proportions by sampling from the stationary distribution via a random walk.

The mixing time of a random walk on a graph determines the number of steps required such that the probability distribution over states is close to the stationary distribution in total variation distance. We can see that for $G(w)$ graphs satisfying (ϵ, δ)-concentration with a large enough minimum degree, mixing times will indeed be fast.

Lemma 4. *For $G(w)$ graphs satisfying (ϵ, δ)-concentration and with $w_n \geq \frac{1}{\epsilon}$, the mixing time of a random walk to within ϵ total variation distance to the stationary distribution is $O(\log(n/\epsilon))$. Further, the largest connected component in A contains $n(1-\exp(-O(1/\epsilon))$ vertices in expectation.*

If a random walk on our graph has some mixing time t to an approximation of the stationary distribution, we can simply record our location after every t steps to generate a sample. Using standard results on learning discrete distributions (see e.g. [12]), $O\left(\frac{n+\log(1/\delta)}{\epsilon^2}\right)$ samples from ϵ-approximate stationary distributions suffice to approximate w^* within ℓ_1 distance of $O(\epsilon\|w^*\|)$ with probability $1-\delta$, directly giving us the desired ℓ_2 bound. Joining this with Lemma 4, our random walk takes a total of $O\left(\frac{n+\log(1/\delta)}{\epsilon^2}\log\left(\frac{n}{\epsilon}\right)\right)$ steps (and thus queries) to obtain our target intervention, starting from an arbitrary vertex in the largest connected component.

[9] The stationary distribution of a random walk on a simple connected graph is $\frac{d_i}{\sum_j d_j}$ for all vertices i, where d_i is the degree. While $G(w)$ graphs may fail to be connected, in many cases the vast majority of vertices will belong to a single component, and we can focus exclusively on that component. We show this in Lemma 4.

4.3 Matrix Reconstruction in SBM Graphs

There is a fair amount of literature on estimation techniques for stochastic block model graphs, often focused on cases where group membership is unknown [1,27–29]. The estimation of eigenvectors is discussed in [5], where they consider stochastic block model graphs as a limit of a convergent sequence of "graphons". Our interest is primarily in recovering eigenvector centrality efficiently from sampling, and we will make the simplifying assumption that group labels are visible for all vertices. This is reasonable in many cases where a close proxy of one's primary group identifier (e.g. location, job, field of study) is visible but connections are harder to map.

In contrast to the $G(w)$ case, degree estimates no longer suffice for estimating the first eigenvector. We assume that there are m groups and that we know each agent's group. Our aim will be to estimate the relative densities of connection between groups. When there are not too many groups, the parameters of a stochastic block model graph can be estimated efficiently with either edge queries or random neighbor queries, From here, we can construct an approximation of $\overline{A}$ and compute its first eigenvector directly. In many cases, the corresponding intervention is near-optimal.

A key lemma in our analysis shows that the "empirical block matrix" is close to its expectation in spectral norm. We prove this for the case where all groups are of similar sizes, but the approach can be generalized to cover any partition.

Lemma 5. *For a stochastic block model graph generated by $\overline{A}$ with m groups, each of size $O(\frac{n}{m})$, let $\hat{A}$ denote the empirical block matrix of edge frequencies for each group. Each entry per block in $\hat{A}$ will contain the number of edges in that block divided by the size of the block. With probability at least $1-\delta$,*

$$\left\|\overline{A} - \hat{A}\right\| \leq O\left(\max\left(\frac{m\sqrt{\log(n/\delta)}}{\sqrt{n}},\ \log^2(n/\delta)\right)\right).$$

The same bound will then apply to the difference of the first eigenvectors, rescaled by the first eigenvalues (which will also be close). Similar bounds can also be obtained when group sizes may vary, but we stick to this case for simplicity.

Edge Queries. If we are allowed to use edge queries, we can estimate the empirical edge frequency for each of the $O(m^2)$ pairs of groups by repeatedly sampling a vertex uniformly from each group and querying for an edge. This allows reconstruction of the empirical frequencies up to ϵ error for each group pair, with probability $1-\delta$, with $O\left(\frac{m^2}{\epsilon^2}\log(m/\delta)\right)$ samples. For the block matrix $\hat{A}$ of edge frequencies for all group pairs, Lemma 5 implies that this will be close to its expectation when there are not too many groups, and so our estimation will be close to $\overline{A}$ in spectral norm as well. If $\overline{A}$ satisfies (ϵ, δ)-concentration and the bound from Lemma 5 is small compared to the norm of $\overline{A}$, then the first eigenvectors of A, $\overline{A}$, and $\hat{A}$ will all be close, and the corresponding intervention proportional to $v_1(\hat{A})$ will be near-optimal.

When all group pairs may have unique probabilities, this will only provide an advantage over a naive graph reconstruction with $O(m^2)$ queries in the case where $m = o(n)$. If we know that all out-group probabilities are the same across groups, our dependence on m becomes linear, as we can treat all pairs of distinct groups as one large group. If in-group probabilities are the same across groups as well, the dependence on m vanishes, as we only have two probabilities to estimate.

Random Neighbor Queries. We can also estimate the empirical group frequency matrix with random neighbor queries. For each group, the row in $\overline{A}$ corresponding to the edge probabilities with other groups can be interpreted as a distribution of frequencies for each group. $O(\frac{m}{\epsilon^2} \log(\frac{m}{\delta}))$ samples per row suffice to get additive error at most ϵ for all of the relative connection probabilities for our chosen group. This lets us estimate each of the m rows up to scaling, at which point we can use the symmetry of the matrix to recover an estimate of $\overline{A}$ up to scaling by some factor. Again, when (ϵ, δ)-concentration holds and the bound from Lemma 5 is small, the first eigenvector of this estimated matrix will give us a near-optimal intervention.

5 Experiments

Our theoretical results require graphs to be relatively large in order for the obtained bounds to be nontrivial. It is natural to ask how well the heuristic interventions we describe will perform on relatively small random graphs, as well as on real-world graphs which do not come from a simple generative model (and may not have independent edges). Here, we evaluate our described interventions on real and synthetic network data, by adjusting b_i values and computing the resulting welfare at equilibrium, and find that performance can be quite good even on small graphs. Our experimental results on synthetic networks are deferred to the full paper [10].

5.1 Real Networks

To test the usefulness of our results for real-world networks which we expect to be "well-behaved" according to our requirements, we simulate the intervention process using network data collected from villages in South India, for purposes of targeted microfinance deployments, from [6]. In this context, we can view actions a_i as indicating levels of economic activity, which we wish to stimulate by increasing individual propensities for spending and creating network effects. The dataset contains many graphs for each village using different edge sets (each representing different kinds of social connections), as well as graphs where nodes are households rather than individuals. We use the household dataset containing the union of all edge sets. These graphs have degree counts ranging from 77 to 365, and our experiments are averaged over 20 graphs from this dataset.

We plot competitive ratios while varying C (scaled by network size) and the spectral radius of βA, fixing $b_i = 1$ for each agent.

The expected degree intervention is replaced by an intervention proportional to exact degree. We also fit a stochastic block model to graphs using a version of the approach described in Sect. 4.3, using exact connectivity probabilities rather than sampling. Our group labels are obtained by running the Girvan-Newman clustering algorithm [21] on the graph, pruning edges until there are either at least 10 clusters with 5 or more vertices or 50 clusters total. We evaluate the intervention proportional to the first eigenvector of the reconstructed block matrix. All interventions are compared to a baseline, where no change is applied to b, for demonstrating the relative degree in social welfare change.

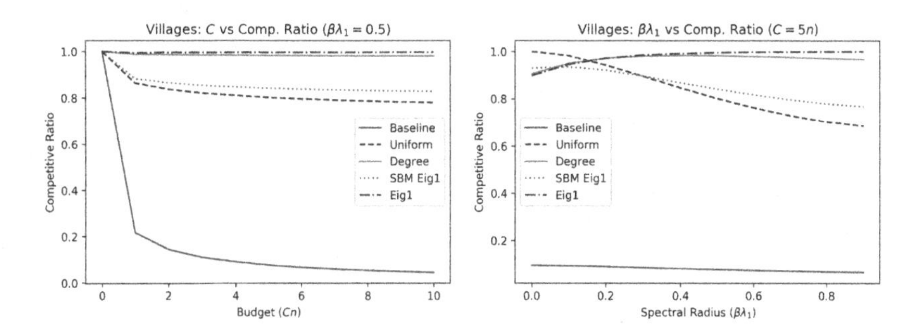

Fig. 1. Intervention in village graphs

In Fig. 1, we find that degree interventions perform quite well, and are only slightly surpassed by first eigenvector interventions. The stochastic block model approach performs better than uniform when the spectral radius sufficiently large, but is still outperformed by the degree and first eigenvector interventions. Upon inspection, the end result of the stochastic block model intervention was often uniform across a large subgraph, with little or no targeting for other vertices, which may be an artifact of the clustering method used for group assignment. On the whole, we observe that minimal-information approaches can indeed perform quite well on both real and simulated networks.

Acknowledgments. We thank Ben Golub, Yash Kanoria, Tim Roughgarden, Christos Papadimitriou, and anonymous reviewers for their invaluable feedback.

References

1. Abbe, E.: Community detection and stochastic block models: recent developments. J. Mach. Learn. Res. **18**(177), 1–86 (2018)
2. Aiello, W., Chung, F., Lu, L.: A random graph model for massive graphs. In: Proceedings of the Thirty-Second Annual ACM Symposium on Theory of Computing, STOC 2000, pp. 171–180. Association for Computing Machinery, New York (2000)

3. Akbarpour, M., Malladi, S., Saberi, A.: Diffusion, seeding, and the value of network information. In: Proceedings of the 2018 ACM Conference on Economics and Computation, EC 2018, p. 641. ACM, New York (2018)
4. Avella-Medina, M., Parise, F., Schaub, M.T., Segarra, S.: Centrality measures for graphons. CoRR abs/1707.09350 (2017). http://arxiv.org/abs/1707.09350
5. Avella-Medina, M., Parise, F., Schaub, M.T., Segarra, S.: Centrality measures for graphons (2017). http://arxiv.org/abs/1707.09350
6. Banerjee, A., Chandrasekhar, A.G., Duflo, E., Jackson, M.O.: The diffusion of microfinance. Science **341**(6144), 1236498 (2013). https://doi.org/10.1126/science.1236498
7. Banerjee, A., Chandrasekhar, A.G., Duflo, E., Jackson, M.O.: Using gossips to spread information: theory and evidence from two randomized controlled trials. Rev. Econ. Stud. **86**(6), 2453–2490 (2019)
8. Banerjee, A., G a, A., Duflo, E., Jackson, M.: Gossip: identifying central individuals in a social network, July 2014
9. Breza, E., Chandrasekhar, A.G., McCormick, T.H., Pan, M.: Using aggregated relational data to feasibly identify network structure without network data (2017)
10. Brown, W., Patange, U.: Targeted intervention in random graphs (2020). http://arxiv.org/abs/2007.06445
11. Candogan, O., Bimpikis, K., Ozdaglar, A.: Optimal pricing in networks with externalities. Oper. Res. **60**(4), 883–905 (2012)
12. Canonne, C.L.: A short note on learning discrete distributions (2020)
13. Chung, F., Lu, L.: Connected components in random graphs with given expected degree sequences. Ann. Comb. **6**(2), 125–145 (2002)
14. Chung, F., Lu, L., Vu, V.: Eigenvalues of random power law graphs. Ann. Comb. **7**(1), 21–33 (2003)
15. Chung, F., Radcliffe, M.: On the spectra of general random graphs. Electr. J. Comb. **18**, P215 (2011)
16. Dasaratha, K.: Distributions of centrality on networks. Papers, September 2017. arXiv.org
17. Dasgupta, A., Kumar, R., Sarlos, T.: On estimating the average degree. In: Proceedings of the 23rd International Conference on World Wide Web, WWW 2014, pp. 795–806. Association for Computing Machinery, New York (2014)
18. Demange, G.: Optimal targeting strategies in a network under complementarities. Games Econ. Behav. **105**, 84–103 (2017)
19. Eckles, D., Esfandiari, H., Mossel, E., Rahimian, M.A.: Seeding with costly network information (2019)
20. Galeotti, A., Golub, B., Goyal, S.: Targeting interventions in networks (2017)
21. Girvan, M., Newman, M.E.J.: Community structure in social and biological networks. Proc. Nat. Acad. Sci. **99**(12), 7821–7826 (2002). https://doi.org/10.1073/pnas.122653799
22. Goldreich, O., Ron, D.: Approximating average parameters of graphs. Random Struct. Algorithms **32**(4), 473–493 (2008)
23. Huang, J., Mani, A., Wang, Z.: The value of price discrimination in large random networks. In: Proceedings of the 2019 ACM Conference on Economics and Computation (2019)
24. Leskovec, J., Faloutsos, C.: Sampling from large graphs. In: Proceedings of the 12th ACM SIGKDD International Conference on Knowledge Discovery and Data Mining, KDD 2006, pp. 631–636. Association for Computing Machinery, New York (2006)

25. Parise, F., Ozdaglar, A.: Graphon games: a statistical framework for network games and interventions (2018)
26. Ruggeri, N., Bacco, C.D.: Sampling on networks: estimating eigenvector centrality on incomplete graphs (2019)
27. Schaub, M.T., Segarra, S., Tsitsiklis, J.N.: Blind identification of stochastic block models from dynamical observations (2019)
28. Tabouy, T., Barbillon, P., Chiquet, J.: Variational inference for stochastic block models from sampled data (2017)
29. Yun, S.Y., Proutiere, A.: Optimal sampling and clustering in the stochastic block model. In: Wallach, H., Larochelle, H., Beygelzimer, A., d'Alché-Buc, F., Fox, E., Garnett, R. (eds.) Advances in Neural Information Processing Systems, vol. 32, pp. 13422–13430. Curran Associates, Inc. (2019)

A New Lower Bound for Deterministic Truthful Scheduling

Yiannis Giannakopoulos, Alexander Hammerl, and Diogo Poças(✉)

TU Munich, Munich, Germany
{yiannis.giannakopoulos,alexander.hammerl,diogo.pocas}@tum.de

Abstract. We study the problem of truthfully scheduling m tasks to n selfish unrelated machines, under the objective of makespan minimization, as was introduced in the seminal work of Nisan and Ronen [NR99]. Closing the current gap of $[2.618, n]$ on the approximation ratio of deterministic truthful mechanisms is a notorious open problem in the field of algorithmic mechanism design. We provide the first such improvement in more than a decade, since the lower bounds of 2.414 (for $n = 3$) and 2.618 (for $n \to \infty$) by Christodoulou et al. [CKV07] and Koutsoupias and Vidali [KV07], respectively. More specifically, we show that the currently best lower bound of 2.618 can be achieved even for just $n = 4$ machines; for $n = 5$ we already get the first improvement, namely 2.711; and allowing the number of machines to grow arbitrarily large we can get a lower bound of 2.755.

Keywords: Mechanism design · Scheduling unrelated machines · Makespan minimization · Truthfulness

1 Introduction

Truthful scheduling of unrelated parallel machines is a prototypical problem in *algorithmic mechanism design*, introduced in the seminal paper of Nisan and Ronen [NR99] that essentially initiated this field of research. It is an extension of the classical combinatorial problem for the makespan minimization objective (see, e.g., [Vaz03, Ch. 17] or [Hal97, Sec. 1.4]), with the added twist that now machines are rational, *strategic* agents that would not hesitate to *lie* about their actual processing times for each job, if this can reduce their personal cost, i.e., their own completion time. The goal is to design a scheduling mechanism, using payments as incentives for the machines to *truthfully* report their true processing costs, that allocates all jobs in order to minimize the makespan, i.e., the maximum completion time across machines.

Supported by the Alexander von Humboldt Foundation with funds from the German Federal Ministry of Education and Research (BMBF). Yiannis Giannakopoulos is an associated researcher with the Research Training Group GRK 2201 "Advanced Optimization in a Networked Economy", funded by the German Research Foundation (DFG). A full version of this paper is available at https://arxiv.org/abs/2005.10054.

T. Harks and M. Klimm (Eds.): SAGT 2020, LNCS 12283, pp. 226–240, 2020.
https://doi.org/10.1007/978-3-030-57980-7_15

Nisan and Ronen [NR01] showed right away that no such truthful deterministic mechanism can achieve an approximation better than 2 to the optimum makespan; this is true even for just $n = 2$ machines. It is worth emphasizing that this lower bound is not conditioned on *any* computational complexity assumptions; it is purely a direct consequence of the added truthfulness requirement and holds even for mechanisms that have unbounded computational capabilities. It is interesting to compare this with the classical (i.e., non-strategic) algorithmic setting where we do know [LST90] that a 2-approximate polynomial-time algorithm does exist and that it is NP-hard to approximate the minimum makespan within a factor smaller than $\frac{3}{2}$. On the positive side, it is also shown in [NR01] that the mechanism that myopically allocates each job to the machine with the fastest reported time for it, and compensates her with a payment equal to the report of the second-fastest machine, achieves an approximation ratio of n (where n is the number of machines); this mechanism is truthful and corresponds to the paradigmatic VCG mechanism (see, e.g., [Nis07]).

Based on these, Nisan and Ronen [NR01, Conjecture 4.9] made the bold conjecture that their upper bound of n is actually the *tight* answer to the approximation ratio of deterministic scheduling; more than 20 years after the first conference version of their paper [NR99] though, very little progress has been made in closing their gap of $[2, n]$. Thus, the *Nisan-Ronen conjecture* remains up to this day one of the most important open questions in algorithmic mechanism design. Christodoulou et al. [CKV07] improved the lower bound to $1 + \sqrt{2} \approx 2.414$, even for instances with only $n = 3$ machines and, soon after, Koutsoupias and Vidali [KV07] showed that by allowing $n \to \infty$ the lower bound can be increased to $1 + \phi \approx 2.618$. The journal versions of these papers can be found at [CKV09] and [KV13], respectively. In our paper we provide the first improvement on this lower bound in well over a decade.

Another line of work tries to provide better lower bounds by imposing further assumptions on the mechanism, in addition to truthfulness. Most notably, Ashlagi et al. [ADL12] were actually able to resolve the Nisan-Ronen conjecture for the important special case of *anonymous* mechanisms, by providing a lower bound of n. The same can be shown for mechanisms with strongly-monotone allocation rules [MS18, Sec. 3.2] and for mechanisms with additive or local payment rules [NR01, Sec. 4.3.3].

Better bounds have also been achieved by modifying the scheduling model itself. For example, Lavi and Swamy [LS09] showed that if the processing times of all jobs can take only two values ("high" and "low") then there exists a 2-approximate truthful mechanism; they also give a lower bound of $\frac{11}{10}$. Very recently, Christodoulou et al. [CKK20] showed a lower bound of $\Omega(\sqrt{n})$ for a slightly generalized model where the completion times of machines are allowed to be submodular functions (of the costs of the jobs assigned to them) instead of additive in the standard setting.

Although in this paper we focus exclusively on deterministic mechanisms, randomization is also of great interest and has attracted a significant amount of attention [NR01, MS18, Yu09], in particular the two-machine case [LY08b, LY08a,

Lu09,CDZ15,KV19]. The currently best general lower bound on the approximation ratio of randomized (universally) truthful mechanisms is $2-\frac{1}{n}$ [MS18], while the upper one is $0.837n$ [LY08a]. For the more relaxed notion of *truthfulness in expectation*, the upper bound is $\frac{n+5}{2}$ [LY08b]. Related to the randomized case is also the fractional model, where mechanisms (but also the optimum makespan itself) are allowed to split jobs among machines. For this case, [CKK10] prove lower and upper bounds of $2-\frac{1}{n}$ and $\frac{n+1}{2}$, respectively; the latter is also shown to be tight for task-independent mechanisms.

Other variants of the strategic unrelated machine scheduling problem that have been studied include the Bayesian model [CHMS13,DW15,GK17] (where job costs are drawn from probability distributions), scheduling without payments [Kou14,GKK19] or with verification [NR01,PV14,Ven14], and strategic behaviour beyond (dominant-strategy) truthfulness [FRGL19]. The *related* machines model, which is essentially a single-dimensional mechanism design variant of our problem, has of course also been well-studied (see, e.g., [AT01,DDDR11,APPP09]) and a deterministic PTAS exists [CK13].

1.1 Our Results and Techniques

We present new lower bounds on the approximation ratio of deterministic truthful mechanisms for the prototypical problem of scheduling unrelated parallel machines, under the makespan minimization objective, introduced in the seminal work of Nisan and Ronen [NR01]. Our main result (Theorem 2) is a bound of $\rho \approx 2.755$, where ρ is the solution of the cubic equation (6). This improves upon the lower bound of $1+\phi \approx 2.618$ by Koutsoupias and Vidali [KV13] which appeared well over a decade ago [KV07]. Similar to [KV13], we use a family of instances with the number of machines growing arbitrarily large ($n \to \infty$).

Furthermore, our construction (see Sect. 3.4) provides improved lower bounds also *pointwise*, as a function of the number of machines n that we are allowed to use. More specifically, for $n = 3$ we recover the bound of $1+\sqrt{2} \approx 2.414$ by [CKV09]. For $n = 4$ we can already match the 2.618 bound that [KV13] could achieve only in the limit as $n \to \infty$. The first strict improvement, namely 2.711, comes from $n = 5$. As the number of machines grows, our bound converges to 2.755. Our results are summarized in Table 1.

A central feature of our approach is the formulation of our lower bound as the solution to a (non-linear) optimization programme (NLP); we then provide optimal, analytic solutions to it for all values of $n \geq 3$ (Lemma 3). It is important to clarify here that, in principle, just giving *feasible* solutions to this programme would still suffice to provide valid lower bounds for our problem. However, the fact that we pin down and use the actual *optimal* ones gives rise to an interesting implication: our lower bounds are provably the best ones that can be derived using our construction.

There are two key elements that allow us to derive our improved bounds, compared to the approach in previous related works [CKV09,KV13]. First, we deploy the weak-monotonicity (Theorem 1) characterization of truthfulness in a slightly more delicate way; see Lemma 1. This gives us better control and flexibility in considering deviating strategies for the machines (see our case-analysis in Sect. 3). Secondly, we consider more involved instances, with two auxiliary parameters (namely r and a; see, e.g., (3) and (4)) instead of just one. On the one hand, this increases the complexity of the solution, which now has to be expressed in an implicit way via the aforementioned optimization programme (NLP). But at the same time, fine-tuning the optimal choice of the variables allows us to (provably) push our technique to its limits. Finally, let us mention that, for a small number of machines ($n = 3, 4, 5$) we get $r = 1/a$ in an optimal choice of parameters. Under $r = 1/a$, we end up with a as the only free parameter, and our construction becomes closer to that of [CKV09, KV13]; in fact, for 3 machines it is essentially the same construction as in [CKV09] (which explains why we recover the same lower bound). However, for $n \geq 6$ machines we need a more delicate choice of r.

Due to space constraints, the proofs of Lemmas 1 to 3 are omitted; they can be found in the full version of our paper [GHP20].

Table 1. Lower bounds on the approximation ratio of deterministic truthful scheduling, as a function of the number of machines n, given by our Theorem 2 (bottom line). The previous state-of-the-art is given in the line above and first appeared in [CKV07] ($n = 3$) and [KV07] ($n \geq 4$). The case with $n = 2$ machines was completely resolved in [NR99], with an approximation ratio of 2.

n	3	4	5	6	7	8	...	∞
Previous work	2.414	2.465	2.534	2.570	2.590	2.601	...	2.618
This paper	2.414	2.618	2.711	2.739	2.746	2.750	...	2.755

2 Notation and Preliminaries

Before we go into the construction of our lower bound (Sect. 3), we use this section to introduce basic notation and recall the notions of mechanism, truthfulness, monotonicity, and approximation ratio. We also provide a technical tool (Lemma 1) that is a consequence of weak monotonicity (Theorem 1); this lemma will be used several times in the proof of our main result.

2.1 Unrelated Machine Scheduling

In the unrelated machine scheduling setting, we have a number n of machines and a number m of tasks to allocate to these machines. These tasks can be performed in any order, and each task has to be assigned to exactly one machine; machine i requires t_{ij} units of time to process task j. Hence, the complete description of a problem instance can be given by a $n \times m$ *cost matrix* of the values t_{ij}, which we denote by $\boldsymbol{t}$. In this matrix, row i, denoted by $\boldsymbol{t}_i$, represents the processing times for machine i (on the different tasks) and column j, denoted by $\boldsymbol{t}_j$, represents the processing times for task j (on the different machines). These values t_{ij} are assumed to be nonnegative real quantities, $t_{ij} \in \mathbb{R}_+$.

Applying the methodology of mechanism design, we assume that the processing times for machine i are known only by machine i herself. Moreover, machines are selfish agents; in particular, they are not interested in running a task unless they receive some compensation for doing so. They may also lie about their processing times if this would benefit them. This leads us to consider the central notion of (direct-revelation) *mechanisms*: each machine reports her values, and a mechanism decides on an allocation of tasks to machines, as well as corresponding payments, based on the reported values.

Definition 1 (Allocation rule, payment rule, mechanism). *Given n machines and m tasks,*

- *a (deterministic)* allocation rule *is a function that describes the allocation of tasks to machines for each problem instance. Formally, it is represented as a function $\boldsymbol{a} : \mathbb{R}_+^{n\times m} \to \{0,1\}^{n\times m}$ such that, for every $\boldsymbol{t} = (t_{ij}) \in \mathbb{R}_+^{n\times m}$ and every task $j = 1, \dots, m$, there is exactly one machine i with $a_{ij}(\boldsymbol{t}) = 1$, that is,*

$$\sum_{i=1}^{n} a_{ij}(\boldsymbol{t}) = 1; \tag{1}$$

- *a* payment rule *is a function that describes the payments to machines for each problem instance. Formally, it is represented as a function $p : \mathbb{R}_+^{n\times m} \to \mathbb{R}^n$;*
- *a (direct-revelation, deterministic)* mechanism *is a pair $(\boldsymbol{a}, p)$ consisting of an allocation and payment rules.*

We let $\mathbb{A}$ denote the set of feasible allocations, that is, matrices $\boldsymbol{a} = (a_{ij}) \in \{0,1\}^{n\times m}$ satisfying (1). Given a feasible allocation $\boldsymbol{a}$, we let $\boldsymbol{a}_i$ denote its row i, that is, the allocation to machine i. Similarly, given a payment vector $\boldsymbol{p} \in \mathbb{R}^n$, we let p_i denote the payment to machine i; note that the payments represent an amount of money given to the machine, which is somewhat the opposite situation compared to other mechanism design frameworks (such as auctions, where payments are done by the agents to the mechanism designer).

2.2 Truthfulness and Monotonicity

Whenever a mechanism assigns an allocation $\boldsymbol{a}_i$ and a payment p_i to machine i, this machine incurs a quasi-linear *utility* equal to her payment minus the sum of processing times of the tasks allocated to her,

$$p_i - \boldsymbol{a}_i \cdot \boldsymbol{t}_i = p_i - \sum_{j=1}^{m} a_{ij} t_{ij}.$$

Note that the above quantity depends on the machine's *true* and *reported* processing times, which in principle might differ. As already explained, machines behave selfishly. Thus, from the point of view of a mechanism designer, we wish to ensure a predictable behaviour of all parties involved. In particular, we are only interested in mechanisms that encourage agents to report their true valuations.

Definition 2 (Truthful mechanism). *A mechanism* $(\boldsymbol{a}, p)$ *is* truthful *if every machine maximizes their utility by reporting truthfully, regardless of the reports by the other machines. Formally, for every machine* i*, every* $\boldsymbol{t}_i, \boldsymbol{t}'_i \in \mathbb{R}^m_+$*,* $\boldsymbol{t}_{-i} \in R^{(n-1)\times m}_+$*, we have that*

$$p_i(\boldsymbol{t}_i, \boldsymbol{t}_{-i}) - \boldsymbol{a}_i(\boldsymbol{t}_i, \boldsymbol{t}_{-i}) \cdot \boldsymbol{t}_i \geq p_i(\boldsymbol{t}'_i, \boldsymbol{t}_{-i}) - \boldsymbol{a}_i(\boldsymbol{t}'_i, \boldsymbol{t}_{-i}) \cdot \boldsymbol{t}_i. \quad \text{(TR)}$$

In (TR), we "freeze" the reports of all machines other than i. The left hand side corresponds to the utility achieved by machine i when her processing times correspond to $\boldsymbol{t}_i$ and she truthfully reports $\boldsymbol{t}_i$. The right hand side corresponds to the utility achieved if machine i lies and reports $\boldsymbol{t}'_i$.

The most important example of a truthful mechanism in this setting is the VCG mechanism, that assigns each task independently to the machine that can perform it fastest, and paying that machine (for that task) a value equal to the second-lowest processing time. This is somewhat the equivalent of second-price auctions (that sell each item independently) for the scheduling setting.

A fundamental result in the theory of mechanism design is the very useful property of truthful mechanisms, in terms of "local" monotonicity of the allocation function with respect to single-machine deviations.

Theorem 1 (Weak monotonicity [NR01,LS09]). *Let* $\boldsymbol{t}$ *be a cost matrix,* i *be a machine, and* $\boldsymbol{t}'_i$ *another report from machine* i*. Let* $\boldsymbol{a}_i$ *be the allocation of* i *for cost matrix* $\boldsymbol{t}$ *and* $\boldsymbol{a}'_i$ *be the allocation of* i *for cost matrix* $(\boldsymbol{t}'_i, \boldsymbol{t}_{-i})$*. Then, if the mechanism is truthful, it must be that*

$$(\boldsymbol{a}_i - \boldsymbol{a}'_i) \cdot (\boldsymbol{t}_i - \boldsymbol{t}'_i) \leq 0. \quad \text{(WMON)}$$

As a matter of fact, (WMON) is also a *sufficient* condition for truthfulness, thus providing an exact *characterization* of truthfulness [SY05]. However, for our purposes in this paper we will only need the direction in the statement of Theorem 1 as stated above. We will make use of the following lemma, which exploits the notion of weak monotonicity in a straightforward way. The second part of this lemma can be understood as a refinement of a technical lemma that appeared before in [CKV09, Lemma 2] (see also [KV13, Lemma 1]).

Lemma 1. *Suppose that machine* i *changes her report from* $\boldsymbol{t}$ *to* $\boldsymbol{t}'$*, and that a truthful mechanism correspondingly changes her allocation from* $\boldsymbol{a}_i$ *to* $\boldsymbol{a}'_i$*. Let* $\{1, \ldots, m\} = S \cup T \cup V$ *be a partition of the tasks into three disjoint sets.*

1. *Suppose that (a) the costs of* i *on* V *do not change, that is,* $\boldsymbol{t}_{i,V} = \boldsymbol{t}'_{i,V}$ *and (b) the allocation of* i *on* S *does not change, that is,* $\boldsymbol{a}_{i,S} = \boldsymbol{a}'_{i,S}$*. Then*

$$(\boldsymbol{a}_{i,T} - \boldsymbol{a}'_{i,T}) \cdot (\boldsymbol{t}_{i,T} - \boldsymbol{t}'_{i,T}) \leq 0.$$

2. *Suppose additionally that (c) the costs of* i *strictly decrease on her allocated tasks in* T *and strictly increase on her unallocated tasks in* T*. Then her allocation on* T *does not change, that is,* $\boldsymbol{a}_{i,T} = \boldsymbol{a}'_{i,T}$*.*

2.3 Approximation Ratio

One of the main open questions in the theory of algorithmic mechanism design is to figure out what is the "best" possible truthful mechanism, with respect to the objective of makespan minimization. This can be quantified in terms of the approximation ratio of a mechanism.

If $\boldsymbol{t}$ is a problem instance (with n machines and m tasks) and $\boldsymbol{a}$ is a feasible allocation, its *makespan* is defined as the quantity

$$\mathrm{makespan}(\boldsymbol{a}, \boldsymbol{t}) = \max_{i=1,\dots,n} \sum_{j=1}^{m} a_{ij} t_{ij}.$$

The *optimal makespan* for a problem instance $\boldsymbol{t}$ is defined as the quantity

$$\mathrm{OPT}(\boldsymbol{t}) = \min_{a \in \mathbb{A}} \mathrm{makespan}(\boldsymbol{a}, \boldsymbol{t}).$$

We say that an allocation rule $\boldsymbol{a}$ has *approximation ratio* $\rho \geq 1$ if, for any problem instance $\boldsymbol{t}$, we have that

$$\mathrm{makespan}(\boldsymbol{a}(\boldsymbol{t}), \boldsymbol{t}) \leq \rho \mathrm{OPT}(\boldsymbol{t});$$

if no such quantity ρ exists, we say that $\boldsymbol{a}$ has *infinite approximation ratio.*

As shown in [NR01], the VCG mechanism has an approximation ratio of n, the number of machines. The long-standing conjecture by Nisan and Ronen states that this mechanism is essentially the best one; any truthful mechanism is believed to attain a worst-case approximation ratio of n (for sufficiently many tasks). In this paper, we prove lower bounds on the approximation ratio of any truthful mechanism (Table 1 and Theorem 2), which converge to 2.755 as $n \to \infty$.

3 Lower Bound

To prove our lower bound, from here on we assume $n \geq 3$ machines, since the case $n = 1$ is trivial and the case $n = 2$ is resolved by [NR99] (with an approximation ratio of 2). Our construction will be made with the choice of two parameters r, a. For now we shall simply assume that $a > 1 > r > 0$. Later we will optimize the choices of r and a in order to achieve the best lower bound possible by our construction.

We will use L_n to denote the $n \times n$ matrix with 0 in its diagonal and ∞ elsewhere,

$$L_n = \begin{bmatrix} 0 & \infty & \cdots & \infty \\ \infty & 0 & \cdots & \infty \\ \vdots & \vdots & \ddots & \vdots \\ \infty & \infty & \cdots & 0 \end{bmatrix}.$$

We should mention here that allowing $t_{ij} = \infty$ is a technical convenience. If only finite values are allowed, we can replace ∞ by an arbitrarily high value. We also follow the usual convention, and use an asterisk * to denote a full or partial allocation. Our lower bound begins with the following cost matrix for n machines and $2n-1$ tasks:

$$A_0 = \left[\begin{array}{c|cccccc} & {}^*1 & 1 & a^{-1} & a^{-2} & \cdots & a^{-n+3} \\ & 1 & 1 & a^{-1} & a^{-2} & \cdots & a^{-n+3} \\ & \infty & 1 & \infty & \infty & \cdots & \infty \\ L_n & \infty & \infty & a^{-1} & \infty & \cdots & \infty \\ & \infty & \infty & \infty & a^{-2} & \ddots & \infty \\ & \vdots & \vdots & \vdots & \ddots & \ddots & \vdots \\ & \infty & \infty & \infty & \infty & \cdots & a^{-n+3} \end{array}\right]. \tag{2}$$

The tasks of cost matrix A_0 can be partitioned in two groups. The first n tasks (i.e., the ones corresponding to the L_n submatrix) will be called *dummy* tasks. Machine i has a cost of 0 for dummy task i and a cost of ∞ for all other dummy tasks. The second group of tasks, numbered $n+1, \ldots, 2n-1$, will be called *proper* tasks. Notice that machines 1 and 2 have the same costs for proper tasks; they both need time 1 to execute task $n+1$ and time a^{-j+2} to execute task $n+j$, for all $j = 2, \ldots n-1$. Finally for $i \geq 3$, machine i has a cost of a^{-i+3} on proper task $n+i-1$ and ∞ cost for all other proper tasks.

In order for a mechanism to have a finite approximation ratio, it must not assign any tasks with unbounded costs. In particular, each dummy task must be assigned to the unique machine that completes it in time 0; and proper task $n+1$ must be assigned to either machine 1 or 2. Since the costs of machines 1 and 2 are the same on all proper tasks, we can without loss assume that machine 1 receives proper task $n+1$. Hence, the allocation on A_0 should be as (designated by an asterisk) in (2).

Next, we reduce the costs of all proper tasks for machine 1, and get the cost matrix

$$A_1 = \left[\begin{array}{c|cccccc} & r & a^{-1} & a^{-2} & a^{-3} & \cdots & a^{-n+2} \\ & 1 & 1 & a^{-1} & a^{-2} & \cdots & a^{-n+3} \\ & \infty & 1 & \infty & \infty & \cdots & \infty \\ L_n & \infty & \infty & a^{-1} & \infty & \cdots & \infty \\ & \infty & \infty & \infty & a^{-2} & \ddots & \infty \\ & \vdots & \vdots & \vdots & \ddots & \ddots & \vdots \\ & \infty & \infty & \infty & \infty & \cdots & a^{-n+3} \end{array}\right]. \tag{3}$$

Under the new matrix A_1, the cost of machine 1 for proper task $n+1$ is reduced from 1 to r; and her cost for any other proper task $n+j$, $j=2,\dots,n-1$, is reduced by a factor of a, that is, from a^{-j+2} to a^{-j+1}. The key idea in this step is the following: we want to impose a constraint on r and a that ensures that *at least one* of the proper tasks $n+1, n+2$ is still allocated to machine 1. Using the properties of truthfulness, namely part 1 of Lemma 1, this can be achieved via the following lemma:

Lemma 2. *Consider a truthful scheduling mechanism that, on cost matrix A_0, assigns proper task $n+1$ to machine* 1. *Suppose also that*

$$1 - r > a^{-1} - a^{-n+2}. \tag{4}$$

Then, on cost matrix A_1, machine 1 *must receive at least one of the proper tasks $n+1, n+2$.*

For the remainder of our construction, we assume that r and a satisfy (4). Next, we split the analysis depending on the allocation of the proper tasks $n+1, \dots 2n-1$ to machine 1 on cost matrix A_1, as restricted by Lemma 2.

3.1 Case 1: Machine 1 Gets *All* Proper Tasks

In this case, we perform the following changes in machine 1's tasks, obtaining a new cost matrix B_1. We increase the cost of dummy task 1, from 0 to 1, and we decrease the costs of all her proper tasks by an arbitrarily small amount. Notice that

- for the mechanism to achieve a finite approximation ratio, it must still allocate the dummy task 1 to machine 1;
- given that the mechanism does not change the allocation on dummy task 1, and that machine 1 only decreases the completion times of her proper tasks, part 2 of Lemma 1 implies that machine 1 still gets all proper tasks.

Thus, the allocation must be as shown below (for ease of exposition, in the cost matrices that follow we omit the "arbitrarily small" amounts by which we change allocated/unallocated tasks):

$$B_1 = \left[\begin{array}{cccccc|ccccc} {}^*1 & \infty & \infty & \infty & \cdots & \infty & {}^*r & {}^*a^{-1} & {}^*a^{-2} & \cdots & {}^*a^{-n+2} \\ \infty & {}^*0 & \infty & \infty & \cdots & \infty & 1 & 1 & a^{-1} & \cdots & a^{-n+3} \\ \infty & \infty & {}^*0 & \infty & \cdots & \infty & \infty & 1 & \infty & \cdots & \infty \\ \infty & \infty & \infty & {}^*0 & \cdots & \infty & \infty & \infty & a^{-1} & \cdots & \infty \\ \vdots & \vdots & \vdots & \vdots & \ddots & \vdots & \vdots & \vdots & \vdots & \ddots & \vdots \\ \infty & \infty & \infty & \infty & \cdots & {}^*0 & \infty & \infty & \infty & \cdots & a^{-n+3} \end{array}\right].$$

This allocation achieves a makespan of $1 + r + a^{-1} + \ldots + a^{-n+2}$, while a makespan of 1 can be achieved by assigning each proper task $n+j$ to machine $j+1$. Hence, this case yields an approximation ratio of at least $1 + r + va^{-1} + \ldots + a^{-n+2}$.

3.2 Case 2: Machine 1 Gets Task $n+1$, But Does *Not* Get All Proper Tasks

That is, at least one of tasks $n+2, \ldots 2n-1$ is not assigned to machine 1. Suppose that task $n+j$ is the lowest indexed proper task that is not allocated to her. We decrease the costs of her *allocated* proper tasks $n+1, \ldots, n+j-1$ to 0, while increasing the cost a^{-j+1} of her (unallocated) proper task $n+j$ by an arbitrarily small amount. By Lemma 1, the allocation of machine 1 on the proper tasks $n+1, \ldots, n+j$ does not change. Hence we get a cost matrix of the form

$$B_2 = \left[\begin{array}{c|cccccc} & {}^*0 & {}^*0 & \cdots & a^{-j+1} & \cdots & a^{-n+2} \\ & 1 & 1 & \cdots & a^{-j+2} & \cdots & a^{-n+3} \\ & \infty & 1 & \cdots & \infty & \cdots & \infty \\ L_n & \vdots & \vdots & \ddots & \vdots & \cdots & \infty \\ & \infty & \infty & \cdots & a^{-j+2} & \ddots & \infty \\ & \vdots & \vdots & \vdots & \ddots & \ddots & \vdots \\ & \infty & \infty & \infty & \infty & \cdots & a^{-n+3} \end{array}\right].$$

Since task $n+j$ is not allocated to machine 1, and the mechanism has finite approximation ratio, it must be allocated to either machine 2 or machine $j+1$. In either case, we increase the cost of the dummy task of this machine from 0 to a^{-j+1}, while decreasing the cost of her proper task $n+j$ by an arbitrarily small amount. For example, if machine 2 got task $n+j$, we would end up with

$$C_2 = \left[\begin{array}{ccccccc|cccccc} {}^*0 & \infty & \infty & \cdots & \infty & \cdots & \infty & 0 & 0 & \cdots & a^{-j+1} & \cdots & a^{-n+2} \\ \infty & {}^*a^{-j+1} & \infty & \cdots & \infty & \cdots & \infty & 1 & 1 & \cdots & {}^*a^{-j+2} & \cdots & a^{-n+3} \\ \infty & \infty & {}^*0 & \cdots & \infty & \cdots & \infty & \infty & 1 & \cdots & \infty & \cdots & \infty \\ \vdots & \vdots & \vdots & \ddots & \vdots & \ddots & \vdots & \vdots & \vdots & \ddots & \vdots & \ddots & \infty \\ \infty & \infty & \infty & \cdots & {}^*0 & \cdots & \infty & \infty & \infty & \cdots & a^{-j+2} & \cdots & \infty \\ \vdots & \vdots & \vdots & \ddots & \vdots & \ddots & \vdots & \vdots & \vdots & \ddots & \vdots & \ddots & \vdots \\ \infty & \infty & \infty & \cdots & \infty & \cdots & {}^*0 & \infty & \infty & \infty & \infty & \cdots & a^{-n+3} \end{array}\right].$$

Similarly to the previous Case 1, the mechanism must still allocate the dummy task to this machine, and given that the allocation does not change on the dummy task, Lemma 1 implies that the allocation must also remain unchanged on the proper task $n+j$. Finally, observe that the present allocation achieves a makespan of at least $a^{-j+1}+a^{-j+2}$, while a makespan of a^{-j+1} can be achieved by assigning proper task $n+j$ to machine 1 and proper task $n+j'$ to machine $j'+1$, for $j'>j$. Hence, this case yields an approximation ratio of at least

$$\frac{a^{-j+1}+a^{-j+2}}{a^{-j+1}}=1+a.$$

3.3 Case 3: Machine 1 Does *Not* Get Task $n+1$

By Lemma 2, machine 1 must receive proper task $n+2$. In this case, we decrease the cost of her task $n+2$, from a^{-1} to 0, while increasing the cost r of her (unallocated) task $n+1$ by an arbitrarily small amount. Since by truthfulness, the allocation of machine 1 for these two tasks does not change, the allocation must be as below:

$$B_3=\left[\begin{array}{c|ccccc} & r & {}^{*}0 & a^{-2} & \cdots & a^{-n+2}\\ & {}^{*}1 & 1 & a^{-1} & \cdots & a^{-n+3}\\ & \infty & 1 & \infty & \cdots & \infty\\ L_n & \infty & \infty & a^{-1} & \cdots & \infty\\ & \vdots & \vdots & \vdots & \ddots & \vdots\\ & \infty & \infty & \infty & \cdots & a^{-n+3}\end{array}\right].$$

Since task $n+1$ is not allocated to machine 1, and the mechanism has finite approximation ratio, it must be allocated to machine 2. We now increase the cost of the dummy task of machine 2 from 0 to $\max\{r,a^{-1}\}$, while decreasing the cost of her proper task $n+1$ by an arbitrarily small amount. Similarly to Cases 1 and 2, the mechanism must still allocate the dummy task to machine 2, and preserve the allocation of machine 2 on the proper task $n+1$. Thus, we get the allocation shown below:

$$C_3=\left[\begin{array}{cccccc|ccccc} {}^{*}0 & \infty & \infty & \infty & \cdots & \infty & r & 0 & a^{-2} & \cdots & a^{-n+2}\\ \infty & {}^{*}\max\{r,a^{-1}\} & \infty & \infty & \cdots & \infty & {}^{*}1 & 1 & a^{-1} & \cdots & a^{-n+3}\\ \infty & \infty & {}^{*}0 & \infty & \cdots & \vdots & \infty & 1 & \infty & \cdots & \infty\\ \infty & \infty & \infty & {}^{*}0 & \cdots & \infty & \infty & \infty & a^{-1} & \cdots & \infty\\ \vdots & \vdots & \vdots & \vdots & \ddots & \vdots & \vdots & \vdots & \vdots & \ddots & \vdots\\ \infty & \infty & \infty & \infty & \cdots & {}^{*}0 & \infty & \infty & \infty & \cdots & a^{-n+3}\end{array}\right].$$

This allocation achieves a makespan of at least $1+\max\{r,a^{-1}\}$, while a makespan of $\max\{r,a^{-1}\}$ can be achieved by assigning proper tasks $n+1,n+2$

to machine 1 and proper task $n+j'$ to machine $j'+1$, for all $j'>2$. Hence, this case yields an approximation ratio of at least

$$\frac{1+\max\{r,a^{-1}\}}{\max\{r,a^{-1}\}}=1+\min\{r^{-1},a\}.$$

3.4 Main Result

The three cases considered above give rise to possibly different approximation ratios; our construction will then yield a lower bound equal to the *smallest* of these ratios. First notice that Case 3 always gives a worse bound than Case 2: the approximation ratio for the former is $1+\min\{r^{-1},a\}$, whereas for the latter it is $1+a$. Thus we only have to consider the minimum between Cases 1 and 3.

Our goal then is to find a choice of r and a that achieves the largest possible such value. We can formulate this as a nonlinear optimization problem on the variables r and a. To simplify the exposition, we also consider an auxiliary variable ρ, which will be set to the minimum of the approximation ratios:

$$\begin{aligned}\rho&=\min\left\{1+r+a^{-1}+\ldots+a^{-n+2},1+\min\{r^{-1},a\}\right\}\\&=\min\left\{1+r+a^{-1}+\ldots+a^{-n+2},1+r^{-1},1+a\right\}.\end{aligned}$$

This can be enforced by the constraints $\rho\leq 1+r+a^{-1}+\ldots+a^{-n+2}$, $\rho\leq 1+r^{-1}$ and $\rho\leq 1+a$. Thus, our optimization problem becomes

$$\begin{aligned}\sup\quad&\rho\\\text{s.t.}\quad&\rho\leq 1+r+a^{-1}+\ldots+a^{-n+2}\\&\rho\leq 1+r^{-1}\\&\rho\leq 1+a\\&0<r<1<a\\&1-r>a^{-1}-a^{-n+2}\end{aligned}\tag{NLP}$$

Notice that *any* feasible solution of (NLP) gives rise to a lower bound on the approximation ratio of truthful machine scheduling. In our next lemma, we characterize the limiting optimal solution of the above optimization problem. Thus, the lower bound achieved corresponds to the best possible lower bound using the general construction in this paper.

Lemma 3. *An optimal solution to the optimization problem given by* (NLP) *is as follows.*

1. *For* $n=3,4,5$, *choose* $\rho=1+a$, $r=\frac{1}{a}$, *and* a *as the positive solution of the equation*

$$\begin{aligned}\frac{2}{a}&=a,\quad\textit{for }n=3;\\\frac{2}{a}+\frac{1}{a^2}&=a,\quad\textit{for }n=4;\\\frac{2}{a}+\frac{1}{a^2}+\frac{1}{a^3}&=a,\quad\textit{for }n=5.\end{aligned}$$

2. *For $n \geq 6$, choose $\rho = 1 + a$, $r = 1 - \frac{1}{a} + \frac{1}{a^{n-2}}$, and a as the positive solution of the equation*

$$1 + \frac{1}{a^2} + \cdots + \frac{1}{a^{n-3}} + \frac{2}{a^{n-2}} = a. \tag{5}$$

Using the above technical lemma, we are able to prove our main result.

Theorem 2. *No deterministic truthful mechanism for unrelated machine scheduling can have an approximation ratio better than $\rho \approx 2.755$, where ρ is the (unique real) solution of equation*

$$(\rho - 1)(\rho - 2)^2 = 1. \tag{6}$$

For a restricted number of machines the lower bounds can be seen in Table 1.

Proof. For n large enough we can use Case 2 of Lemma 3. In particular, taking the limit of (5) as $n \to \infty$, we can ensure a lower bound of $\rho = a + 1$, where a is the (unique) real solution of equation

$$1 + \sum_{i=2}^{\infty} \frac{1}{a^i} = 1 + \frac{1}{a(a-1)} = a.$$

Performing the transformation $a = \rho - 1$, and multiplying throughout by $(\rho - 1)(\rho - 2)$, we get exactly (6).

For a fixed number of machines n, we can directly solve the equations given by either Case 1 ($n = 3, 4, 5$) or Case 2 of Lemma 3 to derive the corresponding value of a, for a lower bound of $\rho = a + 1$. In particular, for $n = 3, 4, 5$ one gets $a = \sqrt{2} \approx 1.414$, $a = \phi \approx 1.618$ (i.e., the *golden ratio*) and $a \approx 1.711$, respectively. The values of ρ for up to $n = 8$ machines are given in Table 1.

References

ADL12. Ashlagi, I., Dobzinski, S., Lavi, R.: Optimal lower bounds for anonymous scheduling mechanisms. Math. Oper. Res. **37**(2), 244–258 (2012). https://doi.org/10.1287/moor.1110.0534

APPP09. Auletta, V., De Prisco, R., Penna, P., Persiano, G.: The power of verification for one-parameter agents. J. Comput. Syst. Sci. **75**(3), 190–211 (2009). https://doi.org/10.1016/j.jcss.2008.10.001

AT01. Archer, A., Tardos, É.: Truthful mechanisms for one-parameter agents. In: Proceedings of the 42nd IEEE Symposium on Foundations of Computer Science (FOCS), pp. 482–491 (2001). https://doi.org/10.1109/sfcs.2001.959924

CDZ15. Chen, X., Du, D., Zuluaga, L.F.: Copula-based randomized mechanisms for truthful scheduling on two unrelated machines. Theory Comput. Syst. **57**(3), 753–781 (2015). https://doi.org/10.1007/s00224-014-9601-5

CHMS13. Chawla, S., Hartline, J.D., Malec, D., Sivan, B.: Prior-independent mechanisms for scheduling. In: Proceedings of the 45th Annual ACM Symposium on Theory of Computing (STOC), pp. 51–60 (2013). https://doi.org/10.1145/2488608.2488616

CK13. Christodoulou, G., Kovács, A.: A deterministic truthful PTAS for scheduling related machines. SIAM J. Comput. **42**(4), 1572–1595 (2013). https://doi.org/10.1137/120866038

CKK10. Christodoulou, G., Koutsoupias, E., Kovács, A.: Mechanism design for fractional scheduling on unrelated machines. ACM Trans. Algorithms **6**(2), 1–18 (2010). https://doi.org/10.1145/1721837.172185410.1145/1721837.1721854

CKK20. Christodoulou, G., Koutsoupias, E., Kovács, A.: On the Nisan-Ronen conjecture for submodular valuations. In: Proceedings of the 52nd Annual ACM SIGACT Symposium on Theory of Computing (STOC), pp.1086–1096 (2020). https://doi.org/10.1145/3357713.3384299

CKV07. Christodoulou, G., Koutsoupias, E., Vidali, A.: A lower bound for scheduling mechanisms. In: Proceedings of the 18th Annual ACM-SIAM Symposium on Discrete Algorithms (SODA), pp. 1163–1170 (2007). https://doi.org/10.5555/1283383.1283508

CKV09. Christodoulou, G., Koutsoupias, E., Vidali, A.: A lower bound for scheduling mechanisms. Algorithmica **55**(4), 729–740 (2009). https://doi.org/10.1007/s00453-008-9165-3

DDDR11. Dhangwatnotai, P., Dobzinski, S., Dughmi, S., Roughgarden, T.: Truthful approximation schemes for single-parameter agents. SIAM J. Comput. **40**(3), 915–933 (2011). https://doi.org/10.1137/080744992

DW15. Daskalakis, C., Weinberg, S.M.: Bayesian truthful mechanisms for job scheduling from bi-criterion approximation algorithms. In: Proceedings of the 26th annual ACM-SIAM Symposium on Discrete Algorithms (SODA), pp. 1934–1952 (2015). https://doi.org/10.1137/1.9781611973730.130

FRGL19. Filos-Ratsikas, A., Giannakopoulos, Y., Lazos, P.: The Pareto frontier of inefficiency in mechanism design. In: Caragiannis, I., Mirrokni, V., Nikolova, E. (eds.) WINE 2019. LNCS, vol. 11920, pp. 186–199. Springer, Cham (2019). https://doi.org/10.1007/978-3-030-35389-6_14

GHP20. Giannakopoulos, Y., Hammerl, A., Poças, D.: A new lower bound for deterministic truthful scheduling. CoRR, abs/2005.10054 (2020). arXiv:2005.10054

GK17. Giannakopoulos, Y., Kyropoulou, M.: The VCG mechanism for Bayesian scheduling. ACM Trans. Econ. Comput. **5**(4), 191–1916 (2017). https://doi.org/10.1145/3105968. arXiv:1509.07455

GKK19. Giannakopoulos, Y., Koutsoupias, E., Kyropoulou, M.: The anarchy of scheduling without money. Theoret. Comput. Sci. **778**, 19–32 (2019). https://doi.org/10.1016/j.tcs.2019.01.022. arXiv:1607.03688

Hal97. Hall, L.A.: Approximation algorithms for scheduling. In: Hochbaum, D.S. (ed.) Approximation Algorithms for NP-Hard Problems, pp. 1–45. PWS Publishing Company (1997)

Kou14. Koutsoupias, E.: Scheduling without payments. Theory Comput. Syst. **54**(3), 375–387 (2013). https://doi.org/10.1007/s00224-013-9473-0

KV07. Koutsoupias, E., Vidali, A.: A lower bound of $1+\phi$ for truthful scheduling mechanisms. In: Proceedings of Mathematical Foundations of Computer Science (MFCS), pp. 454–464 (2007). https://doi.org/10.1007/978-3-540-74456-6_41

KV13. Koutsoupias, E., Vidali, A.: A lower bound of $1+\phi$ for truthful scheduling mechanisms. Algorithmica **66**(1), 211–223 (2013). https://doi.org/10.1007/s00453-012-9634-6

KV19. Kuryatnikova, O., Vera, J.C.: New bounds for truthful scheduling on two unrelated selfish machines. Theory Comput. Syst. **64**(2), 199–226 (2019). https://doi.org/10.1007/s00224-019-09927-x

LS09. Lavi, R., Swamy, C.: Truthful mechanism design for multidimensional scheduling via cycle monotonicity. Games Econ. Behav. **67**(1), 99–124 (2009). https://doi.org/10.1016/j.geb.2008.08.001

LST90. Lenstra, J.K., Shmoys, D.B., Tardos, É.: Approximation algorithms for scheduling unrelated parallel machines. Math. Program. **46**(1), 259–271 (1990). https://doi.org/10.1007/bf01585745

Lu09. Lu, P.: On 2-player randomized mechanisms for scheduling. In: Leonardi, S. (ed.) WINE 2009. LNCS, vol. 5929, pp. 30–41. Springer, Heidelberg (2009). https://doi.org/10.1007/978-3-642-10841-9_5

LY08a. Lu, P., Yu, C.: An improved randomized truthful mechanism for scheduling unrelated machines. In: Proceedings of the 25th International Symposium on Theoretical Aspects of Computer Science (STACS), pp. 527–538 (2008). https://doi.org/10.4230/LIPIcs.STACS.2008.1314

LY08b. Lu, P., Yu, C.: Randomized truthful mechanisms for scheduling unrelated machines. In: Papadimitriou, C., Zhang, S. (eds.) WINE 2008. LNCS, vol. 5385, pp. 402–413. Springer, Heidelberg (2008). https://doi.org/10.1007/978-3-540-92185-1_46

MS18. Mu'alem, A., Schapira, M.: Setting lower bounds on truthfulness. Games Econ. Behav. **110**, 174–193 (2018). https://doi.org/10.1016/j.geb.2018.02.001

Nis07. Nisan, N.: Introduction to mechanism design (for computer scientists). In: Nisan, N., Roughgarden, T., Tardos, É., Vazirani, V. (eds.) Algorithmic Game Theory, Chapter 9. Cambridge University Press (2007). https://doi.org/10.1017/cbo9780511800481.011

NR99. Nisan, N., Ronen, A.: Algorithmic mechanism design (extended abstract). In: The 31st Annual ACM symposium on Theory of Computing (STOC), pp. 129–140 (1999)

NR01. Nisan, N., Ronen, A.: Algorithmic mechanism design. Games Econ. Behav. **35**(1/2), 166–196 (2001). https://doi.org/10.1006/game.1999.0790

PV14. Penna, P., Ventre, C.: Optimal collusion-resistant mechanisms with verification. Games Econ. Behav. **86**, 491–509 (2014). https://doi.org/10.1016/j.geb.2012.09.002

SY05. Saks, M., Yu, L.: Weak monotonicity suffices for truthfulness on convex domains. In Proceedings of the 6th ACM Conference on Electronic Commerce (EC), pp. 286–293 (2005). https://doi.org/10.1145/1064009.1064040

Vaz03. Vazirani, V.V.: Approximation Algorithms. Springer, Heidelberg (2003). https://doi.org/10.1007/978-3-662-04565-7

Ven14. Ventre, C.: Truthful optimization using mechanisms with verification. Theoret. Comput. Sci. **518**, 64–79 (2014). https://doi.org/10.1016/j.tcs.2013.07.034

Yu09. Yu, C.: Truthful mechanisms for two-range-values variant of unrelated scheduling. Theoret. Comput. Sci. **410**(21–23), 2196–2206 (2009). https://doi.org/10.1016/j.tcs.2009.02.001

Modified Schelling Games

Panagiotis Kanellopoulos, Maria Kyropoulou, and Alexandros A. Voudouris(✉)

School of Computer Science and Electronic Engineering, University of Essex, Colchester, UK
{panagiotis.kanellopoulos,maria.kyropoulou, alexandros.voudouris}@essex.ac.uk

Abstract. We introduce the class of *modified Schelling* games in which there are different types of agents who occupy the nodes of a location graph; agents of the same type are friends, and agents of different types are enemies. Every agent is strategic and jumps to empty nodes of the graph aiming to maximize her utility, defined as the ratio of her friends in her neighborhood over the neighborhood size *including* herself. This is in contrast to the related literature on Schelling games which typically assumes that an agent is excluded from her neighborhood whilst computing its size. Our model enables the utility function to capture likely cases where agents would rather be around a lot of friends instead of just a few, an aspect that was partially ignored in previous work. We provide a thorough analysis of the (in)efficiency of equilibria that arise in such modified Schelling games, by bounding the price of anarchy and price of stability for both general graphs and interesting special cases. Most of our results are tight and exploit the structure of equilibria as well as sophisticated constructions.

Keywords: Schelling games · Price of anarchy · Price of stability

1 Introduction

More than 50 years ago, Thomas Schelling [23,24] presented the following simple probabilistic procedure in an attempt to model residential segregation. There are two types of agents who are uniformly at random placed at the nodes of a location graph (such as a line or a grid), and a tolerance threshold parameter $\tau \in (0,1)$. If the neighborhood of an agent consists of at least a fraction τ of agents of her own type, then the agent is happy and remains at her current location. Otherwise, the agent is unhappy and either jumps to a randomly selected empty node of the graph or swaps locations with another randomly chosen unhappy agent. Schelling experimentally showed that this random process can lead to placements such that the graph is partitioned into multiple parts, each containing mostly agents of the same type, even when the agents are tolerant towards having neighbors of the other type (that is, when $\tau < 1/2$).

Since its inception, Schelling's model and interesting variants of it have been studied extensively both experimentally and theoretically from the perspective

T. Harks and M. Klimm (Eds.): SAGT 2020, LNCS 12283, pp. 241–256, 2020.
https://doi.org/10.1007/978-3-030-57980-7_16

of a plethora of different disciplines, including Sociology [12], Economics [22,26], Physics [25], and Computer Science [4,5,9,17]. Most of these works have focused on the analysis of random processes similar to the one proposed by Schelling, either via agent-based simulations or via Markov chains, and have shown that segregation occurs with high probability.

A more recent stream of papers [1,6,10,11,14,16] have considered *Schelling games*, that is, game-theoretic variants of Schelling's model with multiple types of agents and general location graphs. The agents behave strategically and aim to maximize a utility function, which is defined as the minimum between the threshold parameter τ and the ratio of the other agents of the same type within one's neighborhood over the (occupied) neighborhood size. These papers have considered both *jump* games, in which the agents are allowed to jump to empty nodes of the location graph, and *swap* games, in which the agents are only allowed to pairwise swap positions. Among other questions, they have studied the complexity of computing equilibrium assignments (i.e., placements such that no agent wants to jump to an empty node or no pair of agents wants to swap positions), the complexity of maximizing social welfare (i.e., the total utility of the agents), and have shown bounds on the price of anarchy [18] and the price of stability [2].

One limitation of the utility function defined above and used in the related literature on Schelling games, which our model aims to address, is that it does not allow the agents to distinguish between neighborhoods that consist only of agents of their own type, but may vary in size. To give a concrete example, consider a red agent who faces the dilemma of choosing between two empty nodes, one of which is adjacent to one red agent, while the other is adjacent to two red agents. Since the utility is defined as the fraction of red neighbors, both empty nodes offer the same utility of 1 to our agent, which means that she can choose arbitrarily amongst them. However, it is arguably more realistic to assume that the second empty node is more attractive than the first one as it is adjacent to a strictly larger number of red agents, and consequently the agent would normally choose it. To strengthen the ability of the utility function to express preferences of this kind, we redefine it by assuming that the agent considers herself as part of the set of her neighbors, which simply translates to a "+1" term added to the denominator of the ratio; this is similar to fractional hedonic games (see the discussion below). Back to our example, the new *modified* utility function would yield utilities of 1/2 and 2/3 for the two empty nodes, respectively, reflecting the agent's preference for the second node.

Our Setting and Contribution. We introduce the class of modified Schelling games. In such games, there are k types of agents who occupy the nodes of some location graph and aim to maximize their utility, which is defined by the modified function discussed above, by jumping to empty nodes whenever such a move is beneficial; following most of the related work in Schelling games, we assume that $\tau = 1$. Since the modified utility function is able to express preferences over monochromatic neighborhoods of different sizes, a strategic game is induced even when there is a single type of agents. For $k = 1$, we argue that the best-response

Table 1. Overview of our price of anarchy and price of stability bounds. For $k = 1$, the case of balanced games is obviously non-applicable (N/A). For $k \geq 2$, all price of anarchy bounds are for games with at least two agents per type (otherwise, the PoA can be easily seen to be unbounded); the PoA bounds for lines and trees are restricted to balanced games. Unless specified otherwise (like for PoS), the bounds presented are tight.

	PoA				PoS
	Arbitrary	Balanced	Line	Tree	
$k = 1$	$2 - \frac{2}{n}$	N/A	$\frac{4}{3} - \frac{2}{3n}$	$\frac{4}{3} - \frac{2}{3n}$	$\in [\frac{15}{14}, \frac{3}{2}]$
$k \geq 2$	$\frac{2n(n-k)}{n+2}$	$2k$	2 $(k = 2)$ $k + 1/2$ $(k \geq 3)$	$\frac{14}{9}k$ $(k \in \{2, 3\})$ $\frac{2k^2}{k+1}$ $(k \geq 4)$	$\geq \frac{4}{3}$ $(k = 2)$

dynamics always converges to an equilibrium assignment in polynomial time, while this is not generally true for $k \geq 2$. Our main technical contribution is a thorough price of anarchy and price of stability analysis. We distinguish between games on arbitrary location graphs, balanced games in which there is the same number of agents per type (for $k \geq 2$), as well as games with structured location graphs such as lines and trees. We show *tight* bounds on the price of anarchy, by carefully exploiting the structure of equilibrium assignments and the properties of the games we study. We also show lower bounds on the price of stability for $k \in \{1, 2\}$, as well as an upper bound for $k = 1$; to the best of our knowledge, this is the first non-trivial upper bound on the price of stability for general location graphs in the related literature. An overview of our results is given in Table 1. Due to space constraints, many proofs are omitted.

Related Work. We will mainly discuss the related literature on Schelling games. Chauhan *et al.* [11] studied the convergence of the best-response dynamics to an equilibrium assignment in both jump and swap Schelling games with two types of agents and for various values of the threshold parameter τ. They presented a series of positive and negative results depending on the relation of τ to other parameters related to the location graph. Their results were later extended by Echzell *et al.* [14] for more than two types of agents and for two different generalizations of the utility function: one that considers all types in the denominator of the ratio, and one that considers only the type of the agent at hand and the type of maximum cardinality among the remaining types.

Elkind *et al.* [16] considered a variant of jump Schelling games with $k \geq 2$ types of agents who may behave in two different ways: some of them are strategic and aim to maximize their utility, while some others are stubborn and stay at their initial location regardless of the composition of the neighborhood. Elkind *et al.* showed that equilibrium assignments may fail to exist, they proved that the problem of computing an equilibrium or an assignment with high social welfare is intractable, and also showed bounds on the price of anarchy and the price of stability. Furthermore, they discussed several extensions, among which that of *social Schelling games*, where the friendships among agents are specified by a social network. This class of games was further studied by Chan *et al.* [10], who also assumed that the nodes of the location graph can be shared by different agents.

Agarwal *et al.* [1] considered swap Schelling games. Besides studying complexity and price of anarchy questions similar to those of Elkind *et al.*, they also considered related questions for a different objective function over assignments, called *degree of integration*, which aims to capture how diverse an assignment is; this function counts the number of agents who have at least one neighbor of different type. Very recently, Bilò *et al.* [6] performed a refined price of anarchy analysis with respect to the social welfare in the model of Agarwal *et al.* for swap games: they showed improved bounds for $k = 2$, as well as for games with structured location graphs such as cycles, trees, regular graphs, and grids. Furthermore, they initiated the study of games restricted movement, in which the agents can swap positions only with agents within a given radius from their current location (such as their neighbors). In a slightly different context, Massand and Simon [19] studied games that are similar to swap social Schelling games, but with linear utility functions, instead of fractions.

As pointed out by Elkind *et al.*, Schelling games are very similar to hedonic games [8,13], but also quite distinct from them: while one can think of the neighborhoods as coalitions, these coalitions generally overlap depending on the structure of the location graph. Somewhat counter-intuitively, the games studied by almost all the aforementioned papers are analogous to modified fractional hedonic games [15,20,21], where the agents are connected via a weighted social graph and are then partitioned into coalitions; each agent derives a utility which is the total weight of her connections within her coalition divided by the size of the coalition excluding herself. In contrast, the modified Schelling games we study in this paper are analogous to fractional hedonic games [3,7], where the utility of an agent is defined as the total weight of her connections within her coalition divided by the size of the coalition *including* herself.

2 Preliminaries

There are $n \geq 2$ *agents* who are partitioned into $k \geq 1$ *types*. We denote by T_ℓ the set of all agents of type $\ell \in [k]$, and let $n_\ell = |T_\ell|$ such that $n = \sum_{\ell \in [k]} n_\ell$; also, let $\mathbf{n} = (n_\ell)_{\ell \in [k]}$. Agents of the same type are *friends*, and agents of different types are *enemies*. The agents occupy the nodes of a simple *undirected connected* location graph $G = (V, E)$ with $|V| > n$ nodes; following previous work, we refer to this graph as the *topology*. An *assignment* $\mathbf{v} = (v_i)_{i \in [n]}$ is a vector containing the node $v_i \in V$ occupied by each agent $i \in [n]$ such that $v_i \neq v_j$ for $i \neq j$.

For an assignment $\mathbf{v}$, we denote by $N(v|\mathbf{v})$ the set of agents that are adjacent to node $v \in V$. Moreover, let $x(v|\mathbf{v}) = |N(v|\mathbf{v})|$ and denote by $x_\ell(v|\mathbf{v}) = |N(v|\mathbf{v}) \cap T_\ell|$ the number of agents of type $\ell \in [k]$ in the neighborhood of node v. Then, the utility of an agent i of type ℓ who occupies node v_i under assignment $\mathbf{v}$ is defined as

$$u_i(\mathbf{v}) = \frac{x_\ell(v_i|\mathbf{v})}{1 + x(v_i|\mathbf{v})}.$$

To simplify our notation, we will omit $\mathbf{v}$ whenever it is clear from context, and will sometimes use colors to refer to different types.

The agents are strategic and can *jump* to empty nodes of the topology to maximize their utility. An assignment $\mathbf{v}$ is called a *pure Nash equilibrium* (or, simply, *equilibrium*) if no agent prefers to jump to any empty node, that is, $u_i(\mathbf{v}) \geq u_i(v, \mathbf{v}_{-i})$ for every agent i and empty node v, where $(v, \mathbf{v}_{-i})$ is the assignment according to which agent i occupies v and all other agents occupy the same nodes as in $\mathbf{v}$. Let $\text{EQ}(\mathcal{G})$ denote the set of all equilibrium assignments of a *modified k-Schelling game* $\mathcal{G} = (\mathbf{n}, G)$.

The *social welfare* of an assignment $\mathbf{v}$ is the total utility of the agents:

$$\text{SW}(\mathbf{v}) = \sum_{i \in [n]} u_i(\mathbf{v}).$$

For a given game, the maximum social welfare among all possible assignments is denoted by $\text{OPT} = \max_{\mathbf{v}} \text{SW}(\mathbf{v})$. The *price of anarchy* of a modified k-Schelling game $\mathcal{G}$ with $\text{EQ}(\mathcal{G}) \neq \varnothing$ is the ratio of the maximum social welfare achieved by any possible assignment over the minimum social welfare achieved at equilibrium, that is,

$$\text{PoA}(\mathcal{G}) = \frac{\text{OPT}}{\min_{\mathbf{v} \in \text{EQ}(\mathcal{G})} \text{SW}(\mathbf{v})}.$$

Then, the price of anarchy of a class $\mathcal{C}$ of modified k-Schelling games is

$$\text{PoA}(\mathcal{C}) = \sup_{\mathcal{G} \in \mathcal{C}: \text{EQ}(\mathcal{G}) \neq \varnothing} \text{PoA}(\mathcal{G}).$$

Similarly, the *price of stability* of a modified k-Schelling game $\mathcal{G}$ with $\text{EQ}(\mathcal{G}) \neq \varnothing$ is the ratio of the maximum social welfare achieved by any possible assignment over the maximum social welfare achieved at equilibrium, that is,

$$\text{PoS}(\mathcal{G}) = \frac{\text{OPT}}{\max_{\mathbf{v} \in \text{EQ}(\mathcal{G})} \text{SW}(\mathbf{v})},$$

and the price of stability of a class $\mathcal{C}$ of modified k-Schelling games is

$$\text{PoS}(\mathcal{C}) = \sup_{\mathcal{G} \in \mathcal{C}: \text{EQ}(\mathcal{G}) \neq \varnothing} \text{PoS}(\mathcal{G}).$$

Besides general modified k-Schelling games, we will consider *balanced* games in which there are n/k agents of each type $\ell \in [k]$, as well as games in which the topology has a particular set of properties (for instance, it is a line or a tree).

3 One-Type Games

Interestingly, the modified Schelling model that we consider in this paper admits a game even when all agents are of the same type. This is in sharp contrast to the original model in which the utility of any agent who only has neighbors of the same type is always 1, implying that any connected assignment is an equilibrium when there is only one type of agents; see Sect. 1 for a more detailed discussion on the differences between the two utility models. In this section, we focus entirely on the case where there is one type of agents and study the equilibrium properties of the induced strategic games. We start by showing that there always exist equilibrium assignments in such games. The proof follows by defining a suitable potential function.

Theorem 1. *Modified* 1*-Schelling games always admit at least one equilibrium assignment, which can be computed in polynomial time.*

We continue by showing tight bounds on the price of anarchy of modified 1-Schelling games for two cases. The first is the most general one in which the topology can be any arbitrary graph, while the second is for when the topology is a tree.

Theorem 2. *The price of anarchy of modified* 1*-Schelling games on arbitrary graphs is exactly* $2 - \frac{2}{n}$.

Our next result shows that the price of anarchy slightly improves when the topology is more structured.

Theorem 3. *The price of anarchy of modified* 1*-Schelling games on trees and lines is exactly* $\frac{4}{3} - \frac{2}{3n}$.

We now turn our attention to the price of stability. By arguing about the structure of the optimal assignment, and by exploiting the properties of a variant of the best-response dynamics which gives priority to agents of minimum utility, we are able to show an upper bound on the price of stability. We remark that this is the first upper bound on the price of stability in the literature on Schelling games that holds for arbitrary graphs, albeit only when there is a single type of agents.

Theorem 4. *The price of stability of modified* 1*-Schelling games is at most* $3/2$.

Proof. Consider any modified 1-Schelling game, and let $\mathbf{v}^*$ be its optimal assignment. We first claim that if there exists an agent with utility $1/2$ in $\mathbf{v}^*$, then $\mathbf{v}^*$ must be an equilibrium, and thus the price of stability is 1. To see this, suppose otherwise that $\mathbf{v}^*$ is not an equilibrium and there exist agents with utility $1/2$. Since someone can benefit by jumping to an empty node v, it must be the case that there exists an agent i with utility $1/2$ who can increase her utility by jumping to v too. The utility of i will then increase by at least $2/3 - 1/2 = 1/6$, the utility of the agents in $N(v|\mathbf{v}^*)$ will increase by some strictly positive quantity (since the number of their neighbors increases by one), while the utility of i's single neighbor in $\mathbf{v}^*$, who has y neighbors in $\mathbf{v}^*$ (including i), will decrease by $\frac{y}{y+1} - \frac{y-1}{y} = \frac{1}{y(y+1)} \leq \frac{1}{6}$, where the inequality follows since the optimal assignment must form a connected graph, which implies that $y \geq 2$. Since $|x(v|\mathbf{v}^*)| \geq 1$, the jump of i to v leads to a new assignment with strictly larger social welfare than $\mathbf{v}^*$, which contradicts the optimality of $\mathbf{v}^*$. So, it suffices to consider the case where all agents have utility at least 2/3 in the optimal assignment.

We now claim that starting from $\mathbf{v}^*$ the best-response dynamics according to which the agent with the minimum utility jumps in each step, terminates at an equilibrium $\mathbf{v}$ in which there are at most two agents with utility 1/2, while all other agents have utility at least 2/3. This will imply that the maximum social welfare we can achieve at equilibrium is at least $\text{SW}(\mathbf{v}) \geq (n-2)\frac{2}{3} + 1$. Since the optimal social welfare is at most $n-1$, we will obtain an upper bound of $\frac{n-1}{(n-2)\frac{2}{3}+1} \leq \frac{3}{2}$ on the price of stability, as desired.

We use a recursive proof to show that starting with any assignment where the minimum utility among all agents is at least 2/3, we will either reach another assignment with minimum utility 2/3, or an equilibrium where at most two agents have utility 1/2. This is sufficient by the fact that the best response dynamics is guaranteed to terminate to an equilibrium; the proof of Theorem 1 actually uses a potential function to show that the dynamics always converges to an equilibrium.

Let m denote the minimum number of neighbors an agent has in the current assignment. Let a be an agent that has minimum utility $\frac{m}{m+1}$. If $m \geq 3$, then a's jump to an empty node will lead to a new assignment where every agent has at least 2 neighbors, as desired. If $m = 2$, then a's jump leads to at most two agents with utility exactly 1/2 in the new assignment. If this assignment is an equilibrium, then we are done. Otherwise, we distinguish between the following two cases:

Case (1): There are two agents i and j who have utility 1/2 and are connected to each other. According to the best-response dynamics we consider, one of these agents, say i, will jump to an empty node to increase her utility to 2/3. The jump of i will leave j with utility 0, who subsequently will jump to get utility at least 1/2. If j's best response yields her utility exactly 1/2, then there is no empty node adjacent to strictly more than one agents, which implies that the resulting assignment is an equilibrium, in which j is the only agent with utility 1/2. Otherwise, all agents have utility at least 2/3 in the new assignment.

Case (2): There is either only one agent i with utility 1/2, or there is also another agent j with utility 1/2 such that i and j are not neighbors. If i can increase her utility by jumping, then she will no longer have utility 1/2, but such a jump might leave her neighbor with exactly one neighbor (and utility 1/2). However, observe that no other agent can end up with utility 1/2 after i's jump, which means that the number of agents with utility 1/2 in the resulting assignment cannot increase. Again, we distinguish between Cases (1) and (2).

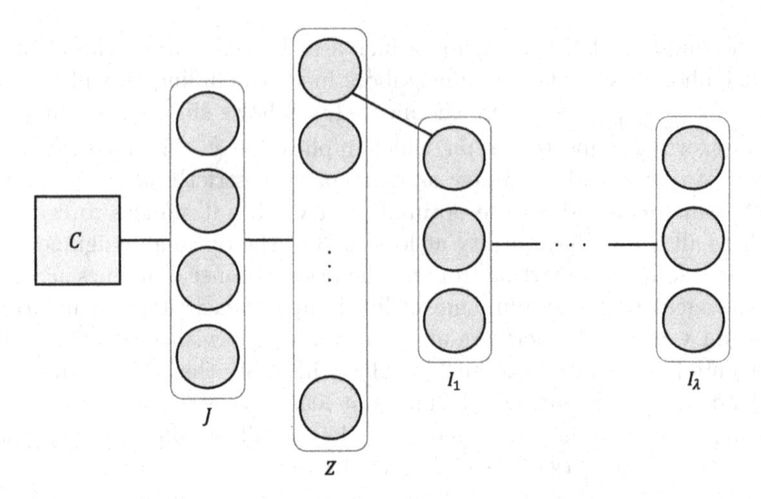

Fig. 1. The topology of the game used for the proof of the price of stability lower bound in Theorem 5. The edges connecting different components indicate that each node of one component is connected to each node of the other one.

Therefore, by starting with the optimal assignment, the process described above will terminate at an equilibrium with at most two agents with utility 1/2, and the bound follows. □

We also show a lower bound on the price of stability, which establishes that even the best equilibrium assignment (in terms of social welfare) is not always optimal.

Theorem 5. *The price of stability of modified* 1*-Schelling games is at least* $15/14 - \varepsilon$, *for any constant* $\varepsilon > 0$.

Proof. Consider a modified 1-Schelling game with $n = 3\lambda + 10$ agents, where λ is a positive integer whose value will be determined later. The topology consists of multiple components: a clique C with 6 nodes, and $\lambda + 2$ independent sets J, Z, $I_1, ..., I_\lambda$ such that $|J| = 4$, $|Z| = 3\lambda$ and $|I_\ell| = 3$ for every $\ell \in [\lambda]$; observe that there are $6\lambda+10$ nodes in total. These components are connected as follows: Every node of C is connected to every node of J; every node of J is connected to every node of Z; one node of Z is connected to one node of I_1; every node of I_ℓ is connected to every node of $I_{\ell+1}$ for $\ell \in [\lambda - 1]$. The topology is depicted in Fig. 1.

The optimal social welfare is at least as high as the social welfare of the assignment according to which the agents occupy all nodes except for those in Z. Since the agents in C have 9 neighbors each, the agents in J have 6, the agents in $I_1 \cup I_\lambda$ have 3, and the agents in $I_2 \cup ... \cup I_{\lambda-1}$ have 6 again, we obtain

$$\begin{aligned}\text{OPT} &\geq 6 \cdot \frac{9}{10} + 4 \cdot \frac{6}{7} + 6 \cdot \frac{3}{4} + 3(\lambda - 2) \cdot \frac{6}{7} \\ &= \frac{18}{7}\lambda + \frac{573}{70}.\end{aligned}$$

Now, consider the assignment $\mathbf{v}$ where the agents are placed at the nodes of $C \cup J \cup Z$. The agents in C have 9 neighbors each, the agents in J have $3\lambda + 6$, and the agents in Z have 4. Since every agent has utility at least 4/5 and would obtain utility at most 1/2 by jumping to any of the empty nodes, $\mathbf{v}$ is an equilibrium. Its social welfare is

$$\begin{aligned}\text{SW}(\mathbf{v}) &= 6 \cdot \frac{9}{10} + 4 \cdot \frac{3\lambda + 6}{3\lambda + 7} + 3\lambda \cdot \frac{4}{5} \\ &= \frac{12}{5}\lambda + \frac{3(47\lambda + 103)}{5(3\lambda + 7)}.\end{aligned}$$

We will now show that $\mathbf{v}$ is the unique equilibrium of this game. Assume otherwise that there exists an equilibrium where at least one agent is at a node in I_ℓ for some $\ell \in [\lambda]$. Let i be an agent occupying a node of I_{ℓ^*}, where ℓ^* is the largest index among all $\ell \in [\lambda]$ such that I_ℓ contains at least one occupied node. Then, the utility of agent i is at most 3/4 (realized in case I_{ℓ^*-1} is fully occupied). Since agent i has no incentive to jump to a node in $C \cup J \cup Z$, it must be the case that either there is no empty node therein, or each of these sets contains at most three occupied nodes. The first case is impossible since $|C \cup J \cup Z| = 3\lambda + 10 = n$ and we have assumed that agent i occupies a node outside this set. Similarly, the second case is impossible since it implies that $C \cup J \cup Z$ should contain at most 9 occupied nodes, but the remaining $n - 9 = 3\lambda + 1$ agents do not fit in the 3λ nodes outside of this set. Therefore, the only possible equilibrium assignments are such that there is no agent outside $C \cup J \cup Z$, which means that $\mathbf{v}$ is the unique equilibrium.

By the above discussion, we have that the price of stability is

$$\frac{\text{OPT}}{\text{SW}(\mathbf{v})} \geq \frac{\frac{18}{7}\lambda + \frac{573}{70}}{\frac{12}{5}\lambda + \frac{3(47\lambda+103)}{5(3\lambda+7)}},$$

which tends to 15/14 as λ becomes arbitrarily large. □

We conclude this section with a result regarding the complexity of computing an assignment with maximum social welfare. Inspired by a corresponding result in [16], we show that, even in the seemingly simple case of modified 1-Schelling games, maximizing the social welfare is NP-hard.

Theorem 6. *Consider a modified* 1*-Schelling game and let* ξ *be a rational number. Then, deciding whether there exists an assignment with social welfare at least* ξ *is NP-complete.*

4 Multi-type Games

In this section, we consider the case of strictly more than one type of agents. We will show bounds on the price of anarchy and the price of stability, both for general games as well as for interesting restrictions on the number of agents per type or on the structure of the topology.

4.1 Arbitrary Topologies

We start by showing tight bounds on the price of anarchy for games on arbitrary graphs when there are at least two agents per type. When there is only one agent per type, any assignment is an equilibrium, and thus the price of anarchy is 1. When there exists a type with at least two agents and one type with a single agent, the price of anarchy can be unbounded: Consider a star topology and an equilibrium assignment according to which the center node is occupied by this lonely agent; then, all agents have utility 0. In contrast, the assignment according to which an agent with at least one friend occupies the center node guarantees positive social welfare.

Theorem 7. *The price of anarchy of modified k-Schelling games with at least two agents per type is exactly* $\frac{2n(n-k)}{n+2}$.

Proof. For the upper bound, consider an arbitrary modified k-Schelling game in which there are $n_\ell \geq 2$ agents of type $\ell \in [k]$. Clearly, the maximum utility that an agent of type ℓ can get is $\frac{n_\ell - 1}{n_\ell}$ when she is connected to all other agents of her type, and only them. Consequently, the optimal social welfare is

$$\text{OPT} \leq \sum_{\ell \in [k]} n_\ell \frac{n_\ell - 1}{n_\ell} = n - k. \tag{1}$$

Now, let $\mathbf{v}$ be an equilibrium assignment, according to which there exists an empty node v which is adjacent to $x_\ell = x_\ell(v)$ agents of type $\ell \in [k]$, such that $x_\ell \geq 1$ for at least one type ℓ; let $x = x(v) = \sum_{\ell \in [k]} x_\ell$. We will now count the contribution of each type ℓ to $\text{SW}(\mathbf{v})$.

- $\underline{n_\ell \geq 3}$. In order to not have incentive to jump to v, every agent of type ℓ must have utility at least $\frac{x_\ell}{x+1}$ if she is not adjacent to v, or $\frac{x_\ell - 1}{x} \geq \frac{x_\ell - 1}{x+1}$ otherwise. Hence, the contribution of all agents of type ℓ to the social welfare is at least

$$(n_\ell - x_\ell)\frac{x_\ell}{x+1} + x_\ell \frac{x_\ell - 1}{x+1} = \frac{(n_\ell - 1)x_\ell}{x+1} \geq \frac{2x_\ell}{x+1}.$$

- $\underline{n_\ell = 2}$. Let i and j be the two agents of type ℓ. If $x_\ell = 2$, then both i and j have utility at least $\frac{1}{x}$ so that they do not have incentive to jump to v. Otherwise, if $x_\ell = 1$ and i is adjacent to v, then i and j must be neighbors, since otherwise they would both have utility 0, and j would want to jump

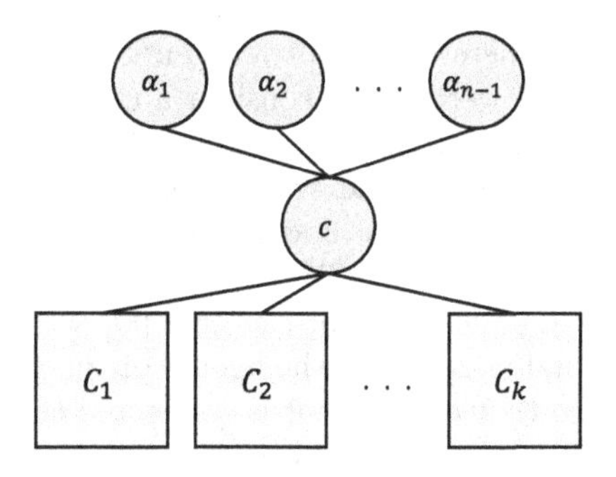

Fig. 2. The topology of the game used for the proof of the lower bound in Theorem 7. The big squares $C_1, ..., C_k$ correspond to cliques such that c is connected only to a single node of each C_ℓ.

to v to increase her utility to positive. Hence, i has utility at least $\frac{1}{n}$ and j has utility at least $\frac{1}{x+1}$. Overall, from both cases, the contribution of the two agents of type ℓ is at least

$$x_\ell \left(\frac{1}{x+1} + \frac{1}{n} \right).$$

Let $\Lambda = \{\ell \in [k] : n_\ell = 2\}$ be the set of all types with exactly two agents. By the above discussion, the social welfare at equilibrium is

$$\begin{aligned}
\mathrm{SW}(\mathbf{v}) &\geq \sum_{\ell \in [k] \setminus \Lambda} \frac{2x_\ell}{x+1} + \sum_{\ell \in \Lambda} x_\ell \left(\frac{1}{x+1} + \frac{1}{n} \right) \\
&= \sum_{\ell \in [k]} \frac{x_\ell}{x+1} + \sum_{\ell \in [k] \setminus \Lambda} \frac{x_\ell}{x+1} + \sum_{\ell \in \Lambda} \frac{x_\ell}{n} \\
&= \frac{x}{x+1} + \sum_{\ell \in [k] \setminus \Lambda} \frac{x_\ell}{x+1} + \sum_{\ell \in \Lambda} \frac{x_\ell}{n}.
\end{aligned}$$

If $\Lambda = \varnothing$, since $x \geq 1$, we obtain

$$\mathrm{SW}(\mathbf{v}) \geq \frac{x}{x+1} + \sum_{\ell \in [k]} \frac{x_\ell}{x+1} = \frac{2x}{x+1} \geq 1.$$

Otherwise, we have

$$\mathrm{SW}(\mathbf{v}) \geq \frac{x}{x+1} + \frac{1}{n} \geq \frac{1}{2} + \frac{1}{n} = \frac{n+2}{2n}.$$

Since $n \geq 2$, it is $\frac{n+2}{2n} \leq 1$, and thus $\mathrm{SW}(\mathbf{v}) \geq \frac{n+2}{2n}$ in any case. By (1), the price of anarchy is at most $\frac{2n(n-k)}{n+2}$.

Observe that the proof of the upper bound implies that the worst case occurs when at equilibrium there exists an empty node that is adjacent to a single agent of some type ℓ such that there are only two agents of type ℓ. Using this as our guide for the proof of the lower bound, consider a modified Schelling game with n agents who are partitioned into k types such that there are $n_1 = 2$ agents of type 1 and $n_\ell \geq 2$ agents of type $\ell \in [k]$. The topology consists of a star with a center node c and $n-1$ leaf nodes $\{\alpha_1, ..., \alpha_{n-1}\}$, as well as k cliques $\{C_1, ..., C_k\}$ such that C_ℓ has size n_ℓ. These subgraphs are connected as follows: c is connected to a single node of C_ℓ for each $\ell \in [k]$; see Fig. 2.

Clearly, in the optimal assignment the agents of type $\ell \in [k]$ are assigned to the nodes of clique C_ℓ so that every agent is connected to all other agents of her type, and only them. Consequently, the optimal social welfare is exactly

$$\sum_{\ell \in [k]} n_\ell \frac{n_\ell - 1}{n_\ell} = n - k.$$

On the other hand however, there exists an equilibrium assignment where c is occupied by one of the agents of type 1 and all other agents occupy the leaf nodes $\alpha_1, ..., \alpha_{n-1}$. Then, only the two agents of type 1 have positive utility, in particular, $1/n$ and $1/2$, respectively. Hence, the price of anarchy is at least

$$\frac{n-k}{\frac{1}{2} + \frac{1}{n}} = \frac{2n(n-k)}{n+2}.$$

This completes the proof. □

From the above theorem it can be easily seen that the price of anarchy can be quite large in general. This motivates the question of whether improvements can be achieved for natural restrictions. One such restriction is to consider balanced games in which the n agents are evenly distributed to the k types, so that there are exactly n/k agents per type. In the following we will focus exclusively on balanced games.

Theorem 8. *The price of anarchy of balanced modified k-Schelling games with at least two agents per type is exactly $2k$.*

We continue by presenting a lower bound on the price of stability for modified 2-Schelling games, which holds even for the balanced case.

Theorem 9. *The price of stability of modified 2-Schelling games is at least $4/3 - \varepsilon$, for any constant $\varepsilon > 0$.*

4.2 Restricted Topologies

We now turn our attention to balanced modified Schelling games on restricted topologies. We start with the case of lines, and show the following statement.

Theorem 10. *The price of anarchy of balanced modified k-Schelling games on a line is exactly 2 for $k = 2$, and exactly $k + 1/2$ for $k \geq 3$.*

Proof. We will only present the proof of the upper bound for $k = 2$. Let there be n agents, with half of them *red* and half of them *blue*. Since the topology is a line, in the optimal assignment the agents of same type are assigned right next to each other and the two types are well-separated by an empty node (which exists). Consequently, for each type, there are two agents with utility 1/2 and $n/2 - 2$ agents with utility 2/3, and thus

$$\text{OPT} = 2 \cdot \left(2 \cdot \frac{1}{2} + \left(\frac{n}{2} - 2\right)\frac{2}{3}\right) = \frac{2(n-1)}{3}. \tag{2}$$

Now, let $\mathbf{v}$ be an equilibrium assignment, and consider an empty node v which, without loss of generality that, is adjacent to a red agent i. We distinguish between three cases:

v is adjacent to another red agent j. Then, $\mathbf{v}$ cannot be an equilibrium. If i and j are the only red agents, they get utility 0 and want to jump to v to get 1/2. Otherwise, there exists a third red agent that gets utility at most 1/2 (by occupying at best the end of a red path) who wants to jump to v to get 2/3.

v is also adjacent to a blue agent j. Since v is connected to a red and a blue agent, every agent must have utility at least 1/3 in order to not want to jump to v. However, observe that v defines two paths that extend towards its left and its right. The two agents occupying the nodes at the end of these paths must be connected to friends and have utility 1/2; otherwise they would have utility 0 and would prefer to jump to v. Therefore, we have two agents with utility exactly 1/2 and $n - 4$ agents with utility at least 1/3; we do not really know anything about the utility of i and j. Putting all these together, we obtain

$$\text{SW}(\mathbf{v}) \geq 2 \cdot \frac{1}{2} + (n-4)\frac{1}{3} = \frac{n-1}{3},$$

and the price of anarchy is at most 2.

v is a leaf or is adjacent to an empty node. Any of the remaining $n/2 - 1$ red agents must have utility at least 1/2 in order to not have incentive to jump to v. So, all red agents are connected only to red agents, which further means that i is also connected to another red agent (otherwise she would be isolated, have utility 0 and incentive to jump), and all blue agents are only connected to other blue agents. Therefore, everyone has utility at least 1/2, yielding price of anarchy at most 4/3.

Hence, the price of anarchy of balanced modified 2-Schelling games is at most 2. □

It should be straightforward to observe that the price of stability of modified k-Schelling games on a line is 1. Indeed, the optimal assignment that allocates agents of the same type next to each other and separates different types with an empty node (if possible) is an equilibrium.

As we showed in Sect. 3, for $k = 1$, the price of anarchy of games on arbitrary trees is the same as the price of anarchy of games on lines. However, this is no longer true when we consider games with $k \geq 2$ types.

Theorem 11. *The price of anarchy of balanced modified k-Schelling games on a tree is exactly* $\frac{14}{9}k$ *for* $k \in \{2, 3\}$, *and exactly* $\frac{2k^2}{k+1}$ *for* $k \geq 4$.

The proof of the theorem follows by distinguishing between cases, depending on the number of agents per type. In particular, we show that the worst case occurs when there are four agents per type for $k \in \{2, 3\}$, and when there are two agents per type for $k \geq 4$.

5 Conclusion and Possible Extensions

We introduced the class of modified Schelling games and studied questions about the existence and efficiency of equilibria. Although we made significant progress in these two fronts, our work leaves many interesting open problems.

In terms of our results, the most interesting and challenging open question is whether equilibria always exist for $k \geq 2$. We remark that to show such a positive result one would have to resort to techniques different than defining a potential function; we can show that it is not possible to define a potential function, even when there are only two types of agents and the topology is a tree. Not being able to argue about the convergence to an equilibrium for $k \geq 2$ further serves as a bottleneck towards proving upper bounds on the price of stability, which we strongly believe that is one of the most challenging questions in Schelling games (not only modified ones). Furthermore, one could also consider bounding the price of anarchy for more special cases such as games on regular or bipartite graphs.

Going beyond our setting, there are many interesting extensions of modified Schelling games that one could consider. For example, when $k \geq 3$, following the work of Echzell *et al.* [14], we could define the utility function of agent i such that the denominator of the ratio only counts the friends of i, the agents of the type with maximum cardinality among all types with agents in i's neighborhood, and herself. Alternatively, following the work of Elkind *et al.* [16], one could focus on social modified Schelling games in which the friendships among the agents are given by a social network.

References

1. Agarwal, A., Elkind, E., Gan, J., Voudouris, A.A.: Swap stability in Schelling games on graphs. In: Proceedings of the 34th AAAI Conference on Artificial Intelligence (AAAI) (2020)
2. Anshelevich, E., Dasgupta, A., Kleinberg, J.M., Tardos, É., Wexler, T., Roughgarden, T.: The price of stability for network design with fair cost allocation. SIAM J. Comput. **38**(4), 1602–1623 (2008)
3. Aziz, H., Brandl, F., Brandt, F., Harrenstein, P., Olsen, M., Peters, D.: Fractional hedonic games. ACM Trans. Econ. Comput. **7**(2), 6:1–6:29 (2019)
4. Barmpalias, G., Elwes, R., Lewis-Pye, A.: Digital morphogenesis via Schelling segregation. In: Proceedings of the 55th IEEE Annual Symposium on Foundations of Computer Science (FOCS), pp. 156–165 (2014)

5. Bhakta, P., Miracle, S., Randall, D.: Clustering and mixing times for segregation models on $\mathbb{Z}^2$. In: Proceedings of the 25th Annual ACM-SIAM Symposium on Discrete Algorithms (SODA), pp. 327–340 (2014)
6. Bilò, D., Bilò, V., Lenzner, P., Molitor, L.: Topological influence and locality in swap Schelling games. In: Proceedings of the 45th International Symposium on Mathematical Foundations of Computer Science (MFCS) (2020)
7. Bilò, V., Fanelli, A., Flammini, M., Monaco, G., Moscardelli, L.: Nash stable outcomes in fractional hedonic games: existence, efficiency and computation. J. Artif. Intell. Res. **62**, 315–371 (2018)
8. Bogomolnaia, A., Jackson, M.O.: The stability of hedonic coalition structures. Games Econ. Behav. **38**(2), 201–230 (2002)
9. Brandt, C., Immorlica, N., Kamath, G., Kleinberg, R.: An analysis of one-dimensional Schelling segregation. In: Proceedings of the 44th Symposium on Theory of Computing Conference (STOC), pp. 789–804 (2012)
10. Chan, H., Irfan, M.T., Than, C.V.: Schelling models with localized social influence: a game-theoretic framework. In: Proceedings of the 19th International Conference on Autonomous Agents and Multiagent Systems (AAMAS), pp. 240–248 (2020)
11. Chauhan, A., Lenzner, P., Molitor, L.: Schelling segregation with strategic agents. In: Deng, X. (ed.) SAGT 2018. LNCS, vol. 11059, pp. 137–149. Springer, Cham (2018). https://doi.org/10.1007/978-3-319-99660-8_13
12. Clark, W., Fossett, M.: Understanding the social context of the Schelling segregation model. Proc. Nat. Acad. Sci. **105**(11), 4109–4114 (2008)
13. Drèze, J.H., Greenberg, J.: Hedonic coalitions: optimality and stability. Econometrica **48**(4), 987–1003 (1980)
14. Echzell, H., et al.: Convergence and hardness of strategic Schelling segregation. In: Caragiannis, I., Mirrokni, V., Nikolova, E. (eds.) WINE 2019. LNCS, vol. 11920, pp. 156–170. Springer, Cham (2019). https://doi.org/10.1007/978-3-030-35389-6_12
15. Elkind, E., Fanelli, A., Flammini, M.: Price of Pareto optimality in hedonic games. In: Proceedings of the 30th AAAI Conference on Artificial Intelligence (AAAI), pp. 475–481 (2016)
16. Elkind, E., Gan, J., Igarashi, A., Suksompong, W., Voudouris, A.A.: Schelling games on graphs. In: Proceedings of the 28th International Joint Conference on Artificial Intelligence (IJCAI), pp. 266–272 (2019)
17. Immorlica, N., Kleinberg, R., Lucier, B., Zadimoghaddam, M.: Exponential segregation in a two-dimensional Schelling model with tolerant individuals. In: Proceedings of the 28th Annual ACM-SIAM Symposium on Discrete Algorithms (SODA), pp. 984–993 (2017)
18. Koutsoupias, E., Papadimitriou, C.H.: Worst-case equilibria. In: Proceedings of the 16th Annual Symposium on Theoretical Aspects of Computer Science (STACS), pp. 404–413 (1999)
19. Massand, S., Simon, S.: Graphical one-sided markets. In: Proceedings of the 28th International Joint Conference on Artificial Intelligence (IJCAI), pp. 492–498 (2019)
20. Monaco, G., Moscardelli, L., Velaj, Y.: Stable outcomes in modified fractional hedonic games. Auton. Agent. Multi-Agent Syst. **34**(1), 1–29 (2019). https://doi.org/10.1007/s10458-019-09431-z
21. Olsen, M.: On defining and computing communities. In: Proceedings of the Conferences in Research and Practice in Information Technology, pp. 97–102 (2012)
22. Pancs, R., Vriend, N.J.: Schelling's spatial proximity model of segregation revisited. J. Public Econ. **91**(1–2), 1–24 (2007)

23. Schelling, T.C.: Models of segregation. Am. Econ. Rev. **59**(2), 488–493 (1969)
24. Schelling, T.C.: Dynamic models of segregation. J. Math. Sociol. **1**(2), 143–186 (1971)
25. Vinković, D., Kirman, A.: A physical analogue of the Schelling model. Proc. Nat. Acad. Sci. **103**(51), 19261–19265 (2006)
26. Zhang, J.: Residential segregation in an all-integrationist world. J. Econ. Behav. Organ. **54**(4), 533–550 (2004)

Race Scheduling Games

Shaul Rosner and Tami Tamir(✉)

School of Computer Science, The Interdisciplinary Center, Herzliya, Israel
shaul.rosner@post.idc.ac.il, tami@idc.ac.il

Abstract. Job scheduling on parallel machines is a well-studied singleton congestion game. We consider a variant of this game in which the jobs are partitioned into *competition sets*, and the goal of every player is to minimize the completion time of his job *relative to his competitors.* Specifically, the primary goal of a player is to minimize the *rank* of its completion time among his competitors, while minimizing the completion time itself is a secondary objective. This fits environments with strong competition among the participants, in which the relative performance of the players determine their welfare.

We define and study the corresponding *race scheduling game* (RSG). We show that RSGs are significantly different from classical job-scheduling games, and that competition may lead to a poor outcome. In particular, an RSG need not have a pure Nash equilibrium, and best-response dynamics may not converge to a NE even if one exists. We identify several natural classes of games, on identical and on related machines, for which a NE exists and can be computed efficiently, and we present tight bounds on the equilibrium inefficiencies. For some classes we prove convergence of BRD, while for others, even with very limited competition, BRD may loop. Among classes for which a NE is not guaranteed to exist, we distinguish between classes for which, it is tractable or NP-hard to decide if a given instance has a NE.

Striving for stability, we also study the Nashification cost of RSGs, either by adding dummy jobs, or by compensating jobs for having high rank. Our analysis provides insights and initial results for several other congestion and cost-sharing games that have a natural 'race' variant.

1 Introduction

Two men are walking through a forest. Suddenly they see a bear in the distance, running towards them. They start running away. But then one of them stops, takes some running shoes from his bag, and starts putting them on. "What are you doing?" says the other man. "Do you think you will run faster than the bear with those?" "I don't have to run faster than the bear," he says. "I just have to run faster than you."

In job-scheduling applications, jobs are assigned on machines to be processed. Many interesting combinatorial optimization problems arise in this setting, which is a major discipline in operations research. A centralized scheduler should assign the jobs in a way that achieves load balancing, an effective use of the system's

T. Harks and M. Klimm (Eds.): SAGT 2020, LNCS 12283, pp. 257–272, 2020.
https://doi.org/10.1007/978-3-030-57980-7_17

resources, or a target quality of service [20]. Many modern systems provide service to multiple strategic users, whose individual payoff is affected by the decisions made by others. As a result, non-cooperative game theory has become an essential tool in the analysis of job-scheduling applications (see e.g., [3,10,12, 19], and a survey in [24]). Job-scheduling is a weighted congestion game [21] with *singleton* strategies, that is, every player selects a single resource (machine).

In traditional analysis of congestion games, the goal of a player is to minimize his cost. We propose a new model denoted *race games* that fits environments with strong competition among the participants. Formally, the players form *competition sets*, and a player's main goal is to do well relative to his competitors. The welfare of a player is not measured by a predefined cost or utility function, but relative to the performance of his competitors. This natural objective arises in many real-life scenarios. For example, in cryptocurrency mining, one needs to be the first miner to build a block. It does not matter how fast a miner builds a block, as long as she is the first to do so. Similarly, when buying event tickets from online vendors, the time spent in the queue is far less important than what tickets are available when it is your turn to buy. Participants' ranking is crucial in numerous additional fields, including auctions with a limited number of winners, where, again, the participants' rank is more important than their actual offer, transplant queues, sport leagues, and even submission of papers to competitive conferences.

In this paper we study the corresponding *race scheduling game* (RSG, for short). We assume that the jobs are partitioned into *competition sets*. The primary goal of a job is to minimize the *rank* of its completion time among its competitors, while minimizing the completion time itself is a secondary objective. As an example, consider a running competition. In order to be qualified for the final, a runner should be faster than other participants in her heat. The runners' ranking is more important than their finish time.

Unfortunately, as we show, even very simple RSGs may not have a NE. We therefore focus on *potentially* more stable instances. In many real-life scenarios, competition is present among agents with similar properties. For example, there is a competition among companies that offer similar services; in sport competitions, the participants are categorized by their sex and age group, or by their weight. Some of our results consider games in which competing players are *homogeneous*. Specifically, we assume that all the jobs in a competition set have the same length.

Our results highlight the differences between RSGs and classical job-scheduling games. We identify classes of instances for which a stable solution exists and can be computed efficiently, we analyze the equilibrium inefficiency, and the convergence of best-response dynamics. We distinguish between different competition structure, and between environments of identical and related machines. In light of our negative results regarding stability existence, we also study the problem of *Nashification*. The goal of Nashification is, given an instance of RSGs, to turn it into an instance that has a stable solution. This is done either by adding dummy jobs, or by compensating jobs for having high rank. We believe that this 'race' model fits many natural scenarios, and should be analyzed for additional congestion and cost-sharing games.

2 Model and Preliminaries

A *race scheduling game* (RSG) is given by $G = \langle \mathcal{J}, \mathcal{M}, \{p(j)\} \; \forall j \in \mathcal{J}, \{d_i\} \; \forall i \in \mathcal{M}, S \rangle$, where $\mathcal{J}$ is a set of n *jobs*, $\mathcal{M}$ is a set of m *machines*, $p(j)$ is the *length* of job j, d_i is the *delay* of machine i, and S is a partition of the jobs into *competition sets*. Specifically, $S = \{S_1, \ldots, S_c\}$ such that $c \leq n$, $\cup_{\ell=1}^{c} S_\ell = \mathcal{J}$, and for all $\ell_1 \neq \ell_2$, we have $S_{\ell_1} \cap S_{\ell_2} = \emptyset$. For every job $j \in S_\ell$, the other jobs in S_ℓ are denoted the *competitors* of j. Let n_ℓ denote the number of jobs in S_ℓ.

Job j is controlled by Player j whose strategy space is the set of machines $\mathcal{M}$. A profile of an RSG is a schedule $s = \langle s_1, \ldots, s_n \rangle \in \mathcal{M}^n$ describing the machines selected by the players[1]. For a machine $i \in \mathcal{M}$, the *load* on i in s, denoted $L_i(s)$, is the total length of the jobs assigned on machine i in s, that is, $L_i(s) = \sum_{\{j|s_j=i\}} p(j)$. When s is clear from the context, we omit it. It takes $p(j) \cdot d_i$ time-units to process job j on machine i. As common in the study of job-scheduling games, we assume that all the jobs assigned on the same machine are processed in parallel and have the same completion time. Formally, the completion time of job j in the profile s is $C_j = L_{s_j}(s) \cdot d_{s_j}$. Machines are called identical if their delays are equal.

Unlike classical job-scheduling games, in which the goal of a player is to minimize its completion time, in race games, the goal of a player is to do well relative to its competitors. That is, every profile induces a ranking of the players according to their completion time, and the goal of each player is to have a lowest possible rank in its competition set. Formally, for a profile s, let $C^s_{S_\ell} = \langle C^s_{\ell_1}, \ldots, C^s_{\ell_{n_\ell}} \rangle$ be a sorted vector of the completion times of the players in S_ℓ. That is, $C^s_{\ell_1} \leq \ldots \leq C^s_{\ell_{n_\ell}}$, where $C^s_{\ell_1}$ is the minimal completion time of a player from S_ℓ in s, etc. The *rank* of Player $j \in S_\ell$ in profile s, denoted $rank_j(s)$ is the rank of its completion time in $C^s_{S_\ell}$. If several players in a competition set have the same completion time, then they all have the same rank, which is the corresponding median value. For example, if $n_\ell = 4$ and $C^s_{S_\ell} = \langle 7, 8, 8, 13 \rangle$ then the players' ranks are $\langle 1, 2.5, 2.5, 4 \rangle$, and if all players in S_ℓ have the same completion time then they all have rank $(n_\ell + 1)/2$. Note that, independent of the profile, $\sum_{j \in S_\ell} rank_j(s) = n_\ell(n_\ell + 1)/2$.

For a profile s and a job $j \in S_\ell$, let $N_{low}(j, s)$ be the number of jobs from S_ℓ whose completion time is lower than $C_j(s)$, and let $N_{eq}(j, s)$ be the number of jobs from S_ℓ whose completion time is $C_j(s)$. We have,

Observation 1. $rank_j(s) = N_{low}(j, s) + \frac{1+N_{eq}(j,s)}{2}$.

The primary objective of every player is to minimize its rank. The secondary objective is to minimize its completion time. Formally, Player j prefers profile s'

[1] In this paper, we only consider *pure* strategies.

over profile s if $rank_j(s') < rank_j(s)$ or $rank_j(s') = rank_j(s)$ and $C_j(s') < C_j(s)$. Note that classic job-scheduling games are a special case of RSGs in which the competition sets are singletons; thus, for every job j, in every profile, s, we have $rank_j(s) = 1$, and the secondary objective, of minimizing the completion time is the only objective.

A machine i is a *best response* (BR) for Player j if, given the strategies of all other players, j's rank is minimized if it is assigned on machine i. Best-Response Dynamics (BRD) is a local-search method where in each step some player is chosen and plays its best improving deviation (if one exists), given the strategies of the other players.

The focus in game theory is on the *stable* outcomes of a given setting. The most prominent stability concept is that of a Nash equilibrium (NE): a profile such that no player can improve its objective by unilaterally deviating from its current strategy, assuming that the strategies of the other players do not change. Formally, a profile s is a NE if, for every $j \in \mathcal{J}$, s_j is a BR for Player j.

Some of our results consider RSGs with *homogeneous* competition sets. We denote by $\mathcal{G}_h$ the corresponding class of games. Formally, $G \in \mathcal{G}_h$ if, for every ℓ, all the jobs in S_ℓ have the same length, p_ℓ. The following example summarizes the model and demonstrates several of the challenges in analyzing RSGs.

Example: Consider a game $G \in \mathcal{G}_h$ on $m = 3$ identical machines, played by $n = 9$ jobs in two homogeneous competition sets. S_1 consists of four jobs having length 4, and S_2 consists of five jobs having length 3 (to be denoted 4-jobs and 3-jobs, respectively). All the machines have the same unit-delay. Figure 1 presents four profiles of this game. The completion times are given above the machines and the jobs are labeled by their ranks. Consider the jobs of S_2 in Profile (a). Their completion times are $C^{(a)}_{S_2} = (7, 12, 12, 12, 12)$. Thus, the 3-job on M_2 has rank 1, and the four jobs on M_3 all have rank $\frac{2+3+4+5}{4} = 3.5$. Profile (a) is a NE. For example, a deviation of a 4-job from M_1 to M_2 leads to Profile (b), and thus involves an increase in the rank of the deviating jobs from 3 to 3.5. It can be verified that other deviations are not beneficial either. This example demonstrates that race games are significantly different from classical job-scheduling games. In particular, a beneficial migration may increase the completion time of a job. For example, the migration of a 3-job that leads from Profile (c) to Profile (a) increases the completion time of the deviating job from 10 to 12, but reduces its rank from 4.5 to 3.5. Moreover, simple algorithms that are known to produce a NE schedule for job-scheduling games without competition need not produce a NE in race games. In our example, Profile (d) is produced by the *Longest Processing Time* (LPT) rule. It is not a NE since the 3-job on M_1 can reduce its rank from 5 to 4 by migration to either M_2 or M_3.

The *social cost* of a profile s, denoted $cost(s)$ is the *makespan* of the corresponding schedule. That is, the maximal completion time of a job, given by $max_i L_i(s) \cdot d_i$. A *social optimum* of a game G is a profile that attains the lowest possible social cost. We denote by $OPT(G)$ the cost of a social optimum profile; i.e., $OPT(G) = \min_s max_i L_i(s) \cdot d_i$.

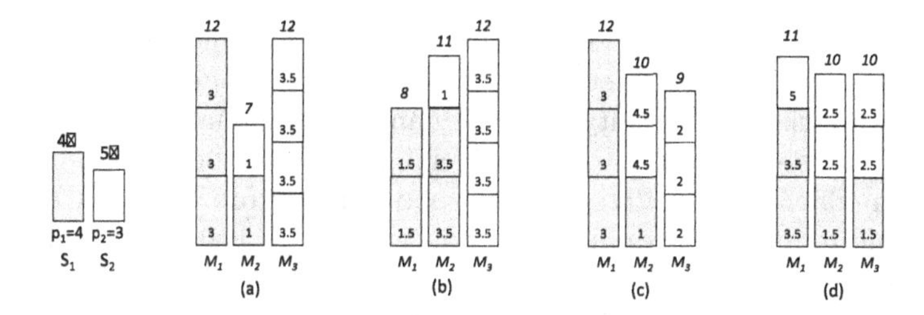

Fig. 1. Jobs are labeled by their ranks. (*a*) A NE profile. (*b*) and (*c*) Deviations from the NE are harmful. (*d*) An LPT schedule.

It is well known that decentralized decision-making may lead to sub-optimal solutions from the point of view of the society as a whole. We quantify the inefficiency incurred due to self-interested behavior according to the *price of anarchy* (PoA) [19] and *price of stability* (PoS) [2,23] measures. The PoA is the worst-case inefficiency of a pure Nash equilibrium, while the PoS measures the best-case inefficiency of a pure Nash equilibrium.

2.1 Related Work

There is wide literature on job scheduling on parallel machines. The minimum makespan problem corresponds to the centralized version of our game in which all jobs obey the decisions of one entity. This is a well-studied NP-complete problem. For identical machines, the simple greedy List-scheduling (LS) algorithm [15] provides a $(2-\frac{1}{m})$-approximation to the minimum makespan problem. A slightly better approximation-ratio of $(\frac{4}{3}-\frac{1}{3m})$ is guaranteed by the Longest Processing Time (LPT) algorithm [16], and A PTAS is given in [17]. For related machines, with various speeds, LS algorithm provides a $\theta(m)$-approximation [9], and a PTAS is given in [18].

Congestion games [21] consist of a set of resources and a set of players who need to use these resources. Players' strategies are subsets of resources. Each resource has a latency function which, given the load generated by the players on the resource, returns the cost of the resource. In *singleton* congestion games players' strategies are single resources. In *weighted* congestion games, each player j has a *weight* $p(j)$, and its contribution to the load of the resources he uses as well as its cost are multiplied by $p(j)$ [7].

The special case of symmetric weighted singleton congestion games corresponds to the setting of job-scheduling: the resources are machines and the players are jobs that need to be processed by the machines. A survey of results of job-scheduling games appears in [24]. For identical machines, it is known that LPT-schedules are NE schedules [14], and that the price of anarchy, which corresponds to the makespan approximation, is $2-\frac{2}{m+1}$ [13]. For uniformly related machines, the price of anarchy is bounded by $\frac{\log m}{\log \log \log m}$ [10]. For two machines, a bound of $\frac{1+\sqrt{5}}{2}$ is given in [19].

Other related work studies additional models in which players' objective involves *social preferences.* In standard game theoretic models, players' objective is to maximize their own utility, while in games with social preferences, players have preferences over vectors of all players' utilities. For example, [25] studies a model in which the *mental state* of a player is a score based on all players' utilities, and in a mental equilibrium, players can not deviate and improve this score. The main difference from our setting is that in their model, optimizing one's utilization has the highest priority, thus, every NE is also a mental equilibrium, which is not the case in race games. Other models that capture preferences based on emotions such as empathy, envy, or inequality aversion are presented and studied in [4,6,11]. A lot of attention has been given to such models in behavioral game theory. We are not aware of previous work that analyzes competition in the framework of congestion games. Other social effect, such as altruism and spite were studied, e.g., in [1,5,8].

2.2 Our Results

We show that competition dramatically impacts job-scheduling environments that are controlled by selfish users. RSGs are significantly different from classical job-scheduling games; their analysis is unintuitive, and known tools and techniques fail even on simple instances. We start by analyzing RSGs on identical machines. We show that an RSG need not have a NE, and deciding whether a game instance has a NE is a NP-complete problem. This is valid even for instances with only two pairs of competing jobs and two machines, and for instances with homogeneous competition sets. Moreover, even in cases where a NE exists, BRD may not converge. On the other hand, we identify several non-trivial classes of instances for which a NE exists and can be calculated efficiently. Each of these positive results is tight in a sense that a slight relaxation of the class characterization results in a game that may not have a NE. Specifically, we present an algorithm for calculating a NE for games with unit-length jobs, for games in $\mathcal{G}_h$ with a limited number of competition sets and machines, or with limited competition-set size, and games in $\mathcal{G}_h$ in which the job lengths form a divisible sequence (e.g., powers of 2).

We then provide tight bounds on the equilibrium inefficiency with respect to the minimum makespan objective. For classical job-scheduling, it is known that $PoS = 1$ and $PoA = 2 - \frac{2}{m+1}$ [13]. We show that for RSGs on identical machines, $PoS = PoA = 3 - \frac{6}{m+2}$. This result demonstrates the 'price of competition'. The fact that $PoS > 1$ implies that even if the system has full control on the initial job assignment, the best stable outcome may not be optimal. Moreover, since $PoA = PoS$, in the presence of competition, having control on the initial job assignment may not be an advantage at all.

For related machines, we start with a negative result showing that even the seemingly trivial case of unit-length jobs is tricky, and a NE may not exist, even if all jobs are in a single competition set. For this class of games, however, it is possible to decide whether a game has a NE, and to calculate one if it exists.

Without competition, for unit-jobs and related machines, a simple greedy algorithm produces an optimal schedule. Moreover, $PoA = PoS = 1$. We show that for RSGs with unit jobs and related machines, $PoS = PoA = 2$. We then move to study games on related machines and arbitrary-length jobs. Striving for positive results, we focus on two machines and homogeneous instances. We present an algorithm for calculating a NE, and prove that any application of BRD converges to a NE. We then bound the equilibrium inefficiency for arbitrary competition structure. Specifically, for RSGs on two related machines, $PoS = PoA = 2$. The PoS lower bound is achieved already with homogeneous competition sets. Note that for classical job-scheduling game on two related machines, it holds that $PoS = 1$ and $PoA = \frac{1+\sqrt{5}}{2}$ [19], thus, again, we witness the harmful effects of a competition.

In light of the negative results regarding equilibrium existence, we discuss possible strategies of the system to modify an RSG instance or the players' utilization, such that the resulting game has a NE. We consider two approaches for *Nashification*. The first is addition of dummy jobs, and the second is compensation of low-rank players. Our hardness results imply that min-cost Nashification is also hard. For both approaches, we present tight bounds on the Nashification cost, in general and for unit-jobs on related machine.

We conclude with a discussion of additional congestion games whose 'race' variant is natural and interesting. We show that some of our results and techniques can be adopted to other games, and suggest some directions for future work. A full version that includes all the proofs is available in [22].

3 Identical Machines - Equilibrium Existence

In this section we assume that all the machines have the same unit-delay, that is, for all $i \in \mathcal{M}, d_i = 1$. The following example demonstrates that even very simple RSGs may not have a NE. Consider an instance with two machines and two competing jobs of lengths $p_1 < p_2$. If the jobs are on different machines, then the long job has a higher completion time and can reduce its rank by joining the short one, so they both have the same completion time and therefore the same rank. If the jobs are on the same machine, then the short job can reduce its rank by escaping to the empty machine. Thus, no profile is a NE.

Hoping for positive results, we turn to consider the class $\mathcal{G}_h$ of RSGs with homogeneous competition sets. Recall that $G \in \mathcal{G}_h$ if, for every $1 \leq \ell \leq c$, all the jobs in S_ℓ have the same length, p_ℓ.

Unfortunately, as demonstrated in Fig. 2, games in this class, even with only three sets and three machines, may not admit a NE. Moreover, as demonstrated in Fig. 3, even if a NE exists, it may be the case that a BRD does not converge.

The next natural question is whether there is an efficient way to decide, given a game $G \in \mathcal{G}_h$, whether G has a NE. We answer this question negatively:

Theorem 2. *Given an instance of RSG with homogeneous competition sets, it is NP-complete to decide whether the game has a NE profile.*

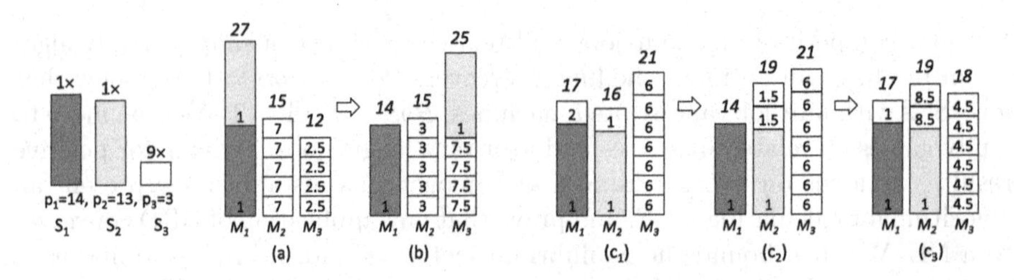

Fig. 2. An example of an RSG with homogeneous competition sets, that has no NE. Jobs are labeled by their ranks. Profiles (a)–(b) show that big jobs must be on different machines. Profiles $(c_1)-(c_2)-(c_3)-(c_1)$ loop when big jobs are on different machines.

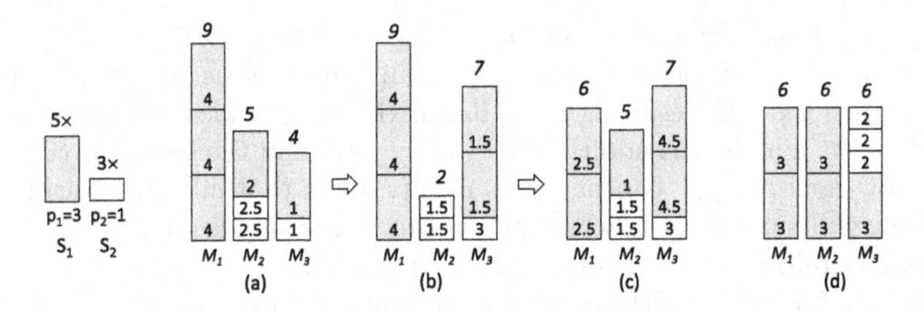

Fig. 3. An example of an RSG with homogeneous competition sets in which $c = 2$, $p_1|p_2$, and BRD may loop (profiles (a)–(b)–(c)–(a)). A NE exists (profile (d)). Jobs are labeled by their ranks.

In light of the above negative results, we would like to characterize instances in which a NE is guaranteed to exist. One such class includes instances of unit-length jobs and arbitrary competition sets.

Theorem 3. *If all jobs have the same length, then a NE exists and can be computed efficiently.*

Another class for which we have a positive result considers instances of only two competition sets and three machines. It is tight in light of the no-NE example given in Fig. 2, in which there are three sets on three machines.

Theorem 4. *If $G \in \mathcal{G}_h$ has $c = 2$ and $m = 3$, then a NE exists and can be computed efficiently.*

Classical job-scheduling games are race games with singleton competition sets. A NE may not exist even if there is just one pair of competing jobs. Also, in the full version we show that it is NP-hard to decide if a NE exists even if there only singletons and two competing pairs. For games with homogeneous competition sets, in which there are only singleton and pairs, we have positive news.

Theorem 5. *If $G \in \mathcal{G}_h$, and for all ℓ, $|S_\ell| \leq 2$ then a NE exists and can be computed efficiently.*

In search of more positive results, we turn to look at games with homogeneous competition sets with divisible lengths. Instances with divisible lengths arise often in applications in which clients can only select several levels of service. Moreover, naturally, in such settings, clients with similar service requirements compete with each other. Let $\mathcal{G}_{div}$ be the class of RSGs with homogeneous competition sets in which the job lengths form a divisible sequence. Formally, let $p_1 > p_2 > \ldots > p_c$ denote the different job lengths in $\mathcal{J}$, then $S_\ell = \{j|p(j) = p_\ell\}$, and for every $\ell_1 > \ell_2$, it holds that $p_{\ell_1}|p_{\ell_2}$. For example, if all job lengths are powers-of-2 and $S_\ell = \{j|p(j) = 2^{c-\ell+1}\}$ then $G \in \mathcal{G}_{div}$.

As demonstrated in Fig. 3, BRD may not converge to a NE even if $G \in \mathcal{G}_{div}$ and $c = 2$. Nevertheless, we prove that a NE can be computed directly for any game $G \in \mathcal{G}_{div}$. In the proof we provide an algorithm for computing a NE for instances in this class.

Theorem 6. *If $G \in \mathcal{G}_{div}$, then a NE exists and can be computed efficiently.*

4 Identical Machines - Equilibrium Inefficiency

In this section we analyze the equilibrium inefficiency of RSGs with respect to the objective of minimizing the maximal cost of a player (equivalent to the makespan of the schedule). For the classical job-scheduling game, the Price of Anarchy is known to be $2 - \frac{2}{m+1}$ for m identical machines, and the Price of Stability is known to be 1. We show that competition causes higher inefficiency. Specifically, both the PoA and the PoS are $3 - \frac{6}{m+2}$. We prove below the upper bound for the PoA and the lower bound for the PoS. In the full version [22], we describe, given $m \geq 3$ and $\epsilon > 0$, a game G, with homogeneous competition sets, for which $PoA(G) = 3 - \frac{6}{m+2} - \epsilon$. Given that we prove a matching PoS bound, the proof in the full version is less interesting. However, it can serve as a warm-up for the lower bound PoS proof, which is more involved.

Theorem 7. *If G is an RSG on m identical machines that has a NE, then $PoA(G) \leq 3 - \frac{6}{m+2}$. For every $m \geq 3$ and $\epsilon > 0$, there exists an RSG G on m identical machines such that G has a NE, and $PoS(G) \geq 3 - \frac{6}{m+2} - \epsilon$.*

Proof. The proof of the PoA upper bound is given in the full version [22]. We describe the lower bound on the PoS. Given $m \geq 3$ and $\epsilon > 0$, we describe an RSG G such that $\text{PoS}(G) = 3 - \frac{6}{m+2} - \epsilon$. Let $E = \{\delta_1, \delta_2, \delta_3\} \cup \{\epsilon_i | 1 \leq i \leq m(m-1)\}$ be a set of $3 + m(m-1)$ small positive numbers such that $\delta_1 < \delta_2 < \delta_3 \leq \frac{\epsilon m}{3}$, $\delta_1 + \delta_2 > \delta_3$, $\sum_{i>0} \epsilon_i < \frac{1}{4}$, and any subset of E with any coefficient in $\{-1, +1\}$ for each element, has a unique sum. That is, $\forall A_1, A_2 \subseteq \{\delta_1, \delta_2, \delta_3, \epsilon_1, \ldots, \epsilon_{m(m-1)}\}$ such that $A_1 \neq A_2$, and any $\gamma_k \in \{-1, +1\}$ we have $\sum_{k \in A_1} \gamma_k \cdot k \neq \sum_{k \in A_2} \gamma_k \cdot k$.

The set of jobs consists of $1 + m(m-1)$ competition sets, $S_0, \ldots, S_{m(m-1)}$:

1. S_0 consists of three jobs, where j_0^i for $i = 1, 2, 3$ has length $m - \delta_i$.
2. For $\ell = 1, \ldots, m(m-1)$, the set S_ℓ consists of two jobs: j_ℓ^1 of length $\frac{1}{4} - \epsilon_\ell$, and j_ℓ^2 of length $\frac{3}{4} + \epsilon_\ell$. Note that $p(j_\ell^1) + p(j_\ell^2) = 1$.

The PoS analysis is based on the fact that in every NE, the three long jobs of S_0 are assigned on the same machine, while an optimal assignment is almost balanced. We first restrict, the possible assignments of the jobs in S_ℓ for all $\ell \geq 0$.

Claim. In every NE, the three jobs of S_0 are assigned on the same machine, and for all $\ell \geq 1$, the two jobs of S_ℓ are assigned on the same machine.

By the above claim, the cost of every NE is at least the load incurred by the three jobs in S_0, that is, $3m - \sum_{i=1}^{3} \delta_i$. We show that a NE of this cost exists. An example for $m = 5$ is given in Fig. 4(a). Assign the jobs of S_0 on M_1. Distribute all remaining jobs such that for all $\ell \geq 1$ the jobs of S_ℓ are on the same machine, and there are m such sets assigned on each machine other than M_1. For all $a \geq 2$ we have $L_a = m$. This assignment is a NE since any migration of a job j from S_ℓ with $\ell \geq 1$ will increase its rank from 1.5 to 2, and any migration of a job j_0^x from S_0, will end up with load at least $2m - \delta_x$ which is more than $2m - \delta_y - \delta_z$, the remaining load on M_1. Thus, such a migration increases the rank of the deviating job from 2 to 3 and is not beneficial.

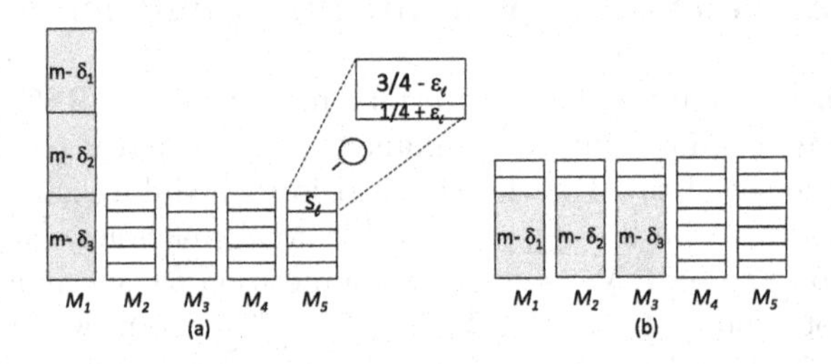

Fig. 4. A tight example for $m = 5$. (a) The only NE profile. (b) an optimal profile.

We turn to describe an optimal assignment. The total load of the jobs is $m(m-1) + 3m - \sum_{i=1}^{3} \delta_i = m(m+2) - \sum_{i=1}^{3} \delta_i$. The maximal length of a job is less than m, and there are $3 \leq m$ long jobs. The remaining jobs can be arranged in unit-length pairs. Thus, an optimal assignment is almost perfectly balanced (up to a gap of δ_3), where the most loaded machine has load $\frac{m(m+2)}{m} = m + 2$. The resulting PoS is $\frac{3m - \sum_{i=1}^{3} \delta_i}{m+2} < 3 - \frac{6}{m+2} - \epsilon$. □

5 Related Machines

In this section we consider RSGs played on related machines. Recall that d_i is the processing delay of machine M_i. Thus, it takes $p(j) \cdot d_i$ time units to process a job of length $p(j)$ on M_i. For a profile s, $C_i(s) = L_i(s) \cdot d_i$ is the completion time of M_i, and the cost of every job assigned on it.

5.1 Unit-Length Jobs

For classical job-scheduling games with unit-jobs and related machines the picture is simple and well understood. For a profile s, let $C_i^+(s) = (L_i(s) + 1) \cdot d_i$ denote the completion time of M_i if one more job would be assigned on it. A simple greedy algorithm that assigns the jobs sequentially, each on a machine minimizing C_i^+, is known to produce a Nash equilibrium profile that also minimizes the makespan. Moreover, every best-response sequence converges to an optimal schedule, thus, without competition, $PoA = PoS = 1$.

Surprisingly, as we show, even this simple setting of RSGs with unit-length jobs may not have a NE. Consider a game with $n = 5$ unit jobs, that form a single competition set. Assume there are three machines with delays $1, 1+\epsilon$ and $1+2\epsilon$. First note that a NE profile must fulfil $L_1 \geq L_2 \geq L_3$, as otherwise, it is clearly beneficial to deviate from a slow machine to a less loaded faster machine. Also note that if $L_1 = 4$, then one of the slower machines is empty and a deviation from M_1 to the empty machine is beneficial. The remaining load vectors are $\{\langle 3,2,0\rangle, \langle 3,1,1\rangle, \langle 2,2,1\rangle\}$. As demonstrated in Fig. 5, none of the corresponding profiles is a NE. Profile (c) is the output of a greedy algorithm. However, a job on M_2 can reduce its rank from 4.5 to 4 by a migration to M_1 (Profile (a)). Once it migrates, it is beneficial for the job on M_3 to join M_2 (Profile (b)), and a migration from M_1 to M_3 brings us back to Profile (c).

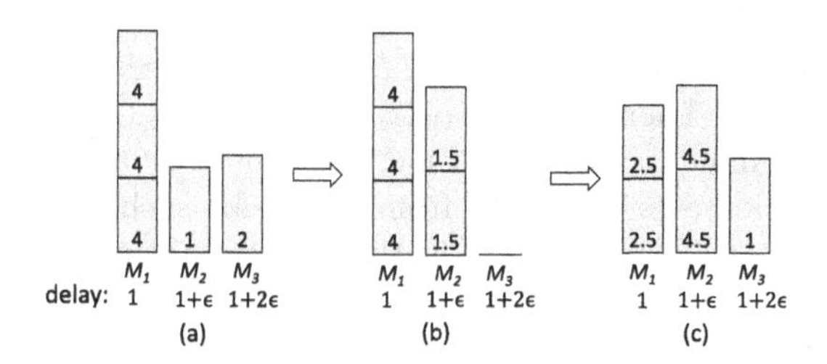

Fig. 5. No NE of an RSG with five competing unit-jobs on three related machines. The profiles loop (a)–(b)–(c)–(a). Jobs are labeled by their ranks.

While a NE may not exist, this class of instances is somewhat simpler. We show that it is possible to decide efficiently whether a given game instance has a NE and to compute one if it exists. Recall that the same task is NP-hard for games in $\mathcal{G}_h$ even on identical machines.

Theorem 8. *Let G be a game with unit-jobs on related machines in which all jobs are in the same competition set ($S_1 = \mathcal{J}$). It is possible to decide efficiently whether G has a NE and to compute one if it exists.*

The above positive result may lead one to expect that it would be possible to modify an instance slightly in order to get a game in which a NE exists. In Sect. 6 we discuss the Nashification of RSGs with unit-jobs by adding dummy jobs, and show that, unfortunately, given n jobs and m related machines, there is no constant k such that a game of $n+k$ jobs on this set of machines is guaranteed to have a NE.

For the equilibrium inefficiency of games with unit-jobs or related machines, we show the following tight bounds.

Theorem 9. *If G is a game on related machines and unit-jobs, for which a NE exists, then $PoA(G) < 2$. Also, for every $\epsilon > 0$ there exists a game for which $PoS(G) = 2 - \epsilon$.*

5.2 Variable-Length Jobs

Our negative results for unit-jobs are clearly valid for variable-length jobs, even with homogeneous competition sets. We are still able to come up with some good news for two machines. We present a linear-time algorithm for calculating a NE, show that any BRD sequence converges to a NE, and provide tight bounds on the equilibrium inefficiency.

Theorem 10. *If $m = 2$ and $G \in \mathcal{G}_h$ then G has a NE, and a NE can be calculated efficiently.*

Theorem 11. *If $m = 2$ and $G \in \mathcal{G}_h$ then BRD converges to a NE.*

Proof. Assume that BRD is performed starting from an arbitrary profile. It is easy to see that a migration from M_a to M_b is never beneficial if $L_a \cdot d_a \leq L_b \cdot d_b$ before the migration. Therefore, the only migrations in the BRD are from the machine with the higher completion time. We denote by a *switching migration*, a beneficial migration of a job $j \in S_\ell$ from M_a to M_b such that $L_a \cdot d_a > L_b \cdot d_b$ but $(L_a - p_i) \cdot d_a < (L_b + p_i) \cdot d_b$, that is, the target of the migration becomes the machine with the higher completion time. Note that a job $j \in S_\ell$ that performs a switching migration has the maximal rank in S_ℓ before the migration, and also the maximal rank in S_ℓ after the migration. The migration is beneficial since the number of jobs from S_ℓ on M_b after the migration is higher than their number on M_a before the migration.

Assume by contradiction that a BRD does not halt. Since the number of profiles is finite, this implies that BRD loops. Let $C_{max} = L_b \cdot d_b$ denote the maximal cost of a machine during the BRD loop, where M_b can be either the fast or the slow machine. Let t be the first time in which C_{max} is achieved during the BRD loop. Since a migration out of the machine with lower completion time is never beneficial, C_{max} is a result of a switching migration into M_b, say of $j \in S_\ell$.

Since BRD loops, a job from S_ℓ migrates back to M_a after time t. We claim that such a migration cannot be beneficial. Before the switching migration, $C_j = (L_a(t) + p_j) \cdot d_a$. The switching migration implies that jobs from S_ℓ have lower rank when their completion time is $C_{max} = L_b(t) \cdot d_b$ compared to their rank (with fewer competitors) on M_a with load $L_a(t) + p_j$. Therefore, a migration of $j' \in S_\ell$ to M_a after time t is beneficial only if the load on M_a is less than $L_a(t)$.

However, this implies that the load on M_b is more than $L_b(t)$, contradicting the choice of C_{max} as the maximal cost during the BRD loop. □

Theorem 12. *If G is an RSG on two related machines, for which a NE exists, then $PoA(G) < 2$. Also, for every $\epsilon > 0$ there exists a game $G \in \mathcal{G}_h$ for which $PoS(G) = 2 - \epsilon$.*

6 Nashification of Race Scheduling Games

In this section we discuss possible strategies of a centralized authority to change the instance or compensate players such that the resulting game has a NE. The first approach we analyze is *addition of dummy jobs.* The cost of such an operation is proportional to the total length of the dummy jobs, as this corresponds to the added load on the system. By Theorem 2, it is NP-hard to identify whether Nashification with budget 0 is possible. Thus, the min-budget problem is clearly NP-hard. We present several tight bounds on the required budget.

Theorem 13. *Let G be an RSG on m identical machines. Let $p_{max} = max_j p(j)$ be the maximal length of a job in $\mathcal{J}$. It is possible to Nashificate G by adding dummy jobs of total length at most $(m-1)p_{max}$. Also, for every m and $\epsilon > 0$ there exists a game G for which jobs of total length $(m-1)p_{max} - \epsilon$ are required to guarantee a NE.*

For related machines and unit-jobs we showed in Sect. 5.1 that a game may not have a NE even with a single competition set. It is tempting to believe that for such simple instances, Nashification may be achieved by an addition of a constant number of dummy jobs. Our next result shows that $m-2$ dummies may be required, and always suffice.

Theorem 14. *For any RSG on m related machines and a single competition set of unit-jobs, it is possible to achieve a NE by adding at most $m-2$ dummy jobs to the instance. Also, for every m there exists an RSG with a single competition set of unit jobs on m machines that requires $m-2$ dummy jobs to be added in order for a NE to exist.*

Proof. For the lower bound, given m, consider a game with $m+4$ unit-jobs and m machines having the following delays: $d_1 = 0.31, d_2 = 0.4, d_3 = 1$, and for all $4 \le i \le m, d_i = 0.5 + i\epsilon$. Figure 6 shows the behaviour of such an instance. A full description of this game, as well as an algorithm that produces a NE by adding at most $m-2$ dummy jobs is provided in the full version [22]. □

A different approach to achieve a NE, is *Nashification by payments.* The cost of a job is $C_j - \gamma_j$ where γ_j is a *compensation* given to the job by the system. A deviating job, will lose the compensation currently suggested to it. The goal is to achieve a NE, while minimizing $\sum_j \gamma_j$. For example, with two competing jobs of length 1 and $1+\epsilon$ on two identical machines, by setting $\gamma_2 = \epsilon$, the optimal schedule is a NE.

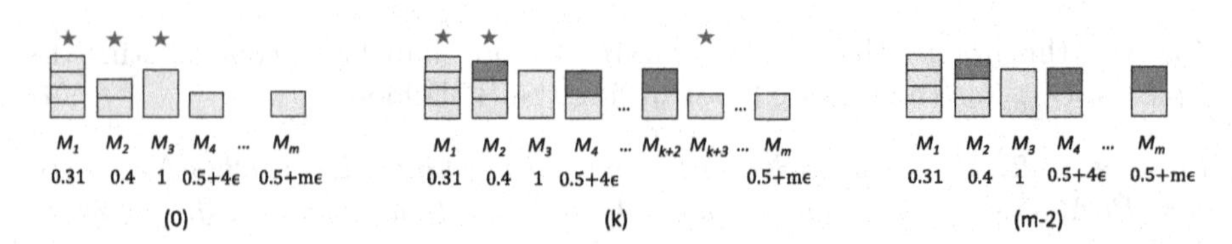

Fig. 6. A game for which an addition of $m - 2$ jobs is inevitable for Nashification. A BRD loop exists on the starred machines. Profile (0) is a dummy-free profile fulfilling simple stability constrains. Profile (k) fulfills the simple stability constraints after k dummy jobs are added. Profile $(m - 2)$ is a NE with $m - 2$ dummy jobs.

Theorem 15. *For any RSG G on identical machines, it is possible to achieve a NE with total compensation less than P, where* $P = \sum_{j \in \mathcal{J}} p(j)$. *Also, for every* m *and* $\epsilon > 0$ *there is a game G for which total compensation* $P - \epsilon$ *is required to achieve a NE.*

7 Conclusions and Directions for Future Work

Our paper suggests a new model for analyzing environments with strong competition. *Race games* are congestion games in which players' welfare depends on their relative performance. The main objective of a player is to perform well relative to his competitors, while minimizing his cost is a minor objective. A profile is a NE if no player can improve her rank, or reduce her cost while keeping her rank.

We analyzed job-scheduling race games on parallel machines. Having an additional constraint for stability, race games are less stable than classical load-balancing games, thus our results for general games are mostly negative. In particular, for all the classes of instances we considered, we showed that $PoS = PoA$, while the same competition-free game has a lower PoA and $PoS = 1$. Practically, it means that competition may lead to a poor outcome even if the system can control the initial players' strategies. Striving for stability, we also studied the cost of Nashification, by either adding dummy jobs to the instance, or compensating jobs for having high rank. While in the general case, Nashification may involve balancing all the machines or jobs' cost, in some natural classes it can be achieved in cheaper ways. Min-cost Nashification of a given instance is NP-complete. We leave open the corresponding approximation problem.

Race games can be studied in various additional settings. In fact, every congestion game in which players are associated with a utility has its 'race' variant. In the full version we list several examples. Additional questions may refer to the structure of the competition-sets, for example, competition sets may overlap, or may be defined according to the players' strategy space (symmetric competition sets). The study of coordinated deviations is another intriguing direction. In the presence of competition, coalitions may be limited to include only members of different competition sets. On the other hand, temporal collaboration may be fruitful even for competing players. Thus, there are many different interesting variants of coordinated deviations in race games.

References

1. Anagnostopoulos, A., Becchetti, L., de Keijzer, B., Schäfer, G.: Inefficiency of games with social context. Theory Comput. Syst. **57**(3), 782–804 (2014). https://doi.org/10.1007/s00224-014-9602-4
2. Anshelevich, E., Dasgupta, A., Kleinberg, J., Tardos, E., Wexler, T., Roughgarden, T.: The price of stability for network design with fair cost allocation. SIAM J. Comput. **38**(4), 1602–1623 (2008)
3. Avni, G., Tamir, T.: Cost-sharing scheduling games on restricted unrelated machines. Theoret. Comput. Sci. **646**, 26–39 (2016)
4. Aumann, R.J.: Rule-Rationality Versus Act-Rationality. Discussion Paper Series DP497, The Hebrew University's Center for the Study of Rationality (2008)
5. Bilò, V., Celi, A., Flammini, M., Gallotti, V.: Social context congestion games. In: Proceedings of 18th SIROCCO, pp. 282–293 (2011)
6. Bolton, G.E., Ockenfels, A.: ERC: a theory of equity, reciprocity, and competition. Am. Econ. Rev. **90**(1), 166–193 (2000)
7. Bhawalkar, K., Gairing, M., Roughgarden, T.: Weighted congestion games: the price of anarchy, universal worst-case examples, and tightness. ACM Trans. Econ. Comput. **2**(4), Article 14 (2014)
8. Chen, P.A., de Keijzer, B., Kempe, D., Schäfer, G.: The robust price of anarchy of altruistic games. In: Proceedings of 7th WINE, pp. 383–390 (2011)
9. Cho, Y., Sahni, S.: Bounds for list schedules on uniform processors. SIAM J. Comput. **9**(1), 91–103 (1980)
10. Czumaj, A., Vöcking, B.: Tight bounds for worst-case equilibria. ACM Trans. Algorithms **3**(1), 4:1–4:17 (2007)
11. Fehr, E., Schmidt, K.M.: A theory of fairness, competition, and cooperation. Q. J. Econ. **114**(3), 817–868 (1999)
12. Fiat, A., Kaplan, H., Levi, M., Olonetsky, S.: Strong price of anarchy for machine load balancing. In: Proceedings of 34th ICALP (2007)
13. Finn, G., Horowitz, E.: A linear time approximation algorithm for multiprocessor scheduling. BIT Numer. Math. **19**(3), 312–320 (1979)
14. Fotakis, D., Kontogiannis, S., Koutsoupias, E., Mavronicolas, M., Spiraklis, P.: The structure and complexity of Nash equilibria for a selfish routing game. In: Proceedings of 29th ICALP, pp. 510–519 (2002)
15. Graham, R.L.: Bounds for certain multiprocessing anomalies. Bell Syst. Tech. J. **45**, 1563–1581 (1966)
16. Graham, R.L.: Bounds on multiprocessing timing anomalies. SIAM J. Appl. Math. **17**, 263–269 (1969)
17. Hochbaum, D.S., Shmoys, D.B.: Using dual approximation algorithms for scheduling problems: practical and theoretical results. J. ACM **34**(1), 144–162 (1987)
18. Jansen, K., Klein, K.M., Verschae, J.: Closing the gap for makespan scheduling via sparsification techniques. In: Proceedings of 43rd ICALP, pp. 1–13 (2016)
19. Koutsoupias, E., Papadimitriou, C.: Worst-case equilibria. Comput. Sci. Rev. **3**(2), 65–69 (2009)
20. Pinedo, M.: Scheduling: Theory, Algorithms, and Systems. Springer, Heidelberg (2008). https://doi.org/10.1007/978-3-642-46773-8_5

21. Rosenthal, R.W.: A class of games possessing pure-strategy Nash equilibria. Int. J. Game Theory **2**, 65–67 (1973)
22. Rosner, S., Tamir, T.: Race Scheduling Games. https://cs.idc.ac.il/~tami/Papers/RSG-full.pdf
23. Schulz, A.S., Stier Moses, N.: On the performance of user equilibria in traffic networks. In: Proceedings of 43rd SODA, pp. 86–87 (2003)
24. Vöcking, B.: Selfish load balancing. In: Algorithmic Game Theory. Chapter 20, Cambridge University Press (2007)
25. Winter, E., Méndez-Naya, L., García-Jurado, I.: Mental equilibrium and strategic emotions. Manage. Sci. **63**(5), 1302–1317 (2017)

Social Choice and Cooperative Games

Line-Up Elections: Parallel Voting with Shared Candidate Pool

Niclas Boehmer[1(✉)], Robert Bredereck[1], Piotr Faliszewski[2], Andrzej Kaczmarczyk[1], and Rolf Niedermeier[1]

[1] Algorithmics and Computational Complexity, TU Berlin, Berlin, Germany
{niclas.boehmer,robert.bredereck,a.kaczmarczyk, rolf.niedermeier}@tu-berlin.de
[2] AGH University, Krakow, Poland
faliszew@agh.edu.pl

Abstract. We introduce the model of line-up elections which captures parallel or sequential single-winner elections with a shared candidate pool. The goal of a line-up election is to find a high-quality assignment of a set of candidates to a set of positions such that each position is filled by exactly one candidate and each candidate fills at most one position. A score for each candidate-position pair is given as part of the input, which expresses the qualification of the candidate to fill the position. We propose several voting rules for line-up elections and analyze them from an axiomatic and an empirical perspective using real-world data from the popular video game FIFA.

Keywords: Single-winner voting · Multi-winner voting · Assignment problem · Axiomatic analysis · Empirical analysis

1 Introduction

Before the start of the soccer World Cup 2014, Germany's head coach Joachim Löw had problems to find an optimal team formation. Due to several injuries, Löw was stuck without a traditional striker. He decided to play with three offensive midfielders instead, namely, Müller, Özil, and Götze. However, he struggled to decide who of the players should play in the center, on the right, and on the left.[1] At the final coaching meeting, he surveyed the opinions of ten coaching assistants asking for each of the candidates, "Is this candidate suitable to play on the left/in the center/on the right?". Coaching assistants were allowed to approve an arbitrary subset of candidate-position pairs. He got answers resulting in the following numbers of approvals for each candidate-position pair:

[1] This story and the opinions of the coaches are fictional. However, Löw really faced the described problem before the World Cup 2014.

N. Boehmer—Supported by the DFG project MaMu (NI 369/19).
P. Faliszewski—Supported by a Friedrich Wilhelm Bessel Award from the Alexander von Humboldt Foundation.
A. Kaczmarczyk—Supported by the DFG project AFFA (BR 5207/1 and NI 369/15).

T. Harks and M. Klimm (Eds.): SAGT 2020, LNCS 12283, pp. 275–290, 2020.
https://doi.org/10.1007/978-3-030-57980-7_18

Candidate	Position		
	Left	Center	Right
Müller	5	10	9
Özil	3	8	5
Götze	4	7	4

After collecting the results, some of the coaches argued that Müller must play in the center, as everyone agreed with this. Others argued that Müller should play on the right, as otherwise this position would be filled by a considerably less suitable player. Finally, someone pointed out that Müller should play on the left, as this was the only possibility to fill the positions such that every position gets assigned a player approved by at least half of the coaches.

The problem of assigning Müller, Özil, and Götze can be modeled as three parallel single-winner elections with a shared candidate pool, where every candidate can win at most one election and each voter is allowed to cast different preferences for each election. In our example, the coaches are the voters, the players are the candidates and the three locations on the field are the positions. Classical single-winner voting rules do not suffice to determine the winners in such settings, as a candidate may win multiple elections. Also multi-winner voting rules cannot be used, as a voter can asses the candidates differently in different elections. Other examples of parallel single-winner elections with a shared candidate pool include a company that wants to fill different positions after an open call for applications, a cooperation electing an executive board composed of different positions, and a professor who assigns students to projects.

In this paper, we introduce a framework for such settings: In a *line-up election*, we get as input a set of candidates, a set of positions, and for each candidate-position pair a score expressing how suitable this candidate is to win the election for this position. The goal of a line-up election is to find a "good" assignment of candidates to positions such that each position gets assigned exactly one candidate and each candidate is assigned at most once. There exist multiple possible sources of the scores. For instance, a variety of single-winner voting rules aggregate preference profiles into single scores for each candidate and then select the candidate with the highest score as the winner of the given election. Examples of such rules include Copeland's voting rule, where the score of a candidate is the number of her pairwise victories minus the number of her pairwise defeats, positional scoring rules, or Dodgson's voting rule. Thus, line-up elections offer a flexible framework that can be built upon a variety of single-winner voting rules.

Our Contributions. This paper introduces *line-up elections*—parallel single-winner elections with a shared candidate pool—and initiates a study thereof. After stating the problem formally, we propose two classes of voting rules, sequential and OWA-rules. Sequential rules fill positions in some order—which may depend on the scores—and select the best still available candidate for a given position. In the versatile class of OWA-rules, a rule aims at maximizing some ordered weighted average (OWA) of the scores of the assigned candidate-position pairs. We highlight seven rules from these two classes. Subsequently, inspired by work on voting, we describe several desirable axioms for line-up voting rules and provide a comprehensive and diverse picture of their axiomatic properties. We complement this axiomatic analysis by empirical investigations on data from the popular soccer video game FIFA [9] and synthetic data.

As our model considers multiple, parallel single-winner elections, it can be seen as an extension of single-winner elections; indeed, we can view the scores of candidates for a position as obtained from some voting rule [2,8,28]. It reduces to multi-winner voting [12] if every voter casts the same vote in all elections. Most of our proposed axioms are generalizations of axioms studied in those settings [4,10,28]. Previously, committee elections where the committee consists of different positions were rarely considered. Aziz and Lee [5] studied multi-winner elections where a given committee is partitioned into different sub-committees and each candidate is only suitable to be part of some of these sub-committees.

Due to lack of space, we defer most of the proofs from our axiomatic analysis and several details of the conducted experiments to the full version [7].

2 Our Model

In a *line-up election* E, we are given a set of m candidates $C = \{c_1, \dots, c_m\}$ and q positions $P = \{p_1, \dots, p_q\}$ with $m \geq q$, together with a score matrix $\mathbf{S} \in \mathbb{Q}^{m \times q}$. For $i \in [m]$ and $j \in [q]$, $\mathbf{S}_{i,j}$ is the score of candidate c_i for position p_j, which we denote as $\text{score}_{p_j}(c_i)$. An outcome of E is an assignment of candidates to positions, where each position is assigned exactly one candidate and each candidate gets assigned to at most one position. We call an outcome a *line-up* π and write, for a position $p \in P$, $\pi_p \in C$ to denote the candidate that is assigned to position p in π. We write a line-up π as a q-tuple $\pi = (\pi_{p_1}, \dots \pi_{p_q}) \in C^q$ with pairwise different entries.

For a candidate-position pair $(c, p) \in C \times P$, we say that (c, p) is assigned in π if $\pi_p = c$ and write $(c, p) \in \pi$. Moreover, for an outcome π and a candidate c, let $\pi(c) \in P \cup \{\square\}$ be the position that c is assigned to in π; that is, $\pi(c) = p$ if $\pi_p = c$ and $\pi(c) = \square$ if c does not occur in π. We write $c \in \pi$ if $\pi(c) \neq \square$, and $c \notin \pi$ otherwise. For a line-up π and a subset of positions $P' \subseteq P$, we write $\pi|_{P'}$ to denote the tuple π restricted to positions P' and $\pi_{P'}$ to denote the set of candidates assigned to positions P' in π. For a position $p \in P$ and line-up π, we write $\text{score}_p(\pi)$ to denote $\text{score}_p(\pi_p)$. Moreover, we refer to $(\text{score}_{p_1}(\pi), \dots, \text{score}_{p_q}(\pi))$ as the *score vector* of π. A *line-up voting rule* f maps a line-up election E to a set of winning line-ups, where we use $f(E) \subseteq 2^{C^q}$ to denote the set of winning line-ups returned by rule f applied to election E.

3 Line-Up Elections as Assignment Problems

It is possible to interpret line-up elections as instances of the Assignment Problem, which aims to find a (maximum weight) matching in a bipartite graph.

The assignment graph G of a line-up election $(\mathbf{S}, C, P)$ is a complete weighted bipartite graph $G = (C \uplus P, E, w)$ with edge set $E := \{\{c, p\} \mid c \in C \wedge p \in P\}$ and weight function $w(c, p) := \text{score}_p(c)$ for $\{c, p\} \in E$. Every matching in the assignment graph which matches all positions induces a valid line-up.

The Assignment Problem and its generalizations have been mostly studied from an algorithmic and fairness perspective [16–18,20,22,23]. For instance, Lesca et al. [22] studied finding assignments with balanced satisfaction from an algorithmic perspective. They utilized ordered weighted average operators and proved that finding assignments that maximize an arbitrary non-decreasing ordered weighted average is NP-hard (see next section for definitions). One generalization of the Assignment Problem is the Conference Paper Assignment Problem (CPAP), which tries to find a many-to-many assignment of papers to reviewers with capacity constraints on both sides [19]. Focusing on egalitarian considerations, Garg et al. [17] studied finding outcomes of CPAP which are optimal for the reviewer that is worst off, where they break ties by looking at the next worst reviewer. They proved that this task is computationally hard in this generalized setting. In contrast to our work and the work by Lesca et al. [22], Lian et al. [23] employed OWA-operators in the context of CPAP on a different level. Focusing on the satisfaction of reviewers, they studied finding assignments maximizing the ordered weighted average of the values a reviewer gave to her assigned papers and conducted experiments where they compare different OWA-vectors.

In contrast to previous work on the Assignment Problem, we look at the problem through the eyes of voting theorists. We come up with several axiomatic and quantitative properties that are desirable to fulfill by a mechanism if we assume that the Assignment Problem is applied in the context of an election.

4 Voting Rules

As we aim at selecting an individually-excellent line-up, a straightforward approach is to maximize the social welfare, which is determined by the scores of the assigned candidate-position pairs. However, it is not always clear which type of social welfare may be of interest. For example, the overall performance of a line-up may depend on the performance of the worst candidate. This may apply to team sports. Sometimes, however, the performance of a line-up is proportional to the sum of the scores and it does not hurt if some positions are not filled by a qualified candidate. OWA-operators provide a convenient way to express both these goals, as well as a continuum of middle-ground approaches [26].

OWA-Rules $\mathcal{F}^\Lambda$. For a tuple $a = (a_1, \ldots, a_k)$ and $i \in [k]$, let $a[i]$ be the i-th largest entry of a. We call $\Lambda := (\lambda_1, \ldots, \lambda_k)$ an ordered weighed-average vector (OWA-vector) and define the ordered weighted average of a under Λ as $\Lambda(a_1, \ldots, a_k) := \sum_{i \in [k]} \lambda_i \cdot a[i]$ [26]. For a line-up π, we define $\Lambda(\pi)$ as the ordered weighted average of the score vector of π under Λ. That is, $\Lambda(\pi) := \Lambda(\text{score}_{p_1}(\pi), \ldots, \text{score}_{p_q}(\pi))$. The score of a line-up π assigned by an OWA-rule f^Λ is $\Lambda(\pi)$. Rule f^Λ chooses (possibly tied) line-ups with the highest score.

Among this class of rules, we focus on the following four natural ones, quite well studied in other contexts, such as finding a collective set of items [11,25]:

- *Utilitarian rule* f^{ut}: $\Lambda^{\mathrm{ut}} := (1, \ldots, 1)$. This corresponds to computing a maximum weight matching in the assignment graph. It is computable in $\mathcal{O}(m^3)$ time for m candidates [20].
- *Egalitarian rule* f^{eg}: $\Lambda^{\mathrm{eg}} := (0, \ldots, 0, 1)$. This corresponds to solving the Linear Bottleneck Assignment Problem, which can be done in $\mathcal{O}(m^2)$ time for m candidates [16].
- *Harmonic rule* f^{har}: $\Lambda^{\mathrm{har}} := (1, \frac{1}{2}, \frac{1}{3}, \ldots, \frac{1}{q})$. The computational complexity of finding a winning line-up under this rule is open. In our experiments, we compute it using Integer Linear Programming (ILP).
- *Inverse harmonic rule* f^{ihar}: $\Lambda^{\mathrm{ihar}} := (\frac{1}{q}, \frac{1}{q-1}, \ldots, \frac{1}{2}, 1)$. While computing the winning line-up for an arbitrary non-decreasing OWA-vector is NP-hard [22], the computational complexity of computing this specific rule is open. We again use an ILP to compute a winning line-up.

Sequential-Rules $\mathcal{F}^{\mathrm{seq}}$. OWA-rules require involved algorithms and cannot be applied by hand easily. In practice, humans tend to solve a line-up election in a simpler way, for instance, by determining the election winners one by one. This procedure results in a class of quite intuitive sequential voting rules. A sequential rule is defined by some function g that, given a line-up election $E = (\mathbf{S}, C, P)$ and a set of already assigned candidates C_{as} and positions P_{as}, returns the next position to be filled. This position is then filled by the remaining candidate $C \setminus C_{\mathrm{as}}$ with the highest score on this position. Sequentializing decisions which partly depend on each other has also proven to be useful in other voting-related problems, such as voting in combinatorial domains [21], or in the House Allocation problem in form of the well-known mechanism of serial dictatorship. We focus on the following three linear-time computable sequential rules.

Fixed-Order Rule $f^{\mathrm{seq}}_{\mathrm{fix}}$. Here, the positions are filled in a fixed order (for simplicity, we assume the order in which the positions appear in the election). The fixed-order sequential rule is probably the simplest way to generalize single-winner elections to the line-up setting and it enables us to make the decisions separately. Moreover, it is not necessary to evaluate all candidates for all positions, which is especially beneficial if evaluating the qualification of a candidate on a position comes at some (computational) cost.

Max-First Rule $f^{\mathrm{seq}}_{\mathrm{max}}$. In the max-first rule, at each step, the position with the highest still available score is filled. That is, $g(\mathbf{S}, C, P, C_{\mathrm{as}}, P_{\mathrm{as}}) := \arg\max_{p \in P \setminus P_{\mathrm{as}}} \max_{c \in C \setminus C_{\mathrm{as}}} \mathrm{score}_p(c)$. This is equivalent to adding at each step the remaining candidate-position pair with the highest score to the line-up. Max-first is intuitively appealing because a candidate who is outstanding at a position is likely to be assigned to it. Notably, this rule is an approximation of the utilitarian rule, as it corresponds to solving the Maximum Weight Matching problem in the assignment graph by greedily selecting the remaining edge with the highest

Table 1. Overview of axiomatic properties of all studied rules. For each of the axioms, we indicate whether this axiom is strongly satisfied (S), only weakly satisfied (W), or not satisfied at all (–). For the egalitarian rule, all entries marked with a † can be improved if a variant of this rule that selects the egalitarian outcome with the highest summed score is used.

	non wasteful	Pareto optimal	reasonable satisfaction	score consistent	mono-tonicity	line-up enlargement
f^{ut}	S	S	–	S	S	S
f^{har}	S	S	–	–	–	–
f^{ihar}	S	S	–	–	–	–
f^{eg}	W†	W†	–	–	–†	W†
$f^{\mathrm{seq}}_{\mathrm{fix}}$	S	W	–	W	S	S
$f^{\mathrm{seq}}_{\mathrm{max}}$	S	W	S	–	S	S
$f^{\mathrm{seq}}_{\mathrm{min}}$	S	–	–	–	–	–

weight. For every possible tie-breaking, this approach is guaranteed to yield a $\frac{1}{2}$-approximation of the optimal solution in polynomial time [3].

Min-First Rule $f^{\mathrm{seq}}_{\mathrm{min}}$. In the min-first rule, the position with the lowest score of the most-suitable remaining candidate is filled next. That is, $g(\mathbf{S}, C, P, C_{\mathrm{as}}, P_{\mathrm{as}}) := \arg\min_{p \in P \setminus P_{\mathrm{as}}} \max_{c \in C \setminus C_{\mathrm{as}}} \mathrm{score}_p(c)$. The reasoning behind this is that the deciders focus first on filling critical positions where all candidates perform poorly.

5 Axiomatic Analysis of Voting Rules

We propose several axioms and properties that serve as a starting point to characterize and compare the introduced voting rules. We checked all introduced voting rules against all the axioms and collected the results in Table 1. The underlying proofs can be found in the full version [7]. We will introduce two efficiency and one fairness axiom for line-ups. These definitions extend to voting rules as follows. A voting rule f *strongly (weakly) satisfies* a given axiom if for each line-up election E, $f(E)$ contains only (some) line-ups satisfying the axiom.

Efficiency Axioms. As our goal is to select individually-excellent outcomes, we aim at selecting line-ups in which the score of each position is as high as possible. Independent of conflicts between positions, there exist certain outcomes which are suboptimal. For example, it is undesirable if there exists an unassigned candidate that is more suitable for some position than the currently assigned one.

Axiom 1 ***Non-wastefulness:*** *In a line-up election* $(\mathbf{S}, C, P)$, *a line-up* π *is non-wasteful if there is no unassigned candidate* $c \notin \pi$ *and a position* $p \in P$ *such that* $\mathrm{score}_p(c) > \mathrm{score}_p(\pi)$.

This axiom implies that in the special case of a single position, the candidate with the highest score for the position wins the election. In the context of non-wastefulness, we only examine whether a line-up can be improved by assigning unassigned candidates. However, it is also possible to consider arbitrary rearrangements. This results in the notion of score Pareto optimality.

Axiom 2 ***Score Pareto optimality:*** *In a line-up election* $(\mathbf{S}, C, P)$*, a line-up* π *is score Pareto dominated if there exists a line-up* π' *such that for all* $p \in P$*,* $\text{score}_p(\pi') \geq \text{score}_p(\pi)$ *and there exists a position* $p \in P$ *with* $\text{score}_p(\pi') > \text{score}_p(\pi)$*. A line-up is score Pareto optimal if it is not score Pareto dominated.*

While all OWA-rules with an OWA-vector containing no zeros are clearly strongly non-wasteful and strongly score Pareto optimal, the egalitarian rule satisfies both axioms only weakly, as this rule selects a line-up purely based on its minimum score. All sequential rules naturally satisfy strong non-wastefulness. Yet, by breaking ties in a suboptimal way, all of them may output line-ups that are not score Pareto optimal. While for the fixed-order rule and the max-first rule at least one winning outcome is always score Pareto optimal, slightly counterintuitively, there exist instances where the min-first rule does not output any score Pareto optimal line-ups.

Fairness Axioms. Another criterion to judge the quality of a voting rule is to assess whether positions and candidates are treated in a fair way. The underlying assumption for fairness in this context is that every position should have the best possible candidate assigned. Similarly, from candidates' perspective, one could argue that each candidate deserves to be assigned to the position for which the candidate is most suitable. In the following, we call a candidate or position for which fairness is violated dissatisfied. Unfortunately, line-ups in which all positions and candidates are satisfied simultaneously may not exist. That is why we consider a restricted fairness property, where we call a candidate-position pair $(c, p) \in \pi$ reasonably dissatisfied in π if candidate c has a higher score for p than π_p and c is either unassigned or c's score for p is higher than for $\pi(c)$.

Axiom 3 ***Reasonable satisfaction:*** *A line-up* π *is reasonably satisfying if there are no two positions* p *and* p' *such that*

$$\text{score}_p(\pi_{p'}) > \text{score}_p(\pi_p) \text{ and } \text{score}_p(\pi_{p'}) > \text{score}_{p'}(\pi_{p'})$$

and there is no candidate $c \notin \pi$ *and position* p *such that* $\text{score}_p(c) > \text{score}_p(\pi_p)$*.*

It is straightforward to prove that all winning line-ups under the max-first rule are reasonably satisfying; hence, a reasonably satisfying outcome always exists. Note that it is also possible to motivate reasonable satisfaction as a notion of stability if we assume that candidates and positions are allowed to leave their currently assigned partner to pair up with a new candidate or position. Thus, reasonable satisfaction resembles the notion of stability in the context of the Stable Marriage problem [15].

However, fulfilling reasonable satisfaction may come at the cost of selecting a line-up that is suboptimal for various notions of social welfare. For some

$\epsilon > 0$, consider an election with $P = \{p_1, p_2\}$, $C = \{a, b\}$, $\text{score}_{p_1}(a) = 2$, $\text{score}_{p_2}(a) = \text{score}_{p_1}(b) = 2 - \epsilon$, and $\text{score}_{p_2}(b) = 0$. We see that the outcome (b, a) of utilitarian welfare $4 - 2\epsilon$ maximizes utilitarian social welfare, while the outcome (a, b) of utilitarian welfare 2 is the only outcome fulfilling reasonable satisfaction. Therefore, it is interesting to measure the price of reasonable satisfaction in terms of utilitarian welfare. Analogous to the price of stability [1], we define this as the maximum utilitarian social welfare achievable by a reasonably satisfying outcome, divided by the maximum achievable utilitarian social welfare. The example from above already implies that this price is upper bounded by $\frac{1}{2}$. In fact, this bound is tight, as the max-first rule that only outputs reasonably satisfying line-ups is a $\frac{1}{2}$-approximation of the utilitarian outcome.

Voting Axioms. We now formulate several axioms that are either closely related to axioms from single-winner [28] or multi-winner voting [10]. We define all axioms using two parts, (a) and (b). On a high level, condition (a) imposes that certain line-ups should be winning after modifying a line-up election, while condition (b) demands that no other line-ups become (additional) winners after the modifications. If a voting rule only fulfills condition (a), then we say that it *weakly satisfies* the corresponding axiom. If it fulfills both conditions, then we say that it *strongly satisfies* the corresponding axiom.

In single-winner and multi-winner voting, the consistency axiom requires that if a voting rule selects the same outcome in two elections (over the same candidate set), then this outcome is also winning in the combined election [27]. We consider a variant of this axiom, adapted to our setting.

Axiom 4 ***Score consistency:*** *For two line-up elections* $(\mathbf{S}, C, P)$ *and* $(\mathbf{S}', C, P)$ *with* $f(\mathbf{S}, C, P) \cap f(\mathbf{S}', C, P) \neq \emptyset$ *it holds that: a)* $f(\mathbf{S}, C, P) \cap f(\mathbf{S}', C, P) \subseteq f(\mathbf{S} + \mathbf{S}', C, P)$ *and b)* $f(\mathbf{S}, C, P) \cap f(\mathbf{S}', C, P) \supseteq f(\mathbf{S} + \mathbf{S}', C, P)$.

The utilitarian rule is the only OWA-rule that satisfies weak (and even strong) score consistency. The fixed-order rule is the only other considered rule that satisfies weak score consistency. For all other rules, it is possible to construct simple two-candidates two-positions line-up elections where this axiom is violated.

Besides focusing on consistency related considerations, it is also important to examine how the winning line-ups change if the election itself is modified. We start by considering a variant of monotonicity [13].

Axiom 5 ***Monotonicity:*** *Let* $(\mathbf{S}, C, P)$ *be a line-up election with a winning line-up* π*. Let* $(\mathbf{S}', C, P)$ *be the line-up election obtained from* $(\mathbf{S}, C, P)$ *by increasing* $\text{score}_p(\pi_p)$ *for some* p*. Then, it holds that (a)* π *is still a winning line-up, that is,* $\pi \in f(\mathbf{S}', C, P)$*, and (b) no new winning line-ups are created, that is, for all* $\pi' \in f(\mathbf{S}', C, P)$ *it holds that* $\pi' \in f(\mathbf{S}, C, P)$*.*

While the utilitarian, fixed-order, and max-first rule all satisfy strong monotonicity, all other rules fail even weak monotonicity. As these negative results are intuitively surprising, we present their proof here.

Proposition 1. *The egalitarian rule* f^{eg}*, the harmonic rule* f^{har}*, and the min-first rule* $f^{\text{seq}}_{\text{min}}$ *all violate weak monotonicity.*

Proof. We present counterexamples for all three rules:

E_1:

	p_1	p_2
a	0	3
b	3	0
c	4	0

E_2:

	p_1	p_2	p_3
a	4	1	0
b	4.75	3	0
c	0	0	2

E_3:

	p_1	p_2
a	2	1
b	0	0

In E_1, (b, a) and (c, a) are winning under f^{eg}. However, after increasing $\text{score}_{p_2}(a)$ to 4, (c, a) has an egalitarian score of 4 and thereby becomes the unique winning line-up. We now turn to election E_2 and voting rule f^{har}. It is clear that c will be assigned to p_3 in every outcome. In fact, (a, b, c) is the winning line-up in E_2, as $\Lambda^{\text{har}}(a, b, c) = 1 \cdot 4 + \frac{1}{2} \cdot 3 + \mathbf{\frac{1}{3}} \cdot \mathbf{2} > 1 \cdot 4.75 + \frac{1}{3} \cdot 1 + \mathbf{\frac{1}{2}} \cdot \mathbf{2} = \Lambda^{\text{har}}(b, a, c)$. After increasing $\text{score}_{p_3}(c)$ to 3, (b, a, c) becomes the unique winning line-up in E_2, as $\Lambda^{\text{har}}(a, b, c) = 1 \cdot 4 + \frac{1}{2} \cdot 3 + \mathbf{\frac{1}{3}} \cdot \mathbf{3} < 1 \cdot 4.75 + \frac{1}{3} \cdot 1 + \mathbf{\frac{1}{2}} \cdot \mathbf{3} = \Lambda^{\text{har}}(b, a, c)$. By the modification, the score of (b, a, c) increases more than the score of (a, b, c), because in (b, a, c), $\text{score}_{p_3}(c)$ is multiplied by a larger coefficient. Lastly, considering $f^{\text{seq}}_{\text{min}}$, (b, a) is the unique winning outcome in E_3. After increasing $\text{score}_{p_2}(a)$ to 3, the ordering in which the positions get assigned changes, and thereby, (a, b) becomes the unique winning line-up. □

In the context of multi-winner voting, an additional monotonicity axiom is sometimes considered: Committee enlargement monotonicity deals with the behavior of the set of winning outcomes if the size of the committee is increased [6,10]. We generalize this axiom to our setting in a straightforward way.

Axiom 6 ***Line-up enlargement monotonicity:*** *Let $(\mathbf{S}, C, P)$ be a line-up election and let $(\mathbf{S}', C, P')$ with $P' = P \cup \{p^*\}$ be an election where position p^* and the scores of candidates for this position have been added. It holds that (a) for all $\pi \in f(\mathbf{S}, C, P)$ there exists some $\pi' \in f(\mathbf{S}', C, P')$ such that $\pi_P \subset \pi'_{P'}$ and (b) for all $\pi' \in f(\mathbf{S}', C, P')$ there exists some $\pi \in f(\mathbf{S}, C, P)$ such that $\pi_P \subset \pi'_{P'}$.*

Note that line-up enlargement monotonicity does not require that the selected candidates are assigned to the same position in the two outcomes π and π'. Despite the fact that this axiom seems to be very natural, neither of the two harmonic rules satisfy it at all. Intuitively, the reason for this is that by introducing a new position, the coefficients in the OWA-vector "shift". Moreover, surprisingly, the min-first rule also violates the weak version of this axiom. The proof for this consists of a rather involved counterexample exploiting the fact that by introducing a new position, the order in which the positions are filled may change. All other rules, apart from the egalitarian one, satisfy strong line-up enlargement monotonicity; the egalitarian rule satisfies line-up enlargement monotonicity only in the weak sense. We conclude with presenting the proof that the utilitarian rule, f^{ut}, satisfies weak line-up enlargement monotonicity, as the proof nicely illustrates how it is possible to reason about this axiom.

Proposition 2. *The utilitarian rule f^{ut} satisfies weak line-up enlargement monotonicity.*

Proof. Let π be a winning line-up of the initial election $E = (\mathbf{S}, C, P)$ and let π' be a winning line-up of the extended election $E' = (\mathbf{S}', C, P \cup \{p^*\})$ such that there exists a candidate $c \in C$ with $c \in \pi$ and $c \notin \pi'$. We claim that it is always possible to construct from π' a winning line-up π^* of the extended election such that all candidates from π appear in π^*: Initially, we set $\pi^* := \pi'$. As long as there exists a candidate $c \in C$ with $c \in \pi$ and $c \notin \pi^*$, we set $\pi^*_{\pi(c)} := c$. Let $\widetilde{P}$ be the set of all positions where π' and π^* differ. Note that none of the replacements can change the candidate assigned to p^*. Thus, it holds that $\widetilde{P} \subseteq P$.

Obviously, all candidates from π appear in π^*. For the sake of contradiction, let us assume that π^* is not a winning line-up of the extended election. Consequently, the summed score of π^* has decreased by the sequence of replacements described above, which implies that the summed scores of candidates on positions from $\widetilde{P}$ is higher in π' than in π: $\Lambda^{\mathrm{ut}}(\pi|_{\widetilde{P}}) < \Lambda^{\mathrm{ut}}(\pi'|_{\widetilde{P}})$. We claim that using this assumption, it is possible to modify π such that its utilitarian score increases, which leads to a contradiction, as we have assumed that π is a winning line-up. Let π_{alt} be a line-up resulting from copying π and then replacing all candidates assigned to positions in $\widetilde{P}$ by the candidates assigned to these positions in π'. This is possible as $p^* \notin \widetilde{P}$. By our assumption, π_{alt} has a higher utilitarian score than π. It remains to argue that π_{alt} is still a valid outcome, that is, every candidate is only assigned to at most one position. This directly follows from the observation that if $p \in \widetilde{P}$ with $\pi'_p \in \pi$, then at some point during the construction of π^*, π'_p is kicked out of the line-up and is assigned to position $\pi(\pi'_p)$ at some later point, which implies that $\pi(\pi'_p) \in \widetilde{P}$. □

Summary. From an axiomatic perspective, the utilitarian rule is probably the most appealing one. Indeed, it satisfies all axioms except weak reasonable satisfaction, which imposes quite rigorous restrictions on every rule fulfilling it. For the egalitarian rule, although this rule is pretty simple, both efficiency axioms are only weakly satisfied and, slightly surprisingly, score consistency and monotonicity are not satisfied at all. As in Proposition 1, most of the counterexamples for the egalitarian rule utilize that the OWA-vector of this rule contains some zeros. If one adapts the egalitarian rule such that the egalitarian outcome with the highest summed score is chosen, strong non-wastefulness, strong score Pareto optimality, strong monotonicity, and strong line-up enlargement monotonicity are additionally satisfied. Note that this variant of the egalitarian rule is still computable in polynomial time.

Both harmonic rules are less appealing from an axiomatic perspective because they do not satisfy any of our considered voting axioms. On a high level, as illustrated in Proposition 1, this is because the corresponding OWA-vectors consist of multiple different entries. Thereby, some modifications change the coefficients by which the scores are multiplied in some undesirable way. Overall, the contrast between the utilitarian rule and the harmonic rule is quite remarkable because they come both from the same class and work pretty similarly.

Turning to sequential rules, apart from reasonable dissatisfaction, the fixed-order rule outperforms the other two. However, a clear disadvantage of the fixed-order rule (and the two other sequential rules) is that a returned winning line-up might not be score Pareto optimal. As the max-first rule fulfills all axioms—except score consistency—at least weakly, and it is the only voting rule that is reasonably satisfying, this rule is also appealing if satisfaction of the candidates or positions is an important criterion. The min-first rule, on the other hand, does not even weakly satisfy any axiom except non-wastefulness. The reason for this is that in some elections modifying the election changes the order in which the positions are filled. The considerable differences between the max-first and min-first rule are quite surprising, as they first appear to be symmetric.

6 Experiments

In this section, we analyze the proposed rules experimentally. We first describe how we generated our data, and then present and analyze the results.

FIFA Data. In the popular video game FIFA 19 [9], 18.207 soccer players have their own avatar. To mimic the quality of a player, experts assessed them on 29 attributes, such as sprint speed, shot power, agility, and heading [14]. From these, the game computes the quality of a player on each possible position, such as left striker or right wing-back etc., in a soccer formation, using a weighted sum of the attribute scores with coefficients depending on the position in question [24]. We used this data to model a coach of a national team $\mathcal{X}$ that wants to find an "optimal" assignment of players with nationality $\mathcal{X}$ to positions in a formation he came up with. This can be modeled as a line-up election.

In soccer, there exist several possible formations consisting of different positions a team can play in. We fixed one formation, that is, a set of ten different positions. In the generated elections, the candidates are some selected number of players of a given nationality with the highest summed score, the positions are the field positions in a selected soccer formation, and the scores are those assigned by FIFA 19 for the player playing on a particular position. We considered 84 national teams (those with over ten field players in FIFA 19).

Synthetic Data. We also generated a synthetic dataset (M2) consisting of 1.000 line-up elections. Here, every candidate c has a ground qualification $\mu_c \in [0.4, 0.7]$ and every position p has a difficulty $\alpha_p \in [1, 2]$ both drawn uniformly at random. For each candidate c and for each position p, we sample a basic score $\beta_{c,p}$ from a Gaussian distribution with mean μ_c and standard deviation 0.05. The score of a candidate-position pair is then calculated as: $\text{score}_p(c) = \beta_{c,p}^{\alpha_p}$. The intuition behind this is that very talented candidates are presumably not strongly affected by the difficulty of a position, whereas weaker candidates may feel completely overburdened by a difficult position.

For both models, we normalized each line-up election by dividing all scores by the maximum score of a candidate-position pair.

Analysis of Experimental Results. We focus on the case of ten candidates on ten positions, as this is the most relevant scenario in the FIFA setting. However, we also conducted experiments for twenty candidates on ten positions and

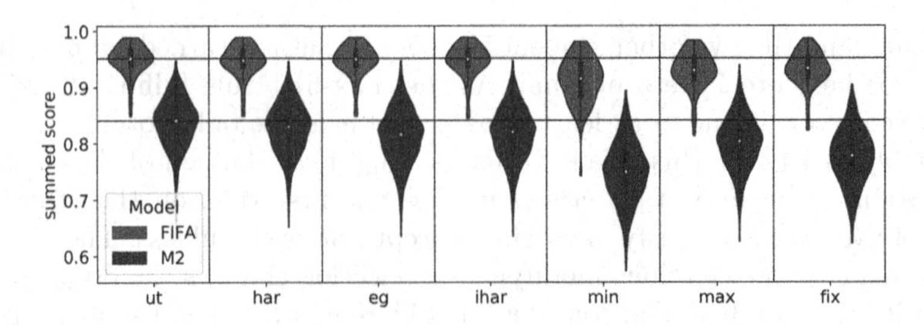

Fig. 1. Summed score of winning line-ups for different voting rules. The black horizontal lines indicate the median for the utilitarian rule on the two datasets.

twenty candidates on twenty positions, where we observed the same trends as in the case presented here. For settings with more candidates than positions, however, the differences between the rules are less visible. In the following, we refer to the (possibly invalid) outcome where every position gets assigned its best candidate as the *utopic* outcome. To visualize our results, we use violin plots. In a violin plot, the white dot represents the median, the thick bar represents the interquartile range, and the thin line represents the range of the data without outliers. Additionally, a distribution interpolating the data is plotted on both sides of the center.

Comparison of Data Models. To compare the datasets, we calculated different metrics designed to measure the amount of "competition" in instances. For example, we calculated the difference between the summed score of the utopic outcome and the summed score of a utilitarian outcome. Generally speaking, the M2 model produces instances with more "competition" than the FIFA data which helps us to make the differences between the rules more pronounced.

Comparison of Voting Rules. We compare the different voting rules by examining the following four metrics: (i) the summed score of the computed winning line-up π normalized by the summed score of the utopic outcome, (ii) the minimum score of a position in the winning line-up, (iii) the Gini coefficient[2] of the score vector, and (iv) the amount of reasonable dissatisfaction measured as the sum of all reasonable dissatisfactions, that is, the difference between the score of a position p in π and the score of a candidate c on p if the candidate-position pair (p, c) is reasonably dissatisfied. For the egalitarian rule, if multiple line-ups are winning, then we always select the line-up with the highest summed score.

Concerning the summed score (see Fig. 1), as expected, all four OWA-rules clearly outperform the three sequential rules. The OWA-rules all behave remarkably similar, especially on the FIFA data. The utilitarian rule produces by definition line-ups with the highest possible summed score, closely followed by the

[2] The Gini coefficient is a metric to measure the dispersion of a probability distribution; it is zero for uniform distributions and one for distributions with a unit step cumulative distribution function.

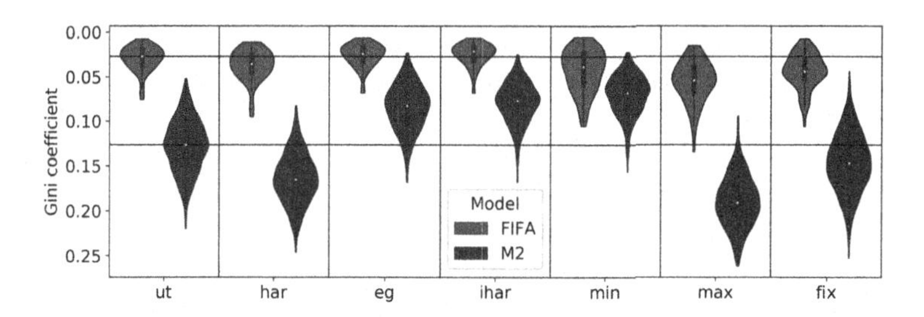

Fig. 2. Gini coefficient of score vector of line-ups for different voting rules.

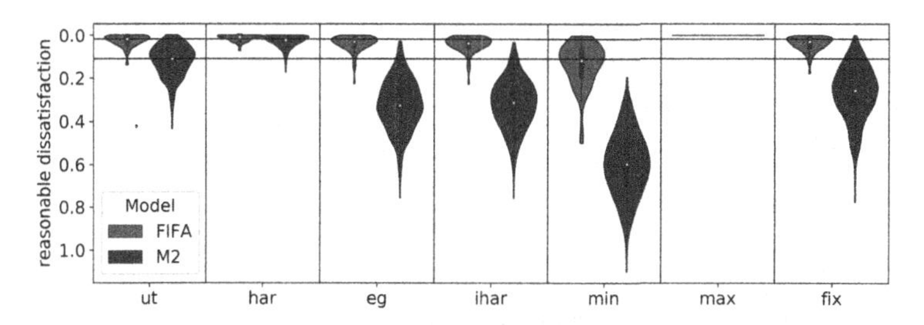

Fig. 3. Reasonable dissatisfaction in winning line-ups for different voting rules.

harmonic rule. Turning to the sequential rules, the min-first rule produces the worst results, while the max-first rule produces the best results.

Turning to the minimum score, the OWA-rules mostly outperform the sequential rules. The two rules producing the highest minimum score are the egalitarian rule and the inverse harmonic rule. The utilitarian and harmonic rule produce slightly worse results on the easier FIFA data and significantly worse results on the more demanding M2 data. Among the sequential rules, the min-first rule performs best, sometimes even outperforming the utilitarian rule, while the max-first rule produces the worst results. For the Gini coefficient (see Fig. 2), the overall picture is quite similar, i.e., rules that produce line-ups with a higher minimum score also produce line-ups that are more balanced in general.

Lastly, considering the amount of reasonable dissatisfaction (see Fig. 3), by definition, the max-first rule does not produce any reasonable dissatisfaction. Among the other rules, the harmonic rule produces the best results followed by the utilitarian rule. The egalitarian, inverse harmonic, and fixed-order rule all produce around double the amount of reasonable dissatisfaction, while the min-first rule produces significantly worse results by an additional factor of two.

Summary. Somewhat surprisingly, all OWA-rules outperform all sequential rules for all quantities, with only two exceptions: The min-first rule produces pretty balanced outcomes and the max-first rule produces no reasonable dissatisfaction. However, even if one aims at optimizing mainly one of these two quantities, it is usually recommendable to use an OWA-rule. Selecting the inverse harmonic rule instead of the min-first rule results in outcomes which are comparably balanced, have significantly higher summed and minimum scores, and have way less reasonable dissatisfaction. Using the harmonic rule instead of the max-first rule will introduce only little reasonable dissatisfaction in exchange for more balanced line-ups with higher summed and minimum scores. Comparing the different OWA-rules to each other, it is possible to differentiate the utilitarian and harmonic rule on the one side, from the egalitarian and inverse harmonic rule (which behave particularly similarly) on the other side: Rules from the former class tend to favor more imbalanced line-ups with lower minimum but higher summed score and less reasonable dissatisfaction.

7 Discussion

Overall, the considered OWA-rules produce better outcomes than the sequential rules. Nevertheless, sequential rules might sometimes be at an advantage, since sequential rules are, generally speaking, more transparent, more intuitive, and easier to explain. If one requires a sequential rule, either the fixed-order rule or the max-first rule should be chosen, as the min-first rule violates nearly all studied axioms and produces undesirable outcomes. Focusing on OWA-rules, the harmonic and inverse harmonic rules are rather to be avoided, as they fail to fulfill all considered voting and fairness axioms. Comparing the utilitarian and egalitarian rule, from an axiomatic perspective, the utilitarian rule is at an advantage, because it satisfies the most axioms among all considered rules. However, choosing between these two rules in practice should depend on the application, as the line-ups produced by these rules maximize different metrics. A rule that could somehow incorporate egalitarian *and* utilitarian considerations is the product rule, which selects outcomes with the highest product of scores.

For future work, it would be interesting to look at line-up elections that take as input the preferences of voters instead of aggregated scores. Analogously to multi-winner voting, new rules for this setting could, for example, focus on selecting proportional and diverse, instead of individually-excellent, line-ups. Such a path would also require developing appropriate axioms. Another possible line of future research could be to run experiments using preference data to examine the influence of the selected single-winner voting rule to aggregate the preferences into scores on the selected line-up. There are also several algorithmic problems that arise from our work. For instance, the computational complexity of computing a winning outcome under the (inverse) harmonic rule and even more generally of computing winning outcomes for arbitrary non-increasing OWA-rules is open.

References

1. Anshelevich, E., Dasgupta, A., Kleinberg, J.M., Tardos, É., Wexler, T., Roughgarden, T.: The price of stability for network design with fair cost allocation. SIAM J. Comput. **38**(4), 1602–1623 (2008)
2. Arrow, K.J., Sen, A., Suzumura, K. (eds.): Handbook of Social Choice and Welfare, vol. 2. Elsevier, Amsterdam (2010)
3. Avis, D.: A survey of heuristics for the weighted matching problem. Networks **13**(4), 475–493 (1983)
4. Aziz, H., Brill, M., Conitzer, V., Elkind, E., Freeman, R., Walsh, T.: Justified representation in approval-based committee voting. Soc. Choice Welf. **48**(2), 461–485 (2017)
5. Aziz, H., Lee, B.E.: Sub-committee approval voting and generalized justified representation axioms. In: AIES 2018, pp. 3–9 (2018)
6. Barberà, S., Coelho, D.: How to choose a non-controversial list with k names. Soc. Choice Welf. **31**(1), 79–96 (2008)
7. Boehmer, N., Bredereck, R., Faliszewski, P., Kaczmarczyk, A., Niedermeier, R.: Line-up elections: parallel voting with shared candidate pool. arXiv preprint arXiv:2007.04960 [cs.GT] (2020)
8. Brams, S.J., Fishburn, P.C.: Voting procedures. In: Handbook of Social Choice and Welfare, chap. 4, pp. 173–236 (2002)
9. EA Sports: FIFA 19. [CD-ROM] (2018)
10. Elkind, E., Faliszewski, P., Skowron, P., Slinko, A.: Properties of multiwinner voting rules. Soc. Choice Welf. **48**(3), 599–632 (2017)
11. Elkind, E., Ismaili, A.: OWA-based extensions of the Chamberlin-Courant rule. In: ADT 2015, pp. 486–502 (2015)
12. Faliszewski, P., Skowron, P., Slinko, A., Talmon, N.: Multiwinner voting: a new challenge for social choice theory. In: Trends in Computational Social Choice, pp. 27–47 (2017)
13. Fishburn, P.C.: Monotonicity paradoxes in the theory of elections. Discrete Appl. Math. **4**(2), 119–134 (1982)
14. Gadiya, K.: FIFA 19 complete player dataset (2019). https://www.kaggle.com/karangadiya/fifa19
15. Gale, D., Shapley, L.S.: College admissions and the stability of marriage. Am. Math. Mon. **69**(1), 9–15 (1962)
16. Garfinkel, R.S.: Technical note - an improved algorithm for the bottleneck assignment problem. Oper. Res. **19**(7), 1747–1751 (1971)
17. Garg, N., Kavitha, T., Kumar, A., Mehlhorn, K., Mestre, J.: Assigning papers to referees. Algorithmica **58**(1), 119–136 (2010)
18. Golden, B., Perny, P.: Infinite order Lorenz dominance for fair multiagent optimization. In: AAMAS 2010, pp. 383–390 (2010)
19. Goldsmith, J., Sloan, R.: The AI conference paper assignment problem. In: MPREF 2007 (2007)
20. Kuhn, H.W.: The Hungarian method for the assignment problem. In: 50 Years of Integer Programming, pp. 29–47 (2010)
21. Lang, J., Xia, L.: Sequential composition of voting rules in multi-issue domains. Math. Soc. Sci. **57**(3), 304–324 (2009)
22. Lesca, J., Minoux, M., Perny, P.: The fair OWA one-to-one assignment problem: NP-hardness and polynomial time special cases. Algorithmica **81**(1), 98–123 (2019)

23. Lian, J.W., Mattei, N., Noble, R., Walsh, T.: The conference paper assignment problem: using order weighted averages to assign indivisible goods. In: AAAI 2018, pp. 1138–1145 (2018)
24. Murphy, R.: FIFA player ratings explained (2018). https://www.goal.com/en-ae/news/fifa-player-ratings-explained-how-are-the-card-number-stats/1hszd2fgr7wgf1n2b2yjdpgynu. Accessed 08 July 2020
25. Skowron, P., Faliszewski, P., Lang, J.: Finding a collective set of items: from proportional multirepresentation to group recommendation. Artif. Intell. **241**, 191–216 (2016)
26. Yager, R.R.: On ordered weighted averaging aggregation operators in multicriteria decisionmaking. IEEE Trans. Syst. Man Cybern. Syst. **18**(1), 183–190 (1988)
27. Young, H.: An axiomatization of Borda's rule. J. Econ. Theory **9**(1), 43–52 (1974)
28. Zwicker, W.S.: Introduction to the theory of voting. In: Handbook of Computational Social Choice, pp. 23–56 (2016)

Recognizing Single-Peaked Preferences on an Arbitrary Graph: Complexity and Algorithms

Bruno Escoffier[1,2(✉)], Olivier Spanjaard[1], and Magdaléna Tydrichová[1]

[1] Sorbonne Université, CNRS, LIP6, 75005 Paris, France
{bruno.escoffier,olivier.spanjaard,magdalena.tydrichova}@lip6.fr
[2] Institut Universitaire de France, Paris, France

Abstract. We study in this paper single-peakedness on arbitrary graphs. Given a collection of preferences (rankings of alternatives), we aim at determining a connected graph G on which the preferences are single-peaked, in the sense that all the preferences are traversals of G. Note that a collection of preferences is always single-peaked on the complete graph. We propose an Integer Linear Programming formulation (ILP) of the problem of minimizing the number of edges in G or the maximum degree of a vertex in G. We prove that both problems are NP-hard in the general case. However, we show that if the optimal number of edges is $m-1$ (where m is the number of candidates) then any optimal solution of the continuous relaxation of the ILP is integer and thus the integrality constraints can be relaxed. This provides an alternative proof of the polynomial time complexity of recognizing single-peaked preferences on a tree. We prove the same result for the case of a path (an axis), providing here also an alternative proof of polynomiality of the recognition problem. Furthermore, we provide a polynomial time procedure to recognize single-peaked preferences on a pseudotree (a connected graph that contains at most one cycle). We also give some experimental results, both on real and synthetic datasets.

1 Introduction

Aggregating the preferences of multiple agents is a primary task in many applications of artificial intelligence, e.g., in preference learning [8,9] or in recommender systems [2,19]. The preferences of agents are often represented as rankings of alternatives, such as cultural products (books, songs, movies...), technological products, candidates for an election, etc. The aim of aggregation is then to produce a single ranking from a collection of rankings (called preference profile).

The preferences are said to be *structured* if they share some common structure [13]. For example, in a political context, it is conventional to assume that each individual preference is decreasing as one moves away from the preferred candidate along a left-right axis on the candidates, axis on which individuals all agree. Such preferences are called *single-peaked* [5]. They have been the subject of

T. Harks and M. Klimm (Eds.): SAGT 2020, LNCS 12283, pp. 291–306, 2020.
https://doi.org/10.1007/978-3-030-57980-7_19

much work in social choice theory. The most well-known result states that if preferences are single-peaked, then one escapes from Arrow's impossibility theorem. We recall that Arrow's theorem states that any unanimous aggregation function for which the pairwise comparison between two alternatives is independent of irrelevant alternatives is dictatorial. Furthermore, from the computational viewpoint, many NP-hard social choice problems (e.g., Kemeny rule and Young rule for rank aggregation [6], Chamberlin-Courant rule for proportional representation [4]) become polynomially solvable if the preferences are single-peaked.

Given the axiomatic and algorithmic consequences, the question of the computational complexity of recognizing single-peaked preferences is thus natural. Bartholdi and Trick [3] have proposed an $O(nm^2)$ algorithm to compute a compact representation of *all* axes on which a collection of n preferences on m candidates are single-peaked, or state that none exists. This complexity can be decreased to $O(nm + m^2)$ if one looks for only *one* possible axis [12].

Several classes of structured preferences have been proposed to generalize the single-peaked domain with respect to an axis, i.e., a path, to more general graphs. Given a set $\mathcal{C} = \{1, \ldots, m\}$ of candidates, a preference order $\succ$ over $\mathcal{C}$ is single-peaked on an undirected graph $\mathcal{G} = (\mathcal{C}, \mathcal{E})$ if it is a *traversal* of $\mathcal{G}$, i.e., for each $j \in \mathcal{C}$ the upper-contour set $\{i \in \mathcal{C} : i \succ j\}$ is connected. A preference profile is then single-peaked on $\mathcal{G}$ if every preference is single-peaked on $\mathcal{G}$. Demange studied single-peakedness on a tree [11]; Peters and Lackner on a circle [22].

Some good axiomatic properties remain valid when preferences are single-peaked on a tree: if the number of voters is odd, such profiles still admit a Condorcet winner (a candidate who is preferred over each other candidate by a majority of voters) [11], and returning this Condorcet winner is a strategyproof voting rule. On the contrary, every majority relation can be realized by a collection of preferences single-peaked on a circle [22], hence single-peaked preferences on a circle do not inherit the good axiomatic properties of single-peakedness on an axis regarding voting rules that are based on the majority relation.

The goal of this paper is to study the recognition problem for single-peaked preferences on arbitrary connected graphs. Although one cannot expect social choice theoretic guarantees from single-peakedness on arbitrary graphs (as it does not result in a domain restriction, any profile being single-peaked on the complete graph), a sparse graph gives some insights on the similarity between candidates/items. This could be used, e.g., in recommendation systems: assume that one discovers that the preferences over movies $\{1, 2, 3, 4, 5\}$ are single-peaked w.r.t. axis $(1, 2, 3, 4, 5)$; if ones knows that an agent likes movies 3 and 5, then it is natural to recommend movie 4. More generally, one can take advantage of single-peakedness on a sparse graph to make recommendations in the neighbourhood of liked items. Thereby, we focus here on determining a graph that minimizes (1) the number of edges or (2) the maximum degree of a vertex. This choice is motivated by the fact that these criteria are measures of sparsity of a graph (the sparser the graph is, the more informative), but also because they generalize known cases such as paths, cycles and trees. Let us indeed emphasize that the mathematical programming approach we propose to identify a graph generalizes the best known instances of the single-peaked recognition problem and provides a uniform treatment of them, leading to simple polynomial time algorithms.

Our Contribution. We propose Integer Linear Programming formulations (ILP) of problems (1) and (2), and we show that both of them are NP-hard (Sect. 2). Nevertheless, if the optimal value for problem (1) is $m - 1$ (where m is the number of candidates), we prove the integrality of the optimal basis solution of the Linear Program (LP) obtained by relaxing the integrality constraint (Sect. 3). This provides an alternative polynomial time method, based on a simple LP solver, to recognize single-peakedness on a tree, as a connected graph with m vertices and $m-1$ edges is a tree. By adding some constraints on the max degree of a vertex, we obtain the same result for the case of paths. As a last theoretical result, we prove that single-peakedness on a *pseudotree* (a connected graph containing at most one cycle) is recognizable in polynomial time (Sect. 4). We also provide some experimental results on real-world and synthetic datasets, where we measure the density of the graphs depending on the diversity of preferences of voters (Sect. 5). All along the article, some proofs are skipped due to lack of space.

Related Work. We briefly describe here some previous contributions that have addressed the concept of single-peakedness on arbitrary graphs, the optimization view of the recognition problem and the use of ILP formulations for computational social choice problems related to structured preferences:

- Nehring and Puppe defined a general notion of single-peaked preferences based on abstract betweenness relations between candidates [18]. In their setting, it is possible to define single-peaked preferences on a graph G by considering the *graphic betweenness relation*: candidate j is between candidates i and k if and only if j lies on a shortest path between i and k in G. A preference profile is then single-peaked on G if for every preference $\succ$, if i^* is the most preferred candidate w.r.t. $\succ$ and j is on a shortest path between i^* and k then $j \succ k$. This definition enables them to state general results regarding strategyproofness on restricted domains of preferences. Note that this definition of single-peakedness on a graph does not coincide with the one we use (see Sect. 2.1).

- Peters and Elkind showed how to compute in polynomial time a compact representation of *all* trees with respect to which a given profile is single-peaked [21]. This structure allows them to find in polynomial time trees that have, e.g., the minimum degree, diameter, or number of internal nodes among all trees with respect to which a given profile is single-peaked. On the contrary, they show that it is NP-hard to decide whether a given profile is single-peaked on a regular tree (where each vertex has degree either 1 or d), or if a profile is single-peaked on a tree isomorphic to a given tree. We provide here alternative proofs for some of the polynomial time results, based on linear programming arguments.

- Peters recently proposed ILP formulations for proportional representation problems, and showed that the binary constraint matrix is totally unimodular if preferences are single-peaked, because the matrix has then the consecutive ones property [20]. We recall that the vertices of a polyhedron defined by a totally unimodular constraint matrix are all integer, thus solving the linear programming relaxation yields an optimal solution to the original ILP problem. We also rely on linear programming for proving the polynomial time complexity of some of the recognition problems we tackle here.

2 ILP Formulation and Complexity

2.1 Problem Definition

We start by recalling some basic terminology of social choice theory. Given a set $\mathcal{C} = \{1, 2, \ldots, m\}$ of candidates and a set $\{v_1, \ldots, v_n\}$ of voters, each voter v_i ranks all candidates from the most to the least preferred one. This ranking is called the *preference* of v_i. It is simply a permutation of $\mathcal{C}$, which can be formally described as an m-tuple $R_i = (\pi_i(1), \ldots, \pi_i(m))$, where $\pi_i(k)$ is the k-th most preferred candidate of voter v_i. The set $\mathcal{P} = \{R_1, \ldots, R_n\}$ of preferences of all voters is called the *profile*. As emphasized in the introduction, several definitions of single-peakedness on an arbitrary graph can be found in the literature. In our study, we are using the following one [13]:

Definition 1 ***Single-peakedness on an arbitrary graph (SP).*** *Let $\mathcal{C}$ be a set of m candidates and $\mathcal{P}$ the profile of preferences of* n *voters. Let $\mathcal{G} = (\mathcal{C}, E)$ be a connected undirected graph. We say that $\mathcal{P}$ is* single-peaked on the graph *$\mathcal{G}$ (SP) if every $R_i \in \mathcal{P}$ is a traversal of $\mathcal{G}$, i.e., for each $R_i \in \mathcal{P}$ and for each $k \in \{1, \ldots m\}$, the subgraph of $\mathcal{G}$ induced by the vertices $\{\pi_i(1), \ldots, \pi_i(k)\}$ is connected.*

This definition coincides with the standard definition on an axis [5]/cycle [22]/tree [24] when $\mathcal{G}$ is a path/cycle/tree. Note that the definition based on shortest paths [18] mentioned earlier does *not* generalize single-peakedness on a circle as defined in [22]. When a profile $\mathcal{P}$ is single-peaked w.r.t. $\mathcal{G}$, for conciseness we say that $\mathcal{P}$ is *compatible* with $\mathcal{G}$ (or that $\mathcal{G}$ is compatible with $\mathcal{P}$).

Example 1. Consider the profile with 4 voters and 5 candidates: $R_1 : (1, 2, 3, 4, 5)$, $R_2 : (1, 3, 4, 2, 5)$, $R_3 : (2, 5, 3, 4, 1)$, and $R_4 : (3, 5, 4, 2, 1)$.

Note that, for R_1, the connectivity constraint applied to the first two candidates makes the edge $\{1, 2\}$ necessary in the graph. The same occurs for $\{1, 3\}$, $\{2, 5\}$ and $\{3, 5\}$. Thus, any solution contains the 4-cycle $(1, 2, 5, 3, 1)$ (in particular the profile is not SP on a tree or on a cycle). One can easily check that adding edge $\{3, 4\}$ makes a graph with 5 edges compatible with the profile, and this is the (unique) optimal solution if we want to minimize the number of edges.

Obviously, any profile is SP on the complete graph. However, this case is not interesting because it does not give any information about the preference structure. That is why we are looking for a *minimal* graph on which the profile is SP. The notion of minimality needs to be made more precise. In our study, we focus essentially on minimizing the number of graph edges. Another criterion we consider is the minimization of the maximum degree of vertices. Put another way, given a preference profile $\mathcal{P}$, we want to determine a graph $\mathcal{G}$ on which the profile is SP, so as to minimize either the number of edges of $\mathcal{G}$, or its (maximum) degree. We emphasize the fact that:

- minimizing the number of edges allows to detect when the profile is compatible with a tree (this occurs iff the minimum number of edges is $m-1$, since $\mathcal{G}$ is necessarily connected);
- minimizing the degree of $\mathcal{G}$ allows to detect when the profile is compatible with a cycle (this occurs iff there exists a graph $\mathcal{G}$ with maximum degree 2);
- combining the objective allows to detect when the profile is compatible with an axis: this occurs iff there is a graph $\mathcal{G}$ with maximum degree 2 and $m-1$ edges.

So the tackled problems generalize the most well known (tractable) recognition problems of single-peakedness.

2.2 ILP Formulation

We now present an ILP formulation of the tackled problems. We are looking for a graph $\mathcal{G}$ with m vertices. For each pair $\{k,l\} \subseteq \{1,\ldots,m\}$ of vertices, we define a binary variable $x_{\{k,l\}}$ which is equal to 1 if such that edge $\{k,l\}$ is present in graph $\mathcal{G}$, and 0 otherwise.

Hence, if we are minimizing the number of graph edges, the objective function $f(x)$ takes the form $f(x) = \sum_{\{k,l\}\subseteq\{1,\ldots,m\}} x_{\{k,l\}}$. If we are minimizing the maximum degree, then $f(x) = \max_{k\in\{1,\ldots,m\}} \sum_{l=1,l\neq k}^{m} x_{\{k,l\}}$. In this latter case, the classical way of linearizing $f(x)$ is to minimize an auxiliary variable z with the constraint $\sum_{l=1,l\neq k}^{m} x_{\{k,l\}} \leq z$, for all $k \in \{1,\ldots,m\}$.

Regardless of the objective function, the other constraints of the problem remain the same. Each $R_i = (\pi_i(1),\ldots,\pi_i(m))$ for $i \in \{1,\ldots,n\}$ has to be a graph traversal. In other words, for each $k \in \{2,\ldots m\}$, $\pi_i(k)$ is connected to at least one of the vertices $\pi_i(1),\ldots,\pi_i(k-1)$. This is formulated as the LP constraints $\sum_{j=1}^{k-1} x_{\{\pi_i(j),\pi_i(k)\}} \geq 1$.

To sum up, the ILP formulation of the tackled problems is

$$\min f(x)$$
$$\text{s.t.} \begin{cases} \sum_{j=1}^{k-1} x_{\{\pi_i(j),\pi_i(k)\}} \geq 1 & \forall i \in \{1,\ldots,n\}, \forall k \in \{2,\ldots,m\} \\ x_{\{k,l\}} \in \{0,1\} & \forall \{k,l\} \subseteq \{1,\ldots,m\} \end{cases}$$

2.3 Minimizing the Number of Edges

In this section, we study the computational complexity of the problem of minimizing the number of edges of $\mathcal{G}$. As a first observation, note that we cannot use the continuous relaxation of this ILP ($x_{\{k,l\}} \in [0,1]$) to solve the problem. The following example indeed shows that the optimal solution (when minimizing the number of edges) of this relaxation is not necessarily integer:

Example 2. Consider the profile with 3 voters and 4 candidates: $R_1 : (1, 2, 4, 3)$, $R_2 : (2, 3, 4, 1)$, and $R_3 : (1, 3, 4, 2)$.

From the two first options of each voter, we see immediately that the edges $\{1, 2\}$, $\{2, 3\}$ and $\{1, 3\}$ are necessarily present in the graph. Then, we observe that vertex 4 needs to be connected to at least one of vertices 1 and 2, at least one of vertices 2 and 3 and finally at least one of vertices 1 and 3. Consequently any integer solution of the problem will be a graph with at least 5 edges. However, there exists a fractional solution of the continuous relaxation with value 4.5:

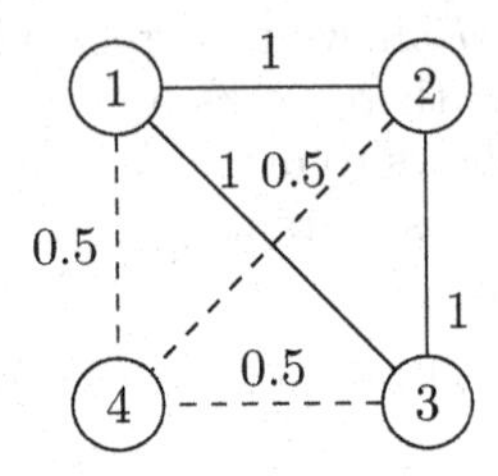

We now show that the problem is actually NP-hard.

Theorem 1. *Given a preference profile $\mathcal{P}$, it is NP-hard to find a graph compatible with $\mathcal{P}$ with a minimum number of edges.*

Proof. We use a polynomial time reduction from the set cover problem, known to be NP-hard [16], where given a finite set $\mathcal{U} = \{e_1, \dots e_n\}$ of elements, a set $\mathcal{S} = \{S_1, \dots, S_m\}$ of subsets of $\mathcal{U}$ and $k \in \mathbb{N}$, the question is to determine if there exists a subset $\mathcal{C} \subseteq \mathcal{S}$ of size k such that $\cup_{S \in \mathcal{C}} S = \mathcal{U}$.

From an instance of set cover, we define a preference profile $\mathcal{P}$ as follows:

(i) Let $\{S_1, \dots, S_m, z\}$ be a set of candidates.
(ii) Let $\{v_1, \dots, v_n\}$ be a set of voters. Let $S_{i_1}, \dots S_{i_l}$ be the subsets in $\mathcal{S}$ containing element $e_i \in \mathcal{U}$, and $S_{i_{l+1}}, \dots, S_{i_m}$ the other subsets in S. Then, the preference of voter v_i is defined as $R_i = (S_{i_1}, \dots, S_{i_l}, z, S_{i_{l+1}}, \dots, S_{i_m})$.
(iii) We add $\frac{m \cdot (m-1)}{2}$ voters $v_{\{i,j\}}, \{i, j\} \subseteq \{1, \dots, m\}$ such that

$$R_{\{i,j\}} = (S_i, S_j, \underbrace{S_1, \dots, S_m}_{\text{except } S_i, S_j}, z).$$

We prove that there exists a set cover of size k if and only if there exists a graph $\mathcal{G}$ compatible with $\mathcal{P}$ that has $\frac{m \cdot (m-1)}{2} + k$ edges.

Let $\mathcal{C}$ be a set cover solution of size k. We generate a graph $\mathcal{G}$ compatible with $\mathcal{P}$ in the following manner:

a) For each $\{i, j\} \in \{1, \dots, m\}^2, i \neq j$, the edge $\{S_i, S_j\}$ is in $\mathcal{G}$ - this is necessary for the preferences of type (*iii*) above to be SP on $\mathcal{G}$.
b) For each $i \in \{1, \dots, m\}$, the edge $\{S_i, z\}$ is in $\mathcal{G}$ if and only if $S_i \in \mathcal{C}$.

Hence, the subgraph formed by vertices $\{S_1, \ldots, S_m\}$ is a clique having $\frac{m\cdot(m-1)}{2}$ edges, and there are exactly k more edges adjacent to z - in total, $\mathcal{G}$ has $\frac{m\cdot(m-1)}{2} + k$ edges. As $k > 0$, the graph is connected and all preferences of type (iii) are SP on $\mathcal{G}$. Let R_i be one of the preferences of type (ii). We need to prove that z is connected to at least one of the vertices $S_{i_1}, \ldots, S_{i_l}$. As the sets $S_{i_1}, \ldots, S_{i_l}$ are the only sets of $\mathcal{S}$ containing the element e_i, and as $\mathcal{C}$ is a solution of the set cover instance, this is true due to b). So, $\mathcal{G}$ is a graph compatible with $\mathcal{P}$ that has $\frac{m\cdot(m-1)}{2} + k$ edges.

To prove the other implication, let $\mathcal{G}$ be a graph compatible with $\mathcal{P}$ that has $\frac{m\cdot(m-1)}{2} + k$ edges. As $\mathcal{G}$ is compatible with $\mathcal{P}$, the subgraph induced by the set of vertices $\{S_1, \ldots S_m\}$ must be a clique so that the preferences $R_{\{i,j\}}$ of type (iii) are SP on $\mathcal{G}$. Hence, this subgraph contains $\frac{m\cdot(m-1)}{2}$ edges, and so, there are exactly k edges adjacent to z. Let us define $\mathcal{C}$ containing S_i iff S_i is adjacent to z in $\mathcal{G}$. As $\mathcal{G}$ is compatible with $\mathcal{P}$, each preference R_i of type (ii) is SP on $\mathcal{G}$. It means that at least one of $S_{i_1}, \ldots, S_{i_l}$ is adjacent to z, so is in $\mathcal{C}$. As all these sets contains e_i, there is an element of $\mathcal{C}$ that covers e_i. The subset $\mathcal{C} \subseteq \mathcal{S}$ is thus a solution of size k of the set cover instance. □

2.4 Minimizing the Maximum Degree

We now consider our second objective function, namely the maximum degree of a vertex in the graph (to be minimized). We come up with similar results.

First, as for the minimization of the number of edges, the ILP formulation we have proposed in Sect. 2.2 is not integer Moreover, here again we show that the problem of minimizing the degree of $\mathcal{G}$ is NP-hard, by a similar reduction.

Theorem 2. *Given a preference profile $\mathcal{P}$, it is NP-hard to find a graph compatible with $\mathcal{P}$ with a minimum degree.*

3 Recognition of Trees and Paths

In this section, we focus on the tree and path recognition - given a profile $\mathcal{P}$, we are looking for a tree (or a path) on which the profile if SP.

Recognizing single-peaked preferences on a tree can be done using the combinatorial algorithm proposed by Trick [24]. As an alternative proof of this result, we show in this section that the continuous relaxation of the ILP formulations given in Sect. 2.2 can be used to solve this recognition problem in polynomial time: in fact, all (optimal) extremal solution are integral (Theorem 3). We show in Theorem 4 a similar result for the recognition of profiles SP on a path.

We start by recalling Trick's procedure [24], as we will use it in the proof of the results of the two theorems mentioned above.

Recognition of Profiles SP on a Tree [24]. Let $\mathcal{P}$ be a profile containing preference lists $\{R_1, \ldots, R_n\}$ of n voters over m candidates. Let k be a candidate placed at the last position by at least one voter. Trick shows that, if preferences are SP on a tree, then k must necessarily be a leaf. More formally, for each $i \in \{1, \dots, n\}$, let us denote by $A(k)_i$ the set of candidates ranked better than k by voter v_i if k is not ranked first by v_i; if k is ranked first by v_i, then $A(k)_i$ is the singleton containing the second most-preferred candidate of v_i. From $A(k) = \bigcap_{i=1}^{n} A(k)_i$, the following conclusions can be drawn:

- if $A(k) = \emptyset$, there does not exist a tree solution.
- Otherwise, $A(k)$ is the set of vertices the leaf k can be connected to.

In the latter case, the algorithm of Trick deletes k from all preferences, and repeats this process on the modified profile with preferences over $m - 1$ candidates.

Using LP to Recognize SP Preferences on a Tree or a Path. Let us consider the following continuous relaxation LP-SP (linear program for single-peakedness) of the ILP introduced in Sect. 2.2:

$$\min \sum_{\{k,l\} \subseteq \{1,\dots,m\}} x_{\{k,l\}}$$
$$\text{s.t.} \begin{cases} \sum_{j=1}^{k-1} x_{\{\pi_i(j),\pi_i(k)\}} \geq 1 \;\; \forall i \in \{1, \dots, n\}, k \in \{2, \dots, m\} \\ x_{\{k,l\}} \in [0, 1] \; \forall \{k, l\} \subseteq \{1, \dots, m\} \end{cases}$$

We show in Theorem 3 that we can use LP-SP to solve in polynomial time the problem to determine, given a profile, whether there exists or not a tree compatible with it.

Theorem 3. *If a profile $\mathcal{P}$ is compatible with a tree, then any extremal optimal solution x of LP-SP is integral, i.e., $x_{\{k,l\}} \in \{0, 1\}$ for any $\{k, l\} \subseteq \{1, \dots, m\}$.*

Proof. The proof is based on two properties of optimal solutions of LP-SP when the profile is compatible with a tree. These two properties allow to come up with a reformulation of the problem as a maximum flow problem, where there is a bijection between the solutions of LP-SP of value $m - 1$ and the (optimal) flows of value $m - 1$. The result then comes from the fact that any extremal solution of the flow problem (with integral capacity) is integral [1].

The first property states that all constraints of LP-SP are tight in a solution of value $m - 1$.

Property 1. If the optimal value of LP-SP is $m - 1$, then all constraints are tight in an optimal solution x^*: $\sum_{j=1}^{k-1} x^*_{\{\pi_i(j),\pi_i(k)\}} = 1$.

Now, let us consider that the profile is SP with respect to a tree. The recognition procedure recalled above starts by identifying a candidate, say m, ranked last in at least one ranking and such that $A(m) \neq \emptyset$. This procedure is then applied recursively, till there is only one candidate. For simplicity, let us assume that the first removed (identified) candidate is m, the second $m - 1$, and so on. Let us now focus on the step when candidate k is identified as a leaf (and then removed from the profile). To avoid confusion, we denote by $B(k)$ the set $A(k)$ at this step, i.e., when considering the profile restricted to the first k candidates.

Property 2. If the profile is SP on a tree, then in an optimal solution of LP-SP, for any candidate $k \geq 2$ we have $\sum_{j \in B(k)} x_{\{j,k\}} = 1$, and $x_{\{j,k\}} = 0$ for any $j \in \{1, \ldots, k-1\} \setminus B(k)$.

Now we reformulate the problem as a flow problem. From $\mathcal{P}$, we build a network (directed graph) R with:

- A source s, a destination t, and for each candidate k two vertices ℓ_k and r_k.
- We have an arc from s to each ℓ_k with capacity 1, and an arc from each r_k to t with capacity ∞.
- For each candidate k, we have an arc (ℓ_k, r_j) for each $j < k$. The capacity of this arc is 1 if $j \in B(k)$, and 0 otherwise.

Let us denote by ϕ a flow on this network, with $\phi(e)$ the flow on edge e. Note that ℓ_1 has no outgoing edge, so the optimal flow is at most $m - 1$.

We show that the correspondence $x_{\{k,j\}} = \phi(k, j)$ (for each $j < k$) is a bijection between solutions of value $m-1$ of LP-SP and (optimal) flows of value $m - 1$ in R.

Let ϕ be a flow of value $m-1$. As there is no flow through ℓ_1, there is a flow of value 1 through each $\ell_k, k > 1$. Since arc (k, j) has capacity 0 if $j \notin B(k)$, by flow conservation we have $\sum_{j \in B(k)} \phi(k, j) = 1$, which means that $\sum_{j \in B(k)} x_{\{k,j\}} = 1$. Now consider a voter v_i for which k is not ranked first. By the procedure of Trick, when k is identified as a leaf, all candidates in $B(k)$ are ranked before k, and the corresponding constraint is satisfied. This is true for all candidates and voters, so x is a feasible solution of LP-SP, of value $m - 1$.

Conversely, let x be a feasible solution of LP-SP of value $m-1$. From Property 2, we have $\sum_{j \in B(k)} x_{\{j,k\}} = 1$ for each candidate $k \geq 2$. This immediately gives a flow of value $m - 1$.

By integrality of extremal flows (any non integral optimal flow is a convex combination of integral flows), any extremal optimal solution of LP-SP is integral (when there exists a tree compatible with $\mathcal{P}$). $\square$

Let us now turn to the recognition of profiles SP on a path. A (connected) graph is a path iff it is a tree with degree at most 2. Hence, we consider the following ILP formulation where we minimize the number of edges and add constraints on the vertex degrees:

$$\min \sum_{\{k,l\} \subset \{1,\ldots,m\}} x_{\{k,l\}}$$

$$\text{s.t.} \begin{cases} \sum_{j=1}^{k-1} x_{\{\pi_i(j),\pi_i(k)\}} \geq 1 & \forall i \in \{1, \ldots, n\}, k \in \{2, \ldots, m\} \\ \sum_{l=1,l\neq k}^{m} x_{\{k,l\}} \leq 2 & \forall k \in \{1, \ldots, m\} \\ x_{\{k,l\}} \in \{0, 1\} & \forall \{k, l\} \subseteq \{1, \ldots, m\} \end{cases}$$

Clearly, a profile is compatible with a path iff the optimal value of the previous ILP is $m-1$. Let us call LP-SP2 the continuous relaxation. By using very similar arguments as above (same reformulation as a flow problem), one can prove the following result.

Theorem 4. *If a profile $\mathcal{P}$ is compatible with a path, then any extremal optimal solution of LP-SP2 is integral, i.e., $x_{\{k,l\}} \in \{0,1\}$ for any $\{k,l\} \subseteq \{1,\ldots,m\}$.*

4 Recognition of Pseudotrees

So far, we have seen that our minimization problem is NP-hard in the general case, but polynomially solvable in the case where the optimal solution is a tree. As a natural extension, we consider the problem to recognize profiles that are single-peaked with respect to a graph with $m-1+k$ edges, for some fixed k, thus allowing k more edges than in a tree. In this section, we consider the case $k=1$. A graph on m vertices with m edges is called a pseudotree. We show that recognizing if there exists a pseudotree compatible with a given profile can be done in polynomial time. We leave as open question the parameterized complexity of the problem when k is the parameter: would the problem be in XP? Or even in FPT?

Let us now deal with the case of pseudotree. Hence, the set of solutions we want to recognize is the class of connected graphs having (at most) m edges. To solve the problem in polynomial time, we devise an algorithm that first identifies the leaves of the pseudotree and then the cycle on the remaining vertices. The second step (cycle recognition) is done using the polynomiality of recognizing single-peakedness on a cycle [22]. For the first step, we need to modify the procedure recalled in Sect. 3. This procedure was able to correctly identify leaves when the profile was compatible with a tree, but it fails to correctly identify leaves when the underlying structure is a pseudotree. With a slight modification though, we obtain in Proposition 1 a necessary and sufficient condition for a candidate to be a leaf in a pseudotree. This is the stepping stone leading to the polynomiality of detecting whether a given profile is compatible with a pseudotree, stated in Theorem 5.

Example 3. Let us consider the profile on 4 voters and 5 candidates given in Example 1, for which there is a (unique) pseudotree compatible with it.

The procedure to detect leaves when looking for a tree focuses on candidates ranked last in some R_i, candidates 1 and 5 here, and $A(1) = A(5) = \emptyset$. Note that the whole profile is not compatible with a cycle, so we need somehow to first detect 4 as a leaf, and then detect that the candidates $1,2,3,5$ are SP with respect to a cycle.

The central property that allows to recognize profiles compatible with a pseudotree is given in the following proposition.

Proposition 1. *Let $\mathcal{P}$ be a preference profile, and suppose that a candidate i is such that $A(i) \neq \emptyset$. Then $\mathcal{P}$ is compatible with a pseudotree if and only if it is compatible with a pseudotree where i is a leaf.*

Proof. Let $\mathcal{G}$ be a pseudotree compatible with $\mathcal{P}$ where i is not a leaf. We transform $\mathcal{G}$ into a pseudo-tree $\mathcal{G}'$ compatible with $\mathcal{P}$ where i is a leaf. Let $j \in A(i)$.

Case 1: $\{i, j\} \in \mathcal{G}$. Let us first consider an easy case, where $\{i, j\} \in \mathcal{G}$. Then we build $\mathcal{G}'$ from $\mathcal{G}$ by simply replacing each edge $\{i, k\}$ (with $k \neq j$) by the edge $\{j, k\}$. Since for each voter either j is ranked before i, or i is first and j second, then this modification creates a graph $\mathcal{G}'$ compatible with all the preferences. Note that $\mathcal{G}'$ has (at most) as many edges as $\mathcal{G}$, so it is a pseudotree (or a tree, and we can add any edge to create a pseudotree).

Case 2: $\{i, j\} \notin \mathcal{G}$. Let us now consider the case where $\{i, j\} \notin \mathcal{G}$. Note that then j is ranked before i in all preferences (otherwise i is first and j is second, and the edge $\{i, j\}$ is forced to be in any compatible graph, a contradiction). Then we transform $\mathcal{G}$ into a graph $\mathcal{G}'$ which is a pseudotree containing the edge $\{i, j\}$, and then Case 1 applies to $\mathcal{G}'$. To do this, let us consider two subcases.

Case 2a. If, in $\mathcal{G}$, in all (simple) paths from j to i the predecessor of i is the same vertex u. Then we create $\mathcal{G}'$ by replacing the edge $\{u, i\}$ by the edge $\{j, i\}$. Consider a voter v. Since j is ranked before i by v, then u is ranked before i by v (the subgraph induced by i and the candidates ranked before him by v is connected and contains i and j, so it contains u). Then the modification does not affect u (it is still connected to one of the candidates ranked before him), and i is now connected to j.

Case 2b. In the other case, in $\mathcal{G}$ there are two simple paths from j to i such that the predecessor of i is u_1 in the first one and $u_2 \neq u_1$ in the second one (note that there cannot be more than 2 since $\mathcal{G}$ is a pseudotree). We build $\mathcal{G}'$ from $\mathcal{G}$ by deleting edges $\{u_1, i\}$ and $\{u_2, i\}$, and adding edges $\{i, j\}$ and $\{u_1, u_2\}$.

Consider a voter v. Since v prefers j to i and the subgraph of $\mathcal{G}$ induced by the candidates up to i in the ranking of v is connected, then u_1 or u_2 is ranked before i by v, say u_1 (we assume wlog that u_1 is preferred to u_2 by v). Then we see that $\mathcal{G}'$ is compatible with the preference of v: indeed, when considering candidates one by one in the order of v, the only modification holds for u_2, which is now connected to u_1 (ranked before him), and for i, which is now connected to j (ranked before him). □

Proposition 2. *If a profile $\mathcal{P}$ is compatible with a pseudotree, then either there exists a candidate i such that $A(i) \neq \emptyset$, or $\mathcal{P}$ is compatible with a cycle.*

Consider now the following procedure DETECT_PSEUDOTREE:

1. Set $E' = \emptyset$
2. While there are at least 4 candidates, and a candidate i such that $A(i) \neq \emptyset$:
   (a) Add edge $\{i, j\}$ to E' for some $j \in A(i)$.
   (b) Remove i from the profile.
3. Detect if there is a cycle C which is compatible with the (remaining) profile.
   (a) If YES: output $E' \cup C$.
   (b) If NO: output NO.

Theorem 5. *Given a preference profile $\mathcal{P}$ on at least 3 candidates, the procedure* DETECT_PSEUDOTREE *is polynomial-time and returns a pseudotree compatible with $\mathcal{P}$ if some exists, or returns NO otherwise.*

Note that the generalization of this polynomiality result to connected graphs with $(n-1+k)$ edges seems to require new techniques (even for fixed k, i.e. to show that the problem is in XP when parameterized by k). Indeed, an enumeration of all subsets of k edges does not allow to reduce the problem to trees. Procedure DETECT_PSEUDOTREE does not seem to generalize either, as it specifically relies on the decomposition of the solution into one cycle and leaves.

5 Experimental Study

We carried out numerical experiments[1] on real and randomly generated instances of the problems tackled in the paper. In the case of real data, we compare the optimal solution of the ILP to that of its continuous relaxation. We also focus on the ability to detect structure in voters' preferences depending on the election context. To go further, we use randomly generated instances to study structural aspects of solutions; we notably study the graph density depending on the number of voters and on the dispersion of their opinions.

5.1 Numerical Tests on Real Data

We used PrefLib data sets available on `www.preflib.org` to perform our numerical tests [17]. While this database offers four different types of data, only the ED (Election Data) type is relevant for our study. Among the ED data sets, we used the complete strict order lists (which correspond to files with `.soc` extension).

At the time we carried out these experiments, 315 data files of this type were available in PrefLib, however, many of them were not adapted to our study for several reasons. The first one is that many elections dealt with only 3 or 4 candidates and a great number of voters, hence the obtained graph was, unsurprisingly, always complete. We also met the opposite problem when there were very few voters, typically 4, so there was no point in looking for some general structure. Thus, in practice, there were 25 real data files usable for our purposes.

The detailed results are given in Fig. 1 The tackled optimization problem was to determine a graph with a *minimal number of edges*.

As for computational issues, the ILP formulation turned out to efficiently solvable for all instances, as for all of them an optimal graph has been obtained in about 40 ms. Moreover, it is noticeable that the (continuous relaxation) linear programming formulation *always* returned an integer solution.

For the results, we note that though no data set is compatible with an axis, most are compatible with a tree or a pseudo-tree. This is true in particular for the Data Set `ED-6` which contains figure skating rankings from various international competitions during the 1998 season. The possible interpretation of these results is that, even though the rankings are based on subjective opinions of the judges, there is something like a "true ranking" behind as some skaters are objectively

[1] All tests were performed on a Intel Core i7-1065G7 CPU with 8 GB of RAM under the Windows OS. We used the IBM Cplex solver for the solution of ILPs.

Set	File	#cand.	#vot.	#edges	Set	File	#cand.	#vot.	#edges
ED-6	3	14	9	13 (tree)	ED-6	34	23	9	22 (tree)
ED-6	4	14	9	13 (tree)	ED-6	35	18	9	17 (tree)
ED-6	7	23	9	22 (tree)	ED-6	36	18	9	17 (tree)
ED-6	8	23	9	22 (tree)	ED-6	37	19	9	18 (tree)
ED-6	11	20	9	20 (ps-tree)	ED-6	44	20	9	19 (tree)
ED-6	12	20	9	20 (ps-tree)	ED-6	46	30	9	30 (ps-tree)
ED-6	18	24	9	23 (tree)	ED-6	48	24	9	23 (tree)
ED-6	21	18	7	17 (tree)	ED-9	1	9	146	8 (tree)
ED-6	22	18	7	17 (tree)	ED-9	2	7	153	6 (tree)
ED-6	28	24	9	23 (tree)	ED-12	1	11	30	25
ED-6	29	19	9	23	ED-14	1	10	5000	45 (clique)
ED-6	32	23	9	23 (ps-tree)	ED-32	2	6	15	7
ED-6	33	23	9	22 (tree)					

Fig. 1. Minimal number of edges (fifth/last column) on real data sets from PrefLib. Specific structures (ps-tree stands for pseudotree) are indicated in parentheses.

better than other ones. Thus, the rankings given by the judges can be viewed as biased observations of the true ranking, so that they are quite close. On the opposite, the instances leading to denser graphs correspond to preferences over T-shirt designs (ED-12) and over sushis (ED-14), consistent with the intuition that there is probably no strong structure behind.

5.2 Experimental Study on Randomly Generated Data

The experimental study on real data revealed some interesting information. Nevertheless, it is limited by the small amount of data available. Here, we conduct experiments on random data in order to study the structure of solutions. As mentioned above, in some contexts we can assume that the voter's preferences are biased observations of a "true" ranking. This idea can be modeled using the Mallows distribution on rankings. In this model, the "true" ranking is called *central permutation* and its probability is the highest one. The probability of other permutations decreases with the Kendall tau distance from the central permutation. Formally, let R_0 be the central permutation. The probability of a permutation R is $P(R) = \frac{\exp(-\theta d(R,R_0))}{\psi(\theta)}$, where $d(.,.)$ is the Kendall tau distance, $\theta \geq 0$ is a dispersion parameter modeling the opinion heterogeneity, and $\psi(\theta)$ is a normalisation constant. If $\theta = 0$, the uniform distribution is obtained. The greater the value of θ, the more the voters agree on the central permutation.

We used the PerMallows R package[2] for generating the random data according to the Mallows model. The number of candidates was set to $m = 20$, the value of θ varied from 0 to 1 by step of 0.1. The number of voters n varied from 20 to 100 by step of 10. For each pair (θ_0, n_0) of parameter values, the results are averaged over

[2] https://cran.r-project.org/web/packages/PerMallows/index.html.

1000 randomly drawn preference profiles. The curves in Fig. 2 show the evolution of the graph density according to these parameter values.

In the best case, the obtained solution is a tree, hence, the density is $(m-1)/\frac{m(m-1)}{2} = \frac{2}{m}$. As we set $m = 20$, this corresponds to a density of 0.1. The function representing the graph density seems indeed to converge to the constant function of value 0.1 while the value of θ increases and the preferences in the profile become similar (the curves get closer and closer to the x-axis). Put another way, the density captures the similarity of voters' preferences, as clearly the higher θ the lower the curve. On the contrary, the graph density becomes of course higher when the number n of voters increases. Nevertheless, note that, even for 100 voters, the graph is still quite far from being complete. Besides, the slope of the curve decreases with n. During our experiments, we plotted functions $1 - \log(\mathit{density})$ and obtained a set of (approximate) straight lines, thus indicating that the convergence towards density 1 (complete graphs) is of the form $1 - e^{-\lambda_\theta n}$, where $\lambda_\theta > 0$ is a parameter decreasing with θ.

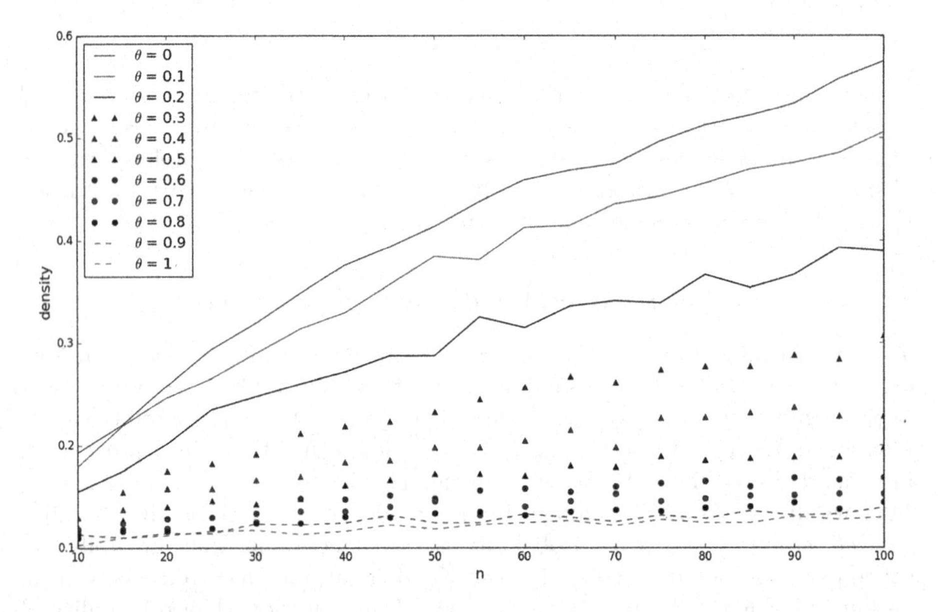

Fig. 2. Density of the graph according to parameters θ and n (with $m = 20$).

6 Concluding Remarks

While the generalization of single-peakedness to arbitrary graphs makes it more plausible to learn some preference structures in applications, an interesting research direction would be to formulate the recognition problem as the determination of a *maximum likelihood* graph (while possibly imposing that the graph is an axis or a tree) on which the preferences are single-peaked. Put another way, one would relax the requirement of perfect compatibility of the graph with the observed preferences in order to facilitate structure learning on real preference data[3]. As shown in the experiments, for an high preference heterogeneity, the graph can indeed become very dense when the number of voters increases (as is often the case in real applications), thus making essential to consider a more flexible view of single-peakedness. Some interesting works in this direction have already been carried out (recognition of nearly structured preferences, e.g., [7,10,14,15]), but a lot remains to be done to make the approach fully operational.

References

1. Ahuja, R.K., Magnanti, T.L., Orlin, J.B.: Network flows. Alfred P. Sloan School of Management, Massachusetts Institute of Technology (1988)
2. Andersen, R., et al.: Trust-based recommendation systems: an axiomatic approach. In: WWW 2008, pp. 199–208. ACM (2008)
3. Bartholdi III, J., Trick, M.A.: Stable matching with preferences derived from a psychological model. Oper. Res. Lett. **5**(4), 165–169 (1986)
4. Betzler, N., Slinko, A., Uhlmann, J.: On the computation of fully proportional representation. J. Artif. Intell. Res. **47**, 475–519 (2013)
5. Black, D.: On the rationale of group decision-making. J. Polit. Econ. **56**(1), 23–34 (1948)
6. Brandt, F., Brill, M., Hemaspaandra, E., Hemaspaandra, L.A.: Bypassing combinatorial protections: polynomial-time algorithms for single-peaked electorates. J. Artif. Intell. Res. **53**, 439–496 (2015)
7. Bredereck, R., Chen, J., Woeginger, G.J.: Are there any nicely structured preference profiles nearby? Math. Soc. Sci. **79**, 61–73 (2016)
8. Cheng, W., Hüllermeier, E.: A new instance-based label ranking approach using the Mallows model. In: Yu, W., He, H., Zhang, N. (eds.) ISNN 2009. LNCS, vol. 5551, pp. 707–716. Springer, Heidelberg (2009). https://doi.org/10.1007/978-3-642-01507-6_80
9. Clémençon, S., Korba, A., Sibony, E.: Ranking median regression: learning to order through local consensus. In: ALT, pp. 212–245 (2018)
10. Cornaz, D., Galand, L., Spanjaard, O.: Bounded single-peaked width and proportional representation. In: ECAI, vol. 12, pp. 270–275 (2012)
11. Demange, G.: Single-peaked orders on a tree. Math. Soc. Sci. **3**(4), 389–396 (1982)
12. Doignon, J., Falmagne, J.: A polynomial time algorithm for unidimensional unfolding representations. J. Algorithms **16**(2), 218–233 (1994)
13. Elkind, E., Lackner, M., Peters, D.: Structured preferences. In: Trends in Computational Social Choice, Chapter 10, pp. 187–207. AI Access (2017)
14. Erdélyi, G., Lackner, M., Pfandler, A.: Computational aspects of nearly single-peaked electorates. J. Artif. Intell. Res. **58**, 297–337 (2017)
15. Faliszewski, P., Hemaspaandra, E., Hemaspaandra, L.A.: The complexity of manipulative attacks in nearly single-peaked electorates. Artif. Intell. **207**, 69–99 (2014)

[3] Note that this problem differs from that studied by Sliwinski and Elkind [23], where voters' preferences are independently sampled from rankings that *are* single-peaked on a given tree, and they manage to identify a maximum likelihood tree.

16. Garey, M.R., Johnson, D.S.: Computers and Intractability: A Guide to the Theory of NP-Completeness. W. H. Freeman & Co., New York (1979)
17. Mattei, N., Walsh, T.: PrefLib: a library for preferences http://www.preflib.org. In: Perny, P., Pirlot, M., Tsoukiàs, A. (eds.) ADT 2013. LNCS (LNAI), vol. 8176, pp. 259–270. Springer, Heidelberg (2013). https://doi.org/10.1007/978-3-642-41575-3_20
18. Nehring, K., Puppe, C.: The structure of strategy-proof social choice-Part I: general characterization and possibility results on median spaces. J. Econ. Theory **135**(1), 269–305 (2007)
19. Pennock, D.M., Horvitz, E., Giles, C.L.: Social choice theory and recommender systems: analysis of the axiomatic foundations of collaborative filtering. In: AAAI/IAAI, pp. 729–734 (2000)
20. Peters, D.: Single-peakedness and total unimodularity: new polynomial-time algorithms for multi-winner elections. In: AAAI, pp. 1169–1176 (2018)
21. Peters, D., Elkind, E.: Preferences single-peaked on nice trees. In: AAAI, pp. 594–600 (2016)
22. Peters, D., Lackner, M.: Preferences single-peaked on a circle. In: AAAI, pp. 649–655 (2017)
23. Sliwinski, J., Elkind, E.: Preferences single-peaked on a tree: sampling and tree recognition. In: IJCAI, August 2019
24. Trick, M.A.: Recognizing single-peaked preferences on a tree. Math. Soc. Sci. **17**(3), 329–334 (1989)

A General Framework for Computing the Nucleolus via Dynamic Programming

Jochen Könemann and Justin Toth(✉)

University of Waterloo, Waterloo, ON N2L 3G1, Canada
{jochen,wjtoth}@uwaterloo.ca

Abstract. This paper defines a general class of cooperative games for which the nucleolus is efficiently computable. This class includes new members for which the complexity of computing their nucleolus was not previously known. We show that when the minimum excess coalition problem of a cooperative game can be formulated as a hypergraph dynamic program its nucleolus is efficiently computable. This gives a general technique for designing efficient algorithms for computing the nucleolus of a cooperative game. This technique is inspired by a recent result of Pashkovich [24] on weighted voting games. However our technique significantly extends beyond the capabilities of previous work. We demonstrate this by applying it to give an algorithm for computing the nucleolus of b-matching games in polynomial time on graphs of bounded treewidth.

Keywords: Combinatorial optimization · Algorithmic game theory · Dynamic programming

1 Introduction and Related Work

Cooperative game theory studies situations in which individual agents form coalitions to work together towards a common goal. It studies questions regarding what sort of coalitions will form and how they will share the surplus generated by their collective efforts. A *cooperative game* is defined by an ordered pair $([n], \nu)$ where $[n]$ is a finite set of players (labelled $1, \ldots, n$), and ν is a function from subsets of $[n]$ to $\mathbb{R}$ indicating the value earned by each coalition.

This paper studies the computational complexity of one of the most classical, deep, and widely applicable solution concepts for surplus division in cooperative games, the *nucleolus*. In particular we study the relationship between the nucleolus, finding the minimum excess of a coalition, congruency-constrained optimization, and dynamic programming. Our first result unifies these areas and provides a general method for computing the nucleolus.

We acknowledge the support of the Natural Sciences and Engineering Research Council of Canada (NSERC). Cette recherche a été financée par le Conseil de recherches en sciences naturelles et en génie du Canada (CRSNG).

T. Harks and M. Klimm (Eds.): SAGT 2020, LNCS 12283, pp. 307–321, 2020.
https://doi.org/10.1007/978-3-030-57980-7_20

Theorem 1. *For any cooperative game (n, ν), if the minimum excess coalition problem on (n, ν) can be solved in time T via an integral dynamic program then the nucleolus of (n, ν) can be computed in time polynomial in T.*

Pashkovich [24] showed how to reduce the problem of computing the nucleolus for weighted voting games to a congruency-constrained optimization problem. Pashkovich then shows how to solve this congruency-constrained optimization problem for this specific class of games via a dynamic program. In Sect. 3 we abstract his reduction to the setting of computing the nucleolus of general combinatorial optimization games.

Our main technical achievement is showing that adding congruency constraints to dynamic programs modelled by a directed acyclic hypergraph model inspired by the work of Campbell, Martin, and Rardin [7] adds only a polynomial factor to the computational complexity. This is the content of Theorem 3, which is instrumental in demonstrating Theorem 1. Our formal model of dynamic programming, where solutions correspond to directed hyperpaths in a directed acyclic hypergraph, is described in Sect. 4. Proving Theorem 3 requires significant new techniques beyond [24]. The series of lemmas in Sect. 4.1 take the reader through these techniques for manipulating directed acyclic hypergraph dynamic programs.

We show how Theorem 1 not only generalizes previous work on computing the nucleolus, but significantly extends our capabilities to new classes of combinatorial optimization games that were not possible with just the ideas in [24]. As we explain in Sect. 1.3, matching games are central to the study of combinatorial optimization games. The problem of computing the nucleolus of weighted matching games was a long-standing open problem [11,16] resolved only recently [20], nearly twenty years after it was first posed. The frontier for the field has now moved to b-matching games, for which computing the nucleolus is believed to be NP-hard in general due to the result in [5] which shows computing leastcore allocations to be NP-hard even in the unweighted, bipartite case with $b \equiv 3$. In Sect. 5 we give a result which significantly narrows the gap between what is known to be tractable and what is known to be intractable in that area.

Theorem 2. *For any cooperative b-matching game on a graph whose treewidth is bounded by a constant, the nucleolus can be computed in polynomial time.*

To achieve this result we give a dynamic program for computing the minimum excess coalition of a b-matching game in Lemma 11 then apply Theorem 1. This dynamic program necessarily requires the use of dynamic programming on hypergraphs instead of just simple graphs, motivating the increased complexity of our model over previous work.

Proofs of our results appear in the full version of the paper [19].

1.1 The Nucleolus

When studying the question of surplus division, it is commonly desirable that shares will be split so all players have an incentive to work together, i.e. that the *grand coalition* forms. A vector $x \in \mathbb{R}^n$ is called an allocation, and if that vector satisfies $x([n]) = \nu([n])$ (efficiency) and $x_i \geq \nu(\{i\})$, for all $i \in [n]$, (individual rationality) we call x an *imputation*. We denote the set of imputations of (n, ν) by $I(n, \nu)$. For any $S \subseteq [n]$ we define $x(S) - \nu(S)$ to be the *excess* of S with respect to allocation x. The following linear program maximizes the minimum excess:

$$\max\{\epsilon : x(S) \geq \nu(S) + \epsilon, \forall S \subseteq [n] \text{ and } x \in I(n, \nu)\} \qquad (P_1)$$

We call this the *leastcore linear program.* For any ϵ be let $P_1(\epsilon)$ denote the set of allocations x such that (x, ϵ) is a feasible solution to (P_1). If we let ϵ_1 denote the optimal value of (P_1) then we call $P_1(\epsilon_1)$ the *leastcore* of the cooperative game.

For an imputation $x \in I(n, \nu)$ let $\theta(x) \in \mathbb{R}^{2^n-2}$ be the vector obtained by sorting the list of excess values $x(S) - \nu(S)$, for each $\emptyset \neq S \subset [n]$, in non-decreasing order.

Definition 1. *The nucleolus is the imputation which lexicographically maximizes $\theta(x)$, formally: the nucleolus is equal to* $\arg lex\max\{\theta(x) : x \in I(n, \nu)\}$.

The nucleolus was first defined by Schmeidler [29]. In the same paper, Schmeidler showed the nucleolus to be a unique allocation and a continuous function of ν. The nucleolus is a classical object in game theory, attracting attention for its geometric beauty [22], and its surprising applications. The most ancient of which is the application of the nucleolus as a bankruptcy division scheme in the Babylonian Talmud [2]. Some notable applications of the nucleolus include but are not limited to water supply management [1], fair file sharing on peer-to-peer networks [23], resource sharing in job assignment [30], and airport pricing [6].

1.2 Computing the Nucleolus

Multiple approaches exist for algorithmically finding the nucleolus of a cooperative game. The most ubiquitous of which is Maschler's Scheme [22] which operates by solving a hierarchy of at most n linear programs, the last of which has the nucleolus as its unique optimal solution. In Sect. 2 we elaborate on Maschler's Scheme and a natural relaxation thereof. An alternative method of computing the nucleolus via characterization sets was proposed independently by Granot, Granot, and Zhu [13] and Reinjerse and Potters [27].

The complexity of computing the nucleolus varies dramatically depending on how the cooperative game (n, ν) is presented as input. If the function ν is presented explicitly, by giving as input the value of $\nu(S)$ for each $S \subseteq [n]$, then Maschler's Scheme can be used to compute the nucleolus in polynomial time. The issue in this case is that the size of the specification of ν is exponential in the number of players and so the computation is trivial. We are interested cooperative games where ν can be determined implicitly via some auxiliary information given as input, which we call a *compact representation* of (n, ν).

One prominent example of a cooperative game with a compact representation is the class of *weighted voting games.* In a weighted voting game, each player $i \in [n]$ is associated with an integer weight $w_i \in \mathbb{Z}$. Additionally a threshold value $T \in \mathbb{Z}$ is given. For each $S \subseteq [n]$ the value of $\nu(S) \in \{0, 1\}$ is 1 if and only if $w(S) \geq T$.

It is not hard to see that (n, ν) is completely determined by (w, T). In this case (w, T) is a compact representation of the weighted voting game (n, ν). Even though they may appear simple at first, weighted voting games can have a lot of modelling power. In fact the voting system of the European Union can be modelled by a combination of weighted voting games [3]. In [10] Elkind, Goldberg, Goldberg, and Wooldridge show that the problem of computing the nucleolus of a weighted voting game is NP-hard, in fact even the problem of testing if there is a point in the leastcore of a weighted voting game that assigns a non-zero payoff to a given player is NP-complete. Pashkovich [24] later followed up with an algorithm based on Maschler's Scheme which solves $O(n)$ linear programs, each in pseudopolynomial time, and thus computes the nucleolus of a weighted voting game in pseudopolynomial time.

Pashkovich's result crucially relies on the existence of a well-structured dynamic program for knapsack cover problems which runs in pseudopolynomial time. Theorem 1 and Sect. 5 place Pahskovich's algorithm in the context of a general framework for computing the nucleolus of cooperative games where a natural associated problem has a dynamic program: the *minimum excess coalition problem.*

Definition 2. *In the minimum excess coalition problem the given input is a compact representation of a cooperative game* (n, ν) *and an imputation* x. *The goal is to output a coalition* $S \subseteq [n]$ *which minimizes excess, i.e.* $x(S) - \nu(S)$, *with respect to* x.

1.3 Combinatorial Optimization Games

A very general class of cooperative games with compact representations comes from the so-called *cooperative combinatorial optimization games.* In games of this class some overarching combinatorial structure is fixed on the players, and for each subset S of players, $\nu(S)$ can be determined by solving an optimization problem on this structure. Many classes of combinatorial optimization games can be defined and the complexity of their nucleoli have been studied leading to polynomial time algorithms, such as fractional matching, cover, and clique games [8], simple flow games [26], assignment games [31] and matching games [16]. Fleiner, Solymosi and Sziklai used the concept of dually essential coalitions [32] to compute the nucleolus of a large class of directed acyclic graph games [33] via the characterization set method. Other cases have led to NP-hardness proofs, such as flow games [9], weighted voting games [10], and spanning tree games [12].

A prominent example of combinatorial optimization games is matching games. In matching games the players are vertices of a graph G and $\nu(S)$ is equal to the size of the largest matching on $G[S]$. The question of computing the nucleolus of weighted matching games in polynomial time was open for a long time [11,16]. Solymosi and Raghavan [31] gave an algorithm for computing the nucleolus of matching games on bipartite graphs. Biró, Kern and Paulusma gave a combinatorial algorithm for computing the nucleolus of weighted matching games with a non-empty core [4]. Recently Koenemann, Pashkovich, and Toth [20] resolved the question by giving a compact formulation for each linear program in Maschler's Scheme for weighted matching games with empty core.

A natural generalization of matching games is to weighted b-matching games. In weighted b-matching games a vector $b \in \mathbb{Z}^{V(G)}$ and vector $w \in \mathbb{R}^{E(G)}$ are given in addition to the graph G. The value of $S \subseteq [n]$ is equal to the maximum w-weight subset of edges in $G[S]$ such that each playere $v \in V(G)$ is incident to at most b_v edges. In [5] they show how to test if an allocation is in the core of b-matching games when $b \leq 2$, and they show that for matching games where $b \equiv 3$ deciding if an allocation is in the core is coNP-complete. This result likely means that computing the nucleolus of $b \equiv 3$-matching games is NP-hard. In [20] they show how to separate over the leastcore of any $b \leq 2$-matching game. The question of computing the nucleolus of b-matching games remains open. By the preceding complexity discussion, it is highly likely that it is necessary to impose some structure on b-matching games to compute their nucleolus in polynomial time. In Theorem 2 we impose the structure of bounded treewidth and use our general framework to give an algorithm which computes the nucleolus of weighted b-matching games on graphs which have bounded treewidth.

2 Maschler's Scheme

The most prominent technique for computing the nucleolus is Maschler's Scheme [22]. To define Maschler's Scheme we need the notion of a fixed set for a polyhedron. For any polyhedron Q, we define the set $\text{Fix}(Q)$ as

$$\text{Fix}(Q) := \{S \subseteq [n] : \exists c \in R \text{ such that } \forall x \in Q, x(S) = c\}.$$

In Maschler's Scheme a sequence of linear programs $(P_1), (P_2), \ldots, (P_N)$ is computed where the ith linear program $(i \geq 2)$ is of the form

$$\max\{\epsilon : x(S) \geq \nu(S) + \epsilon, \forall S \notin \text{Fix}(P_{i-1}(\epsilon_{i-1})) \text{ and } x \in P_{i-1}(\epsilon_{i-1})\} \qquad (P_i)$$

and the first linear program is the leastcore linear program (P_1). The method terminates when the optimal solution is unique (yielding the nucleolus), and this happens after at most n rounds [25], since the dimension of the set of characteristic vectors of sets in $\text{Fix}(P_i(\epsilon_i))$ increases by at least one in each iteration.

Since Maschler's Scheme ends after at most n linear program solves, the run time of the method is dominated by the time it takes to solve (P_i). To use the Ellipsoid Method [17,21] to implement Maschler's Scheme we need be able to separate over the constraints corresponding to all coalitions in $\text{Fix}(P_{i-1}(\epsilon_{i-1}))$ in each iteration. There can be an exponential number of such constraints in general, and some structure on the underlying cooperative game would need to be observed in order to separate these constraints efficiently. This requirement can be relaxed somewhat, and still retain the linear number of iterations required to compute the nucleolus.

2.1 The Relaxed Maschler's Scheme

We will define a sequence of linear programs $Q_1, Q_2, \ldots, Q_N$ where the unique optimal solution to Q_N is the nucleolus of (n, ν). With each linear program Q_i there will be an associated set of vectors V_i contained in the set of incidence vectors of $\text{Fix}(Q_i)$. The feasible solutions to Q_i will lie in $\mathbb{R}^n \times \mathbb{R}$. In keeping with the notion we used for $P_i(\epsilon_i)$, for each linear program Q_i we let $\bar{\epsilon}_i$ be the optimal value of Q_i and let

$$Q_i(\bar{\epsilon}_i) := \{x \in \mathbb{R}^n : (x, \bar{\epsilon}_i) \text{ is feasible for } Q_i\}.$$

We will describe the linear programs $\{Q_i\}_i$ inductively. The first linear program is again the leastcore linear program of (n, ν). That is to say Q_1 is equal to (P_1). Let $V_1 \subseteq \mathbb{R}^n$ be a singleton containing the incidence vector of one coalition in $\text{Fix}(Q_1(\bar{\epsilon}_1))$. Now given Q_{i-1} and V_{i-1} we describe Q_i as follows

$$\max\{\epsilon : x(S) \geq \nu(S) + \epsilon, \forall S : \chi(X) \notin \text{span}(V_{i-1}) \text{ and } x \in Q_{i-1}(\bar{\epsilon}_{i-1}\} \qquad (Q_i)$$

Now we choose $v \in \text{Fix}(Q_i(\bar{\epsilon}_i)) \setminus \text{span}(V_{i-1})$ and set $V_i := V_{i-1} \cup \{v\}$. By the optimality of $\bar{\epsilon}_i$, v always exists as long as $Q_i(\bar{\epsilon}_i)$ has affine dimension at least 1. If $Q_i(\bar{\epsilon}_i)$ has affine dimension 0 we terminate the procedure and conclude that $Q_i(\bar{\epsilon}_i)$ is a singleton containing the nucleolus.

A nice proof of correctness for this scheme is given in [24], where this scheme is used to give a pseudopolynomial time algorithm for computing the nucleolus of weighted voting games.

Lemma 1. *When the Relaxed Maschler Scheme is run on a cooperative game (n, ν) yielding a hierarchy of linear programs $Q_1, \ldots, Q_N$, with optimal values $\bar{\epsilon}_1, \ldots, \bar{\epsilon_N}$ respectively, the set $Q_N(\bar{\epsilon_N})$ is a singleton containing the nucleolus of (n, ν). Moreover N is at most n.*

3 The Linear Subspace Avoidance Problem

Motivated by the desire to design a separation oracle for the constraints of (Q_i) we initiate a general study of combinatorial optimization problems whose feasible region avoids a linear subspace. For our purposes, we say a *combinatorial optimization problem* is an optimization problem of the form

$$\max\{f(x) : x \in \mathcal{X}\} \qquad (P)$$

where $\mathcal{X} \subseteq \{0,1\}^n$ is known as the feasible region, and $f : \mathcal{X} \to \mathbb{R}$ is the objective function. Normally (P) is presented via a compact representation. For example in the shortest path problem on a directed graph, $\mathcal{X}$ is the family of paths in a directed graph D and $f(x)$ is a linear function. The entire feasible set $\mathcal{X}$ is uniquely determined by the underlying directed graph D, and f is determined by weights on the arcs of D. When giving as input D and the arc weights, the problem is completely determined without specifying every one of the exponentially many paths in $\mathcal{X}$.

For compactly represented cooperative games the minimum excess coalition problem can be phrased as a problem of the form (P). Simply take $\mathcal{X}$ to be the set of incidence vectors of subsets of $[n]$ and take $f(x)$ to be $x(S) - \nu(S)$.

Now consider a linear subspace $\mathcal{L} \subseteq \mathbb{R}^E$. For our combinatorial optimization problem (P), the associated *linear subspace avoidance problem* is

$$\max\{f(x) : x \in \mathcal{X}\backslash\mathcal{L}\} \qquad (P_{\mathcal{L}})$$

Even when (P) can be solved in polynomial time with respect to its compact representation and $\mathcal{L}$ is given through a basis, $(P_{\mathcal{L}})$ can be NP-hard.

Lemma 2. *$(P_{\mathcal{L}})$ is NP-hard in general even when (P) can be solved in polynomial time with respect to its compact representation and $\mathcal{L}$ is given through a basis.*

Observe that when we formulate the minimum excess coalition problem for a cooperative game (n, ν) as a problem of the form (P) and we take $\mathcal{L} = \text{span}(V_{i-1})$ then $(P_{\mathcal{L}})$ is the ellipsoid method separation problem for (Q_i), the i-th linear program in the relaxed Maschler Scheme. This discussion yields the following easy lemma

Lemma 3. *If (P) is a minimum excess coalition problem of a cooperative game (n, ν) and one can solve the associated $(P_{\mathcal{L}})$ for any $\mathcal{L}$ in polynomial time then the nucleolus of (n, ν) can be computed in polynomial time.*

3.1 Reducing Linear Subspace Avoidance to Congruency-Constrained Optimization

The goal of this subsection is to show the connection between solving $(P_{\mathcal{L}})$ and solving congruency-constrained optimization. This connection was first drawn in the work of Pashkovich [24] for the special case of weighted voting games. Here we abstract their work to apply it to our more general framework.

By the following lemma, we can restrict our attention from linear independence over $\mathbb{R}$ to linear independence over finite fields.

Lemma 4 *(Pashkovich [24]). Let P be a set of prime numbers such that $|P| \geq \log_2(n!)$ with $n \geq 3$. A set of vectors $v_1, \ldots, v_k \in \{0,1\}^n$ are linearly independent over $\mathbb{R}$ if and only if there exists $p \in P$ such that $v_1, \ldots, v_k$ are linearly independent over $\mathbb{F}_p$.*

Moreover, the set P can be found in $O(n^3)$ time, and each p in P can be encoded in $O(\log(n))$ bits.

This lemma enables us to reduce the problem $(P_{\mathcal{L}})$ to the problem of computing (P) subject to a congruency constraint with respect to a given prime p, $k \in \mathbb{Z}_p$, $v \in Z_p^E$:

$$\max\{f(x) : x \in \mathcal{X}, v^T x = k \mod p\}. \qquad (P_{\mathcal{L},p,v,k})$$

Lemma 5 *If one can solve $(P_{\mathcal{L},p,v,k})$ in time T then one can solve $(P_{\mathcal{L}})$ in time $O(n^6 T)$.*

4 Dynamic Programming

Our goal is to define a class of problems where tractability of (P) can be lifted to tractability of $(P_{\mathcal{L},p,v,k})$ and hence via Lemma 5 to $(P_{\mathcal{L}})$. Our candidate will be problems which have a dynamic programming formulation. The model of dynamic programming we propose is based on the model of Martin, Rardin, and Campbell [7]. The essence of a dynamic programming solution to a problem is a decomposition of a solution to the program into optimal solutions to smaller subproblems. We will use a particular type of hypergraph to describe the structure of dependencies of a problem on its subproblems.

To begin we will need to introduce some concepts. A *directed hypergraph* $H = (V, E)$ is an ordered pair, where V is a finite set referred to as the *vertices* or *nodes* of the hypergraph, and E is a finite set where each element is of the form (v, S) where $S \subseteq V$ and $v \in V \backslash S$. We refer to the elements of E as *edges* or *arcs* of H. For an arc $e = (v, S) \in E$ we call v the *tail* of e and say e is *outgoing* from v. We call S the *heads* of e, call each $u \in S$ a *head* of e, and say e is *incoming* on each $u \in S$. We call vertices with no incoming arcs *sources* and we call vertices with no outgoing arcs *sinks*. For a directed hypergraph H, the set $L(H)$ denotes the set of sinks of H.

For any non-empty strict subset of vertices $U \subset V$, we define the *cut* induced by U, denoted $\delta(U)$, as follows $\delta(U) := \{(v, S) \in E : v \in U \text{ and } S \cap (V \backslash U) \neq \emptyset\}$. We say a directed hypergraph is *connected* if it has no empty cuts.

A *directed hyperpath* is a directed hypergraph P satisfying the following:

- there is a unique vertex $s \in V(P)$ identified as the *start* of P,
- the start s is the tail of at most one arc of P, and the head of no arcs of H,
- every vertex in $V(P) \backslash \{s\}$ is the tail of precisely one arc of H,
- P is connected.

Observe that there is at least one, and potentially many, vertices of a path which have one incoming arc and no outgoing arcs. These vertices we call the *ends* of the path. If there is a path starting from a vertex u and ending with a vertex v then we say u is an *ancestor* to v and v is a *descendant* of u. For any vertex

$v \in V(H)$, the subgraph of H *rooted at* v, denoted H_v, is the subgraph of H induced by the descendants of v (including v).

We say that a directed hypergraph $H = (V, E)$ is *acyclic* if there exists a topological ordering of the vertices of H. That is to say, there exists a bisection $t : V \to [|V|]$ such that for every $(v, S) \in E$, for each $u \in S$, $t(v) < t(u)$.

A common approach to dynamic programming involves a table of subproblems (containing information pertaining to their optimal solutions), and a recursive function describing how to compute an entry in the table based on the values of table entries which correspond to smaller subproblems. The values in the table are then determined in a bottom-up fashion. In our formal model, the entries in the table correspond to vertices of the hypergraph, and each hyperarc (v, S) describes a potential way of computing a feasible solution to the subproblem at v by composing the solutions to the subproblems at each node of S.

Consider a problem of the form (P). That is, we have a feasible region $\mathcal{X} \subseteq \mathbb{R}^n$ and an objective function $f : \mathcal{X} \to \mathbb{R}$ and we hope to maximize $f(x)$ subject to $x \in \mathcal{X}$. We need some language to describe how solutions to the dynamic program, i.e. paths in the directed hypergraph, will map back to solutions in the original problem space. To do this mapping back to the original space we will use an affine function. A function $g : \mathbb{R}^m \to \mathbb{R}^n$ is said to be *affine* if there exists a matrix $A \in \mathbb{R}^{n \times m}$ and a vector $b \in \mathbb{R}^n$ such that for any $x \in \mathbb{R}^m$, $g(x) = Ax + b$.

Oftentimes an affine function g will have a domain R^E indexed by a finite set E. When this happens for any $S \subseteq E$ we use $g(S)$ as a shorthand for $g(\chi(S))$ where $\chi(S)$ is the incidence vector of S. We further shorten $g(\{e\})$ to $g(e)$.

Definition 3. *Let $H = (V, E)$ be a directed acyclic connected hypergraph with set of sources T. Let $\mathcal{P}(H)$ denote the set of paths in H which begin at a source in T and end only at sinks of H. Let $g : \mathbb{R}^E \to \mathbb{R}^n$ be an affine map which we will use to map between paths in $\mathcal{P}(H)$ and feasible solutions in $\mathcal{X}$. Let $c : \mathbb{R}^E \to \mathbb{R}$ be an affine function we will use as an objective function. We say (H, g, c) is a* dynamic programming formulation *for (P) if $g(\mathcal{P}(H)) = \mathcal{X}$, and moreover for any $x \in \mathcal{X}$, $f(x) = \max\{c(P) : P \in g^{-1}(x)\}$.*

In other words, the optimal values of

$$\max\{c(P) : P \in \mathcal{P}(H)\} \qquad (DP)$$

and (P) are equal, and the feasible region of (P) is the image (under g) of the feasible region of (DP). The *size* of a dynamic programming formulation is the number of arcs in $E(H)$.

In [7] the authors show that (DP) has a totally dual integral extended formulation of polynomial size. Thus they show that (DP) can be solved in polynomial time via linear programming. They further show that the extreme point optimal solution of this extended formulation lie in $\{0, 1\}^E$ under a condition which is equivalent to the following *no common descendants* condition: for each $(\ell, J) \in E(H)$ for all $u \neq v \in J$, there does not exist $w \in V(H)$ such that w is a descendant of both u and v. We say that a dynamic programming formulation (H, g, c) of a problem (P) is *integral* if H satisfies the no common descendants condition. By the preceding discussion we have the following lemma

Lemma 6. *If a problem (P) has an integral dynamic programming formulation (H, g, c) then (P) can be solved in time polynomial in the encoding of (H, g, c).*

4.1 Congruency Constrained Dynamic Programming

In this subsection our goal is to show that when a problem of the form (P) has a dynamic programming formulation, then its congruency constrained version $(P_{\mathcal{L},p,v,k})$ has a dynamic programming formulation that is only a $O(p^3)$ factor larger than the formulation for the original problem. This will prove Theorem 3.

We begin with a handy lemma for constructing dynamic programming formulations of combinatorial optimization problems.

Lemma 7. *If (H, g, c) is a dynamic programming formulation for (P) and we have a dynamic programming formulation (H', g', c') for (DP) with respect to hypergraph H and costs c then $(H', g \circ g', c')$ is a dynamic programming formulation for (P).*

Consider a directed hypergraph $H = (V, E)$ and an edge $(u, S) \in E$. For $v \in S$ we define the *hypergraph obtained from the subdivision of* (u, S) *with respect to* v to be the hypergraph $H' = (V', E')$ where $V' = V \dot{\cup} \{b_v\}$ for a new dummy vertex b_v and $E' = (E \backslash \{(u, S)\}) \cup \{(u, \{v, b_v\}), (b_v, S \backslash \{v\})\}$. That is, H' is obtained from H by replacing edge (u, S) with two edges: $(u, \{v, b_v\})$ and $(b_v, S \backslash \{v\})$. We call the edges $(u, \{v, b_v\})$ and $(b_v, S \backslash \{v\})$ the *subdivision* of edge (u, S).

Lemma 8. *Let $H = (V, E)$ be a directed acyclic hypergraph and let $H' = (V', E')$ be the directed acyclic hypergraph obtained via a subdivision of $(u, S) \in E$ with respect to $v \in S$. Then there is an affine function $g : \mathbb{R}^{E'} \to \mathbb{R}^E$, such that for any affine function $c : \mathbb{R}^E \to \mathbb{R}$, there exists an affine function $c' : \mathbb{R}^{E'} \to \mathbb{R}$ such that (H, g, c') is a dynamic programming formulation of the problem (DP) on H with objective c.*

Moreover if H satisfies the "no common descendants" property, this dynamic programming formulation is integral.

For a directed hypergraph $H = (V, E)$ let $\Delta(H) := \max\{|S| : (u, S) \in E\}$ and let $\Gamma(H) := |\{(u, S) \in E : |S| = \Delta(H)\}|$. The following Lemma shows that we may assume the number of heads of any arc in a dynamic programming formulation is constant.

Lemma 9. *Consider a combinatorial optimization problem of the form (P). If there exists a dynamic programming formulation (H, g, c) for (P) then there exists a dynamic programming formulation (H^*, g^*, c^*) for (P) such that $\Delta(H^*) \leq 2$, and $|E(H^*)| = \sum_{u \in V(H)} \sum_{(u,S) \in E(H)} (|S| - 1)$.*

Moreover, if H is integral then H^* is integral.

The next lemma is our main technical lemma. It provides the backbone of our dynamic programming formulation for $(P_{\mathcal{L},p,v,k})$ by showing that we can track the congruency of all hyperpaths rooted at a particular vertex by expanding the size of our hypergraph by a factor of $p^{\Delta(H)} + 1$.

Lemma 10. *Let $H = (V, E)$ be a directed acyclic hypergraph. Let p be a prime. Let $k \in \mathbb{Z}_p$ and let $a \in \mathbb{Z}_p^E$. There exists a directed acyclic hypergraph $H' = (V', E')$ and an affine function $g' : \mathcal{P}(H') \to \mathcal{P}(H)$, $g'(x) = Ax + b$, such that:*

1) $|E'| \leq p^{\Delta(H)+1}|E|$
2) For every $v \in V \backslash L(H)$, for every $k \in \mathbb{Z}_p$, if $\{P \in \mathcal{P}(H_v) : a(P) = k \mod p\} \neq \emptyset$ then there exists $v' \in V(H')$ such that

$$g'(\mathcal{P}(H'_{v'})) = \{P \in \mathcal{P}(H_v) : a(P) = k \mod p\}.$$

Moreover if H satisfies the "no common descendants" property then H' satisfies the "no common descendants" property.

We are now ready to show our main theorem. Theorem 3 with Lemma 6 yield Corollary 1. Via Corollary 1, Lemma 3, and Lemma 5 we obtain Theorem 1.

Theorem 3. *Consider an instance of a combinatorial optimization problem (P). Let p be a prime, let $v \in \mathbb{Z}_p^n$, and let $k \in \mathbb{Z}_p$. Consider the corresponding congruency-constrained optimization problem $(P_{\mathcal{L},p,v,k})$. If (P) has a dynamic programming formulation (H, g, c) then $(P_{\mathcal{L},p,v,k})$ has a dynamic programming formulation (H', g', c') such that $|E(H')| \leq p^3 \cdot |V(H)| \cdot |E(H)|$.*

Moreover if (H, g, c) is integral then (H', g', c') is integral.

Corollary 1. *If (P) has an integral dynamic programming formulation (H, g, c) then for any v, k, p problem $(P_{\mathcal{L},p,v,k})$ can be solved in time polynomial in size of H, the prime p, and the encoding of g, c, v, k,*

5 Applications

In this section we show a couple of applications of Theorem 1 to computing the nucleolus of cooperative games. The first application is to Weighted Voting Games. In [24] a pseudopolynomial time algorithm for computing the nucleolus of Weighted Voting Games was given. We show how the same result can be obtained as a special case of Theorem 1. Recall that a weighted voting game (n, ν) has value function $\nu : 2^{[n]} \to \{0, 1\}$ determined by a vector $w \in \mathbb{Z}^n$ and $T \in Z$, such that for any $S \subseteq [n]$, $\nu(S) = 1$ if and only if $w(S) \geq T$.

We partition $2^{[n]}$ into two classes: $N_0 := \{S \subseteq [n] : w(S) < T\}$ and $N_1 := \{S \subseteq [n] : w(S) \geq T\}$. If we can design a dynamic programming formulation for the minimum excess coalition problem restricted to N_0: $\max\{-x(S) : w(S) \leq T-1, S \subseteq [n]\}$ and a dynamic programming formulation for the minimum excess coalition problem restricted to N_1: $\max\{-x(S) + 1 : w(S) \geq T, S \subseteq [n]\}$, then the dynamic programming formulation which takes the maximum of these two formulations will provide a dynamic programming formulation for the minimum excess coalition problem of the weighted voting game.

If we let $W[k, D]$ denote $\max\{-x(S) : w(S) \leq D, S \subseteq [k]\}$ then we can solve the minimum excess coalition problem restricted to N_0 by computing $W[n, T-1]$

via the following recursive expression

$$W[k, D] = \begin{cases} \max\{W[k-1, D-w_k], W[k-1, D]\}, & \text{if } k > 1 \\ -x_1, & \text{if } k = 1 \text{ and } w_1 \leq D \\ -\infty, & \text{if } k = 1 \text{ and } w_1 > D. \end{cases}$$

It is not hard to construct a dynamic programming formulation (H_0, g_0, c_0) for the minimum excess coalition problem restricted to N_0 by following this recursive expression. The hypergraph H_0 will in fact be a rooted tree (i.e. all heads will have size one), and H_0 will have $O(nT)$ vertices and arcs. Via a similar technique, a dynamic programming formulation (H_1, g_1, c_1) with $O(nT)$ arcs can be constructed for the minimum excess problem restricted to N_1. Then by taking the union these dynamic programming formulations, we obtain an integral dynamic programming formulation of size $O(nT)$. Therefore by Theorem 1 we obtain a short proof that

Theorem 4 *([10, 24]). The nucleolus of a weighted voting game can be computed in pseudopolynomial time.*

In the following subsections we will see how the added power of hyperarcs lets us solve the more complex problem of computing the nucleolus of b-matching games on graphs of bounded treewidth.

5.1 Treewidth

Consider a graph $G = (V, E)$. We call a pair $(T, \mathcal{B})$ a *tree decomposition* [14,28] of G if $T = (V_T, E_T)$ is a tree and $\mathcal{B} = \{B_i \subseteq V : i \in V_T\}$ is a collection of subsets of V, called *bags*, such that 1) $\bigcup_{i \in V_T} B_i = V$, i.e. every vertex is in some bag, 2) for each $v \in V$, the subgraph of T induced by $\{i \in V_T : v \in B_i\}$ is a tree, and 3) for each $uv \in E$, there exists $i \in V_T$ such that $u, v \in B_i$.

The *width* of a tree decomposition is the size of the largest bag minus one, i.e. $\max_{i \in V_T}\{|B_i| - 1\}$. The *treewidth* of graph G, denoted $\text{tw}(G)$, is minimum width of a tree decomposition of G. We may assume that tree decompositions of a graph have a special structure. We say a tree decomposition $(T, \mathcal{B})$ of G is *nice* if there exists a vertex $r \in V_T$ such that if we view T as a tree rooted at r then every vertex $i \in V_T$ is one of the following types:

- **Leaf:** i has no children and $|B_i| = 1$.
- **Introduce:** i has one child j and $B_i = B_j \dot{\cup} \{v\}$ for some vertex $v \in V$.
- **Forget:** i has one child j and $B_i \dot{\cup} \{v\} = B_j$ for some vertex $v \in V$.
- **Join:** i has two children j_1, j_2 with $B_i = B_{j_1} = B_{j_2}$.

Nice tree decompositions can be computed in polynomial time.

Theorem 5 *([18] Lemma 13.1.3). If $G = (V, E)$ has a tree decompostion of width w with n tree vertices then there exists a nice tree decomposition of G of width w and $O(|V|)$ tree vertices which can be computed in $O(|V|)$ time.*

5.2 Dynamic Program for b-Matching Games

We want to show that on graphs of bounded treewidth, the nucleolus of b-matching games can be computed efficiently. Fix a graph $G = (V, E)$, a vector of b-values $b \in \mathbb{Z}^V$, and tree decomposition $(T, \mathcal{B})$ of treewidth w, where T is rooted at r, to be used throughout this section. For $i \in V(T)$ let T_i denote the subtree of T rooted at i, and also let $G_i := G[\bigcup_{j \in V(T_i)} B_j]$. For any $v \in V(G_i)$, let $\delta_i(v) := \{uv \in E(G_i)\}$.

For any $i \in V(T)$, $X \subseteq B_i$, $d \in \{d \in \mathbb{Z}^{B_i} : 0 \leq d \leq \Delta(G)\}$, and $F \subseteq E(B_i)$, we define the combinatorial optimization problem $C[i, X, d, F]$ to be the problem of finding a b-matching M and a set of vertices S such that M uses only edges of G_i, S uses only vertices of G_i, the intersection of M and $E(B_i)$ is F, the number of edges in M adjacent to u is d_u for each u in B_i, and the vertices in S not intersecting an edge in F is X. We define $C[i]$ to be the union over all $C[i, X, d, F]$. A formal definition of $C[i, X, d, F]$ and $C[i]$ is given in the full version of the paper [19].

We will show a dynamic programming formulation (H, g, c) for $C[i]$. Since the feasible region of the minimum excess coalition problem for b-matching games is the image of the feasible region of $C[i]$ under the linear map which projects out M, and $\nu(S) - x(S) = \max_{(M,S) \text{feasible for } C[i]} w(M) - x(S)$, the existence of (H, g, c) will imply the existence a dynamic programming formulation of the minimum excess coalition problem for b-matching games of the same encoding length. Lemma 11, and Theorem 1 we have shown Theorem 2.

Lemma 11 *Let $i \in V(T)$. There exists an integral dynamic programming formulation (H, g, c) for $C[i]$ such that for every $j \in V(T_i)$, $X \subseteq B_i$, $d \in \mathbb{Z}^{B_i}$, and $F \subseteq E(B_i)$, if $C[i, X, d, F]$ has a feasible solution then there exists $a \in V(H)$ such that (H_a, g, c) is an integral dynamic programming formulation for $C[i, X, dF]$. Moreover $|E(H)| \leq |V(T_i)| \cdot w \cdot \Delta(G)^w \cdot w^2$.*

6 Conclusion and Future Work

We have given a formalization of dynamic programming, and shown that adding congruency constraints to this model only increases the complexity by a polynomial factor of the prime modulus. Further, we showed that whenever the minimum excess coalition problem of a cooperative game can be solved via dynamic programming, its nucleolus can be computed in time polynomial in the size of the dynamic program. Using this result we gave an algorithm for computing the nucleolus of b-matching games on graphs of bounded treewidth.

In [15] they show that a generalization of the dynamic programming model in [7] called Branched Polyhedral Systems also has an integral extended formulation. It is natural to wonder how our framework could extend to Branched Polyhedral Systems and if that would enable to computation of the nucleolus for any interesting classes of cooperative games.

References

1. Akbari, N., Niksokhan, M.H., Ardestani, M.: Optimization of water allocation using cooperative game theory (Case study: Zayandehrud basin). J. Environ. Stud. **40**, 10–12 (2015)
2. Aumann, R.J., Maschler, M.: Game theoretic analysis of a bankruptcy problem from the Talmud. J. Econ. Theory (1985). https://doi.org/10.1016/0022-0531(85)90102-4
3. Bilbao, J.M., Fernández, J.R., Jiménez, N., López, J.J.: Voting power in the European Union enlargement. Eur. J. Oper. Res. (2002). https://doi.org/10.1016/S0377-2217(01)00334-4
4. Biró, P., Kern, W., Paulusma, D.: Computing solutions for matching games. Int. J. Game Theory **41**, 75–90 (2012). https://doi.org/10.1007/s00182-011-0273-y
5. Biró, P., Kern, W., Paulusma, D., Wojuteczky, P.: The stable fixtures problem with payments. Games Econ. Behav. **108**, 245–268 (2017). https://doi.org/10.1016/J.GEB.2017.02.002
6. Brânzei, R., Iñarra, E., Tijs, S., Zarzuelo, J.M.: A simple algorithm for the nucleolus of airport profit games. Int. J. Game Theory (2006). https://doi.org/10.1007/s00182-006-0019-4
7. Campbell, B.A., Martin, R.K., Rardin, R.L., Campbell, B.A.: Polyhedral characterization of discrete dynamic programming. Oper. Res. **38**(1), 127–138 (1990). https://doi.org/10.1287/opre.38.1.127
8. Chen, N., Lu, P., Zhang, H.: Computing the nucleolus of matching, cover and clique games. In: AAAI (2012)
9. Deng, X., Fang, Q., Sun, X.: Finding nucleolus of flow game. J. Comb.inatorial Optim. (2009). https://doi.org/10.1007/s10878-008-9138-0
10. Elkind, E., Goldberg, L.A., Goldberg, P.W., Wooldridge, M.: On the computational complexity of weighted voting games. Ann. Math. Artif. Intell. **56**(2), 109–131 (2009). https://doi.org/10.1007/s10472-009-9162-5
11. Faigle, U., Kern, W., Fekete, S.P., Hochstättler, W.: The nucleon of cooperative games and an algorithm for matching games. Math. Program. **83**(1–3), 195–211 (1998). https://doi.org/10.1007/BF02680558
12. Faigle, U., Kern, W., Kuipers, J.: Note computing the nucleolus of min-cost spanning tree games is NP-hard. Int. J. Game Theory **27**(3), 443–450 (1998). https://doi.org/10.1007/s001820050083
13. Granot, D., Granot, F., Zhu, W.R.: Characterization sets for the nucleolus. Int. J. Game Theory (1998). https://doi.org/10.1007/s001820050078
14. Halin, R.: S-functions for graphs. J. Geom. **8**(1–2), 171–186 (1976). https://doi.org/10.1007/BF01917434
15. Kaibel, V., Loos, A.: Branched polyhedral systems. In: Eisenbrand, F., Shepherd, F.B. (eds.) IPCO 2010. LNCS, vol. 6080, pp. 177–190. Springer, Heidelberg (2010). https://doi.org/10.1007/978-3-642-13036-6_14
16. Kern, W., Paulusma, D.: Matching games: the least core and the nucleolus. Math. Oper. Res. **28**(2), 294–308 (2003). https://doi.org/10.1287/moor.28.2.294.14477
17. Khachiyan, L.G.: Polynomial algorithms in linear programming. USSR Comput. Math. Math. Phys. (1980). https://doi.org/10.1016/0041-5553(80)90061-0
18. Kloks, T. (ed.): Treewidth. LNCS, vol. 842. Springer, Heidelberg (1994). https://doi.org/10.1007/BFb0045375
19. Koenemann, J., Toth, J.: A general framework for computing the nucleolus via dynamic programming. arXiv preprint arXiv:2005.10853 (2020)

20. Könemann, J., Pashkovich, K., Toth, J.: Computing the nucleolus of weighted cooperative matching games in polynomial time. Math. Program. 1–27 (2020). https://doi.org/10.1007/s10107-020-01483-4
21. Leung, J., Grotschel, M., Lovasz, L., Schrijver, A.: Geometric algorithms and combinatorial optimization. J. Oper. Res. Soc. (1989). https://doi.org/10.2307/2583689
22. Maschler, M., Peleg, B., Shapley, L.S.: Geometric properties of the kernel, nucleolus, and related solution concepts. Math. Oper. Res. **4**(4), 303–338 (1979). http://www.jstor.org/stable/3689220
23. Militano, L., Iera, A., Scarcello, F.: A fair cooperative content-sharing service. Comput. Netw. (2013). https://doi.org/10.1016/j.comnet.2013.03.014
24. Pashkovich, K.: Computing the nucleolus of weighted voting games in pseudo-polynomial time, pp. 1–12 (2018). http://arxiv.org/abs/1810.02670
25. Paulusma, D.: Complexity Aspects of Cooperative Games. Twente University Press, Enschede (2001)
26. Potters, J., Reijnierse, H., Biswas, A.: The nucleolus of balanced simple flow networks. Games Econ. Behav. **54**(1), 205–225 (2006)
27. Reijnierse, H., Potters, J.: The B-nucleolus of TU-games. Games Econ. Behav. (1998). https://doi.org/10.1006/game.1997.0629
28. Robertson, N., Seymour, P.D.: Graph minors. III. Planar tree-width. J. Comb. Theory Ser. B **36**(1), 49–64 (1984)
29. Schmeidler, D.: The nucleolus of a characteristic function game. SIAM J. Appl. Math. **17**(6), 1163–1170 (1969). https://doi.org/10.1137/0117107
30. Solymosi, T., Raghavan, T.E., Tijs, S.: Computing the nucleolus of cyclic permutation games. Eur. J. Oper. Res. (2005). https://doi.org/10.1016/j.ejor.2003.02.003
31. Solymosi, T., Raghavan, T.E.S.: An algorithm for finding the nucleolus of assignment games. Int. J. Game Theory **23**(2), 119–143 (1994). https://doi.org/10.1007/BF01240179
32. Solymosi, T., Sziklai, B.: Characterization sets for the nucleolus in balanced games. Oper. Res. Lett. (2016). https://doi.org/10.1016/j.orl.2016.05.014
33. Sziklai, B., Fleiner, T., Solymosi, T.: On the core and nucleolus of directed acyclic graph games. Math. Program. 243–271 (2016). https://doi.org/10.1007/s10107-016-1062-y

How Many Freemasons Are There? The Consensus Voting Mechanism in Metric Spaces

Mashbat Suzuki(✉) and Adrian Vetta

McGill University, Montreal, Canada
mashbat.suzuki@mail.mcgill.ca, adrian.vetta@mcgill.ca

Abstract. We study the evolution of a social group when admission to the group is determined via *consensus* or *unanimity voting.* In each time period, two candidates apply for membership and a candidate is selected if and only if all the current group members agree. We apply the *spatial theory of voting* where group members and candidates are located in a metric space and each member votes for its closest (most similar) candidate. Our interest focuses on the expected cardinality of the group after T time periods. To evaluate this we study the geometry inherent in dynamic consensus voting over a metric space. This allows us to develop a set of techniques for lower bounding and upper bounding the expected cardinality of a group. We specialize these methods for two-dimensional Euclidean metric spaces. For the unit ball the expected cardinality of the group after T time periods is $\Theta(T^{\frac{1}{8}})$. In sharp contrast, for the unit square the expected cardinality is at least $\Omega(\ln T)$ but at most $O(\ln T \cdot \ln \ln T)$.

1 Introduction

This paper studies the evolution of social groups over time. In an *exclusive social group*, the existing group members vote to determine whether or not to admit a new member. Familiar examples include the freemasons, fraternities, membership-run sports and social clubs, acceptance to a condominium, as well as academia. To analyze the inherent dynamics we use the model of Alon, Feldman, Mansour, Oren and Tennenholtz [1]. In each time period, two candidates apply for membership and the current members vote to decide if either or none of them is acceptable. The spatial model of voting is used: each candidate is located uniformly at random in a metric space and each group member votes for the candidate closest to them.

Alon et al. [1] analyze social group dynamics in a one-dimensional Euclidean metric space, specifically, the unit interval $[0, 1]$. They examine how outcomes vary under different winner determination rules, in particular, majority voting and consensus voting. In *consensus voting* or *unanimity voting* a candidate is elected if and only if the group members agree unanimously. Equivalently, every member may *veto* a potential candidate.

T. Harks and M. Klimm (Eds.): SAGT 2020, LNCS 12283, pp. 322–336, 2020.
https://doi.org/10.1007/978-3-030-57980-7_21

Our interest lies in the evolution of the group size under consensus voting; that is, what is the expected cardinality of the social group G^T after T time periods? In the one-dimensional setting the answer is quite simple. There, Alon et al. [1] show that under consensus voting *if* a candidate is elected in round t then, with high probability, it is within a distance $\Theta(1/\sqrt{t})$ of one endpoint of the interval. Because the winning candidate must be closer to all group members than the losing candidate, both candidates must therefore be near the endpoints. This occurs with probability $\Theta(1/t)$. As a consequence, in a one-dimensional Euclidean metric space, the expected size of the social group after T time periods is $\mathbb{E}[|G^T|] \cong \ln T$. Here we use the notation $f \cong g$ if both $f \lesssim g$ and $g \lesssim f$, where $f \lesssim g$ if $f \leq c \cdot g$ for some constant $c > 0$.

Bounding the expected group size in higher-dimensional metric spaces is more complex and is the focus of this paper. To do this, we begin in Sect. 2 by examining the geometric aspects of consensus voting in higher-dimensional metric spaces. More concretely, we explain how winner determination relates to the convex hull of the group members and the Voronoi cells formed by the candidates. This geometric understanding enables us to construct, in Sect. 3, a set of techniques, based upon *cap methods* in probability theory, that allow for the upper bounding and lower bounding of expected group size under the *Euclidean metric*. In Sects. 4 and 5, we specialize these techniques to two-dimensional Euclidean spaces for application on the fundamental special cases of the unit square and the unit ball. Specifically, for the unit square we show the following lower and upper bounds on expected group size.

Theorem 1. *The expected cardinality of the social group on the unit square* $\mathbb{H}$ *after* T *periods is bounded by* $\ln T \lesssim \mathbb{E}[|G^T|] \lesssim \ln T \ln\ln T$.

Thus, expected group size for the two-dimensional unit square is comparable to that of the one-dimensional interval. Surprisingly, there is a dramatic difference in expected group size between the unit square and the unit ball. For the unit ball, the expected group size evolves not logarithmically but polynomially with time.

Theorem 2. *The expected cardinality of the social group on the unit ball* $\mathbb{B}$ *after* T *periods is* $\mathbb{E}[|G^T|] \cong T^{\frac{1}{8}}$.

1.1 Background and Related Work

Here we discuss some background on the spatial model and consensus voting. The spatial model of voting utilized in this paper dates back nearly a century to the celebrated work of Hotelling [16]. His objective was to study the division of a market in a duopoly when consumers are distributed over a one-dimension space, but he noted his work had intriguing implications for electoral systems. Specifically, in a two-party system there is an incentive for the political platforms of the two parties to *converge*. This was formalized in the *median voter theorem* of Black [5]: in a one-dimensional ideological space the location of the median voter induces a Condorcet winner[1], given single-peaked voting preferences. The traditional voting assumption in a metric space is *proximity voting* where each voter supports its closest candidate; observe that proximity voting gives single-peaked preferences.

The *spatial model of voting* was formally developed by Downs [10] in 1957, again in a one-dimensional metric space. Davis, Hinich, and Ordeshook [9] expounded on practical necessity of moving beyond just one dimension. Interestingly, they observed that in two-dimensional metric spaces, a Condorcet winner is not guaranteed even with proximity voting. Of particular relevance here is their finding that, in dynamic elections, the order in which candidates are considered can fundamentally affect the final outcome [5,9].

There is now a vast literature on spatial voting, especially concerning the strategic aspects of simple majority voting; see, for example, the books and surveys [11,12,20,21,23]. There has also been a vigorous debate concerning whether voter utility functions in spatial models should be distance-based (such as the standard assumption of proximity voting used here), relational (e.g. directional voting [22]), or combinations thereof [20]. This debate has been philosophical, theoretical and experimental [7,14,17–19,25]. Recently there has also been a large amount of interest in the spatial model by the artificial intelligence community [2,3,6,13,24]. It is interesting to juxtapose these modern potential applications with the original motivations suggested by Black [5], such as the administration of colonies!

Consensus is one of the oldest group decision-making procedures. In addition to exclusive social groups, it is familiar in a range of disparate settings, including judicial verdicts, Japanese corporate governance [26], and even decision making in religious groups, such as the Quakers [15]. From a theoretic perspective, consensus voting in a metric spaces has also been studied by Colomer [8] who highlights the importance the initial set of voters can have on outcomes in a dynamic setting.

2 The Geometry of Consensus Voting

In this section, we present a simple geometric interpretation of a single election using consensus voting in the spatial model. In the subsequent sections, we will apply this understanding, developed for the static case, to study the dynamic model. Specifically, we examine how a group grows over time when admission to the group is via a sequence of consensus elections.

Let $G^0 = \{v_1, \cdots, v_k\}$ denote the initial set of group members[2], selected uniformly and independently from a metric space K. In the consensus voting mechanism, for each round $t \geq 1$, a finite set of candidates $C^t = \{w_1, \cdots, w_n\} \subseteq K$, drawn uniformly and independently from K, applies for membership. Members

[1] A candidate is a *Condorcet winner* if, in a pairwise majority vote, it beats every other candidate.

[2] We may take the cardinality of the initial group to be any constant k. In particular, we may assume $k = 1$.

of the group at the start of round t, denoted G^{t-1}, are eligible to vote. Assuming the *spatial theory of voting*, each group member will vote for the candidate who is closest to her in the metric space. That is, member v_i votes for candidate w_j if and only if $d(v_i, w_j) \leq d(v_i, w_k)$ for every candidate $w_k \neq w_j$. Under the *consensus (unanimity) voting rule*, if *every* group member selects the candidate $w_j \in C^t$ then w_j is accepted to the group and $G^t = G^{t-1} \cup w_j$; otherwise, if the group does not vote unanimously then no candidate wins selection and $G^t = G^{t-1}$.

As stated, to study how group size evolves over time, our first task is to develop a more precise understanding of when a candidate will be selected under consensus voting in a single election. Fortunately, there is a nice geometric characterization for this property in terms of the *Voronoi cells* (regions) formed in the metric space K by the candidates (points) $C = \{w_1, \cdots w_n\}$. Specifically, the Voronoi cell H_i associated with point w_i is $H_i := \{v \in K \mid d(v, w_i) \leq d(v, w_j) \text{ for all } i \neq j\}$. The characterization theorem for the property that a candidate is selected under the consensus voting mechanism is then:

Theorem 3. *Let $C = \{w_1, w_2, \ldots, w_n\}$ be the candidates and let $H_1, H_2, \ldots, H_n$ be the Voronoi cells on K generated by C. Then there is a winner under consensus voting* if and only if *$G \subseteq H_i$ for some candidate w_i.*

Proof. Assume $G \subseteq H_i$ for some candidate $w_i \in C$. Then, for every voter $v_j \in G$, we have $d(v_j, w_i) \leq d(v_j, w_k)$ for any other candidate $w_k \in C$. Hence, every voter prefers candidate w_i over all the other candidates. Thus candidate w_i is selected. Conversely, assume that candidate w_i is selected. Then, by definition of consensus voting, each voter $v_j \in G$ voted for w_i. Thus $d(v_j, w_i) \leq d(v_j, w_k)$ for all $k \neq i$. Ergo, $G \subseteq H_i$. □

The next lemma requires the following definition: let $B(v, w)$ denote the *ball* centred at v with radius $d(v, w)$, that is $B(v, w) := \{u \in K \mid d(v, u) \leq d(v, w)\}$.

Lemma 1. *Let G be current set of group members and $C = \{w_1, w_2, \cdots, w_n\}$ be the candidates. Under consensus, there is a winning candidate if and only if $\exists w_i \in C$ such that*

$$w_i \in \bigcap_{k \in [n] \setminus i} \bigcap_{v_j \in G} B(v_j, w_k) \tag{1}$$

Due to space constraints, the proof of Lemma 1 and several subsequent results in the paper are deferred to the full version.

As discussed, our main focus is on Euclidean metric spaces. In this setting the Voronoi cells are convex. Hence, $H_i = \text{conv}(H_i)$ and so, in Theorem 3, the condition $G \subseteq H_i$ is equivalent to $\text{conv}(G) \subseteq H_i$. Furthermore, when the metric is Euclidean, condition (1) in Lemma 1 is equivalent to $w_i \in \bigcap_{k \in [n] \setminus i} \bigcap_{v_j \in \partial(\text{conv}(G))} B(v_j, w_k)$, where $\partial(\text{conv}(G))$ denotes the extreme points of convex hull of the group members. To see this note that $\partial(\text{conv}(G)) \subseteq G$ so one inclusion follows immediately. For the other direction, suppose $w_i \in \bigcap_{k\in[n]\setminus i} \bigcap_{v_j\in\partial(\mathrm{conv}(G))} B(v_j, w_k)$. Then $d(v_j, w_i) \leq d(v_j, w_k)$ for each $v_j \in \partial(\mathrm{conv}(G))$ and any $k \neq i$. Hence $\partial(\mathrm{conv}(G)) \subseteq H_i$. Given H_i is convex when the metric is Euclidean, taking convex hulls on both sides gives $G \subseteq H_i$ and condition (1) is satisfied.

Following Alon et al. [1], from now on we restrict attention to case of $n = 2$ candidates in each round. The case $n \geq 3$ is not conceptually harder and the basic techniques presented in this paper do extend to that setting, but mathematically the analyses would be even more involved than those that follow.

3 General Tools for Bounding Expected Group Size

In this section we introduce a general approach for obtaining both upper and lower bounds on the expected cardinality of the social group in round t. From this section on we make the natural assumption that the underlying metric is Euclidean. These techniques apply for consensus voting in any convex compact domain K. In the rest of the paper we will specialize these methods for the cases in which K is either a unit ball or a unit square. In particular, lower bounds are provided for these two domains in Sect. 4 and upper bounds in Sect. 5.

Let K be a convex compact set, and let $C^t = \{w_1, w_2\}$ be candidates appearing in round t. Again, we assume each candidate w_i is distributed uniformly and independently on K. We may also assume that $\mathrm{vol}(K) = 1$, as otherwise we can absorb the associated constant factor into our bounds. Note that the expected group size is $\mathbb{E}[|G^T|] = \sum_{t=1}^{T} \Pr[X^t]$, where X^t denotes the event a new candidate wins in round t.

In Euclidean spaces the convex hull of the voters $S^t := \mathrm{conv}(G^t)$ play an important role. By Theorem 3 and noting the metric is Euclidean, we know that a candidate is accepted if and only if $S^t \subseteq H_i$ for some candidate i. We can use this fact to obtain the following:

Corollary 1. *Let $C^t = \{w_1, w_2\}$ be set of candidates. If there is a candidate accepted with $S^t = A$, then the same candidate is also accepted with $S^t = B$ for any convex set $B \subseteq A$.*

Let's first present the intuition behind our approach to upper bounding the probability of selecting a candidate in any round. Recall that, by Theorem 3, given two candidates $\{w_1, w_2\}$ in round $t + 1$, we accept candidate i if and only if $S^t \subseteq H_i(w_1, w_2)$. Now in order for the convex hull to satisfy $S^t \subseteq H_i(w_1, w_2)$, it must be the case that in the previous round (i) $S^{t-1} \subseteq H_i(w_1, w_2)$, and (ii) a new candidate did not get accepted inside the complement $\overline{H_i(w_1, w_2)} = \mathbf{cl}(K \setminus H_i(w_1, w_2))$, where **cl** denotes set closure. Applying this argument recursively with respect to the worst case convex hulls for accepting candidates inside $\overline{H_i(w_1, w_2)}$, we will obtain an upper bound on the probability of accepting a candidate. Such worst case convex hulls can be found by appropriately applying Corollary 1.

To formalize this intuition, we require some more notation. We denote by $Z(H)$ the event that a new candidate is selected inside H. Let $\Pr[Z(H)|S = A]$ denote the probability of selecting a candidate inside H given the convex hull of the group members is A. The shorthand $\Pr[Z(H)|A] = \Pr[Z(H)|S = A]$ will be used when the context is clear. Note, by Theorem 3, the probability of acceptance depends only on the shape of the convex hull of the members, and not on the round. That is, if $S^t = S^{\hat{t}} = A$ for two rounds $t \neq \hat{t}$ then the probabilities of accepting a candidate inside a given region in the rounds $t+1$ and $\hat{t}+1$ are exactly the same.

We say a set A is a *cap* if there exists a half space W such that $A = K \cap W$. We remark that caps have been widely used for studying the convex hull of random points; see the survey article [4] and the references therein. Of particular relevance here is that, in the case of two candidates, the Voronoi regions for the candidates are caps. Furthermore, $\overline{H_1(w_1, w_2)} = H_2(w_1, w_2)$ and vice versa.

Theorem 4. *Let K be convex compact domain and let $f_K(w_1, w_2)$ be any function which satisfies $f_K(w_1, w_2) \leq \min_i \Pr[Z(H_i(w_1, w_2)) \mid \overline{H_i(w_1, w_2)}]$. Then*

$$\Pr[X^{t+1}] \lesssim \int_K \int_K e^{-tf_K(w_1,w_2)}\, dw_1\, dw_2$$

Proof. Observe that, for any cap A, we have the following inequality:

$$\begin{aligned}
\Pr[S^t \subseteq A] &= \left(1 - \Pr[Z(\overline{A}) \mid S^{t-1} \subseteq A]\right) \cdot \Pr[S^{t-1} \subseteq A] \\
&\leq \left(1 - \Pr[Z(\overline{A}) \mid S^{t-1} = A]\right) \cdot \Pr[S^{t-1} \subseteq A] \\
&\leq \left(1 - \Pr[Z(\overline{A}) \mid A\ ]\right)^t \qquad (2)
\end{aligned}$$

Here the first inequality follows from Corollary 1. The second inequality is obtained by repeating the argument inductively for S^{t-1}. Hence:

$$\begin{aligned}
\Pr[X^{t+1}] &= \int\int \Pr[X^{t+1} \mid (w_1, w_2) \text{ are candidates}]\, dw_1\, dw_2 \\
&= \int\int \left(\Pr[S^t \subseteq H_1(w_1, w_2)] + \Pr[S^t \subseteq H_2(w_1, w_2)]\right) dw_1\, dw_2 \\
&\leq \sum_{i=1}^{2} \int\int \left(1 - \Pr[Z(H_i(w_1, w_2)) \mid \overline{H_i(w_1, w_2)}\ ]\right)^t dw_1\, dw_2 \\
&\leq 2\int\int (1 - f_K(w_1, w_2))^t\, dw_1\, dw_2 \\
&\leq 2\int\int e^{-tf_K(w_1,w_2)}\, dw_1\, dw_2
\end{aligned}$$

The second equality holds by Theorem 3. The first inequality follows by combining inequality (2) and the facts $\overline{H_1(w_1, w_2)} = H_2(w_1, w_2)$ and $\overline{H_2(w_1, w_2)} = H_1(w_1, w_2)$. Next, by assumption, $f_K(w_1, w_2) \leq \Pr[Z(H_i(w_1, w_2)) \mid \overline{H_i(w_1, w_2)}]$ for each i; the last two inequalities follow immediately. □

As stated, Theorem 4 allows us to upper bound the expected group size. However, the theorem is not easily applicable on its own. To rectify this, consider the following easier to apply corollary.

Corollary 2. *Let K be compact space. If f_K satisfies the conditions of Theorem 4 then*

$$\Pr[X^{t+1}] \lesssim \frac{1}{t} + \int_0^{\frac{\ln(t)}{t}} te^{-t\lambda} \cdot \Phi(\lambda)\, d\lambda$$

where $\Phi(\lambda) = \int \int \mathbb{I}\left[f_K(w_1, w_2) \leq \lambda\right] dw_1 dw_2$.

As alluded to earlier, when obtaining upper bounds for the unit ball and the unit square, we will apply Corollary 2. Of course, in order to do this, we must find an appropriate function $f_K(w_1, w_2)$ which lower bounds the probability of acceptance inside a Voronoi region $H_i(w_1, w_2)$, given the current convex hull is $\overline{H_i(w_1, w_2)}$. Finding such a function f_K can require some ingenuity, but Lemma 2 below will be useful in assisting in this task. Moreover, as we will see in Sect. 4, this lemma can be used to obtain lower bounds as well as upper bounds on the expected cardinality of the group.

Lemma 2. *Given a two-dimensional convex compact domain K and a cap A. If z_1 and z_2 are the endpoints of the line segment separating A and $\overline{A}$ then*

$$\Pr[Z(A) \mid \overline{A}\,] = 2 \cdot \int_A vol(B(z_1, \xi) \cap B(z_2, \xi) \cap A)\; d\xi$$

4 Lower Bounds on Expected Group Size

In this section we provide lower bounds for the cases where K is either a unit ball $\mathbb{B}$ or a unit square $\mathbb{H}$. Recall, we assumed that $\mathrm{vol}(K) = 1$ but $\mathrm{vol}(\mathbb{B}) = \pi$ for the unit ball. We remark that this is of no consequence as we may absorb the associated constant factor into our bounds.

4.1 Lower Bound for the Unit Ball

For the unit ball $\mathbb{B}$, a *circular segment* is the small piece of the circle formed by cutting along a *chord*. Evidently, this means that every cap of $\mathbb{B}$ is either a circular segment or the complement of a circular segment. Thus, to analyze the case of the unit ball we must study circular segments.

Lemma 3. *Let $\mathbb{B}$ be a unit ball, and J_δ be a circular segment with height $\delta \leq \frac{1}{8}$ then* $\Pr[Z(J_\delta) \mid \overline{J_\delta}\,] \gtrsim \delta^4$.

Note that Lemma 3 can be used to prove lower bound on the expected cardinality of the group. It is also used in later sections to obtain appropriate function f_K when using Corollary 2.

Theorem 5. *For the unit ball $\mathbb{B}$, expected size of the social group after T rounds satisfies $\mathbb{E}[|G^T|] \gtrsim T^{\frac{1}{8}}$.*

Proof. For each t, we construct collection $\{A_1^t, A_2^t, \ldots, A_{N(t)}^t\}$ of *disjoint* circular segments on the unit ball. To do this, let the height of each circular segment in the collection be $\delta(t) = \frac{1}{4t^{\frac{1}{4}}}$. Then we can fit $N(t) = \left\lfloor \pi \cdot t^{\frac{1}{8}} \right\rfloor$ of these segments into $\mathbb{B}$. To see this, observe that a circular segment of height δ has a chord of length $2\sqrt{\delta}\sqrt{2-\delta}$. The central angle of the segment is then $\theta = 2\arctan\left(\frac{\sqrt{\delta}\sqrt{2-\delta}}{1-\delta}\right) \leq 4\sqrt{\delta}$, implying the existence of at least $N = \left\lfloor \frac{\pi}{2\sqrt{\delta}} \right\rfloor$ disjoint circular segments of height δ. Now define $\tau \leq T$ to be the last round for which $\Pr[S^\tau \cap A_i^\tau \neq \emptyset] \geq \frac{1}{2}$. Thus,

$$\mathbb{E}[|G^\tau|] \geq \sum_{i=1}^{N(\tau)} \Pr[S^\tau \cap A_i^\tau \neq \emptyset] = N(\tau) \cdot \Pr[S^\tau \cap A_1^\tau \neq \emptyset] \geq \frac{1}{2} \left\lfloor \pi \tau^{\frac{1}{8}} \right\rfloor \gtrsim \tau^{\frac{1}{8}} \quad (3)$$

Here, the first inequality follows from the observation that if $S^t \cap A_i^t \neq \emptyset$ then there is a least one group member inside A_i^t. The equality is due to symmetry; that is, $\Pr[S^t \cap A_i^t \neq \emptyset] = \Pr[S^t \cap A_j^t \neq \emptyset]$ for each pair $1 \leq i, j \leq N(t)$. The second inequality follows because, by definition, $\Pr[S^\tau \cap A_i^\tau \neq \emptyset] \geq \frac{1}{2}$. Next consider rounds $t > \tau$. For these rounds, by definition of τ, we know $\Pr[S^t \cap A_i^t \neq \emptyset] \leq \frac{1}{2}$ which implies $\Pr[S^t \cap A_i^t = \emptyset] \geq \frac{1}{2}$. Therefore,

$$\begin{aligned}
\Pr[X^{t+1}] &\gtrsim \sum_{i=1}^{N(t)} \Pr\left[Z(A_i^t) \ \wedge \ (S^t \cap A_i^t = \emptyset)\right] \\
&= N(t) \cdot \Pr\left[Z(A_i^t) \mid S^t \cap A_i^t = \emptyset\right] \cdot \Pr[\ S^t \cap A_i^t = \emptyset\ ] \\
&\gtrsim N(t) \cdot \Pr\left[Z(A_i^t) \mid \overline{A_i^t}\right] \qquad (4)
\end{aligned}$$

Where the last inequality follows by Corollary 1. Finally by Lemma 3, we see $\Pr\left[Z(A_i^t) \mid \overline{A_i^t}\right] \gtrsim \frac{1}{t}$, and thus $\Pr[X^{t+1}] \gtrsim t^{-\frac{7}{8}}$ for any $t > \tau$. We may now lower bound the expected group size at the end of round T. Specifically, for $T \geq 4$,

$$\mathbb{E}[|G^T|] = \mathbb{E}[|G^\tau|] + \sum_{t=\tau+1}^{T} \Pr[X^t = 1] \gtrsim \tau^{\frac{1}{8}} + \sum_{t=\tau+1}^{T} t^{-\frac{7}{8}} \gtrsim T^{\frac{1}{8}}$$

The last inequality was obtained using integral bounds. □

4.2 Lower Bound for the Unit Square

For the unit square $\mathbb{H}$ caps are either *right-angled triangles* or *right-angled trapezoids* (trapezoids with two adjacent right angles). We can bound the probability of accepting a point inside a right-angled trapezoid by consideration of the largest inscribed triangle it contains. Thus, it suffices to consider only the case in which the cap forms a triangle.

Lemma 4. *Let $J_{a,b}$ be triangular cap on the unit square with perpendicular side lengths $a \leq b$. Then* $\Pr[Z(J_{a,b}) \mid \overline{J_{a,b}}\,] \geq \frac{1}{2^{11}} a^4 \cdot \left(1 + \ln\left(\frac{b}{a}\right)\right)$.

Similar to the ball case, Lemma 4 is used to prove lower bound on the expected cardinality.

Theorem 6. *For the unit square $\mathbb{H}$, the expected size of the social group after T rounds satisfies $\mathbb{E}[|G^T|] \gtrsim \ln(T)$.*

Proof. For each t, consider the triangle $A^t = \operatorname{conv}\left((0,0), \left(0, \frac{1}{4t^{\frac{1}{4}}}\right), \left(\frac{1}{4t^{\frac{1}{4}}}, 0\right)\right)$. As discussed, the triangle A^t is a cap of the unit square.

$$\begin{aligned}
\Pr[X^{t+1}] &\geq \Pr[Z(A^t) \wedge (S^t \cap A^t = \emptyset)] \\
&= \Pr[Z(A^t) \mid (S^t \cap A^t = \emptyset)\,] \cdot \Pr[\,S^t \cap A^t = \emptyset\,] \\
&\geq \Pr[Z(A^t) \mid \overline{A^t}\,] \cdot \Pr[\,S^t \cap A^t = \emptyset\,] \\
&\gtrsim \frac{1}{t} \cdot \Pr[\,S^t \cap A^t = \emptyset\,]
\end{aligned} \tag{5}$$

Here the second inequality follows from Corollary 1. The third inequality is derived by applying Lemma 4 with respect to the cap A^t, for which $a = b = \frac{1}{4t^{\frac{1}{4}}}$.

Now, let T_1, T_2, T_3 and T_4 be right-angled triangles each containing one of the corners with perpendicular side lengths $a = b = \frac{1}{4}$. By Lemma 4, the probability of accepting a candidate inside triangle T_ℓ, given no member is currently in T_ℓ, is lower bounded by $\frac{a^4}{2^{11}} = \frac{1}{2^{19}}$. Thus

$$\Pr[T_\ell \cap S^i = \emptyset] \leq \left(1 - \frac{1}{2^{19}}\right)^i = k^i \tag{6}$$

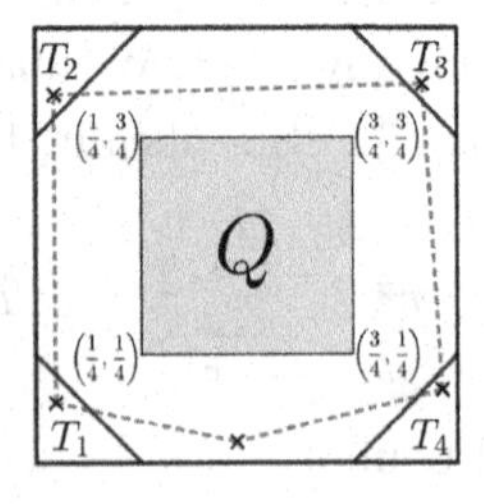

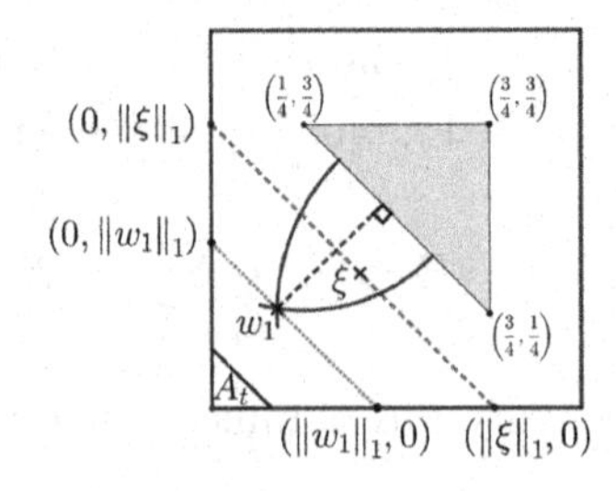

Fig. 1. Figure on the left illustrates that if there is a member in each T_ℓ then $Q \subseteq S_i$. The figure on the right shows if $\xi \in B\left(\left(\frac{1}{4}, \frac{3}{4}\right), w_1\right) \cap B\left(\left(\frac{3}{4}, \frac{1}{4}\right), w_1\right)$ then $\|\xi\|_1 \geq \|w_1\|_1$

Next let Q be the square $\left[\frac{1}{4}, \frac{3}{4}\right]^2$. Observe that if there is a member of S^i selected in each of the four triangles T_1, T_2, T_3 and T_4 then $Q \subseteq S^i$, as illustrated in Fig. 1. Consequently, applying the union bound and (6), for all $1 \leq \ell \leq 4$, we have $\Pr[Q \not\subseteq S^i] \leq 4k^i$. Note, since $k = 1 - \frac{1}{2^{19}}$, there exists a fixed round τ such that $4k^\tau < 1$; hence, $\Pr[Q \subseteq S^\tau] \geq 1 - 4k^\tau$ is lower bounded by a constant. Therefore, for $t \geq \tau + 1$,

$$\Pr[S^t \cap A^t = \emptyset] = \Pr[S^t \cap A^t = \emptyset \mid Q \subseteq S^\tau] \Pr[Q \subseteq S^\tau] \gtrsim \Pr[S^t \cap A^t = \emptyset \mid Q \subseteq S^\tau]$$

To complete the proof, it suffices to show that $\Pr[\, S^t \cap A^t = \emptyset \mid Q \subseteq S^\tau \,]$ is bounded below by some constant. To show this, we use the following basic idea: if the convex hull contains the square Q *and* a candidate is selected inside A^t then it must have been the case that both candidates were in A^t. Recursively, we have

$$\begin{aligned}
\Pr[S^t \cap A^t = \emptyset \mid Q \subseteq S^\tau \,] &= \Pr[\neg Z(A^t) \wedge (S^{t-1} \cap A^t = \emptyset) \mid Q \subseteq S^\tau] \\
&= \Pr[\neg Z(A^t) \mid (S^{t-1} \cap A^t = \emptyset) \wedge (Q \subseteq S^\tau)] \cdot \Pr[\, S^{t-1} \cap A^t = \emptyset \mid Q \subseteq S^\tau] \\
&= \prod_{i=\tau}^{t-1} (1 - \Pr[Z(A^t) \mid S^i \cap A^t = \emptyset \wedge (Q \subseteq S^\tau)]) \cdot \Pr[\, S^\tau \cap A^t = \emptyset \mid Q \subseteq S^\tau] \\
&\gtrsim \prod_{i=\tau}^{t-1} (1 - \Pr[Z(A^t) \mid S^i \cap A^t = \emptyset \wedge (Q \subseteq S^\tau)]) \cdot \Pr[\, S^\tau \cap A^\tau = \emptyset \mid Q \subseteq S^\tau] \\
&\gtrsim \prod_{i=\tau}^{t-1} (1 - \Pr[Z(A^t) \mid S^i \cap A^t = \emptyset \wedge (Q \subseteq S^\tau)]) \\
&\gtrsim \prod_{i=\tau}^{t-1} (1 - \Pr[Z(A^t) \mid Q \,]) \qquad (7)
\end{aligned}$$

Here the first inequality holds because, by definition, $A^t \subseteq A^\tau$ for all $t \geq \tau$. The second inequality holds for any constant τ satisfying $4k^\tau < 1$ (for example, $\tau = 1000000$) because then $\Pr[\, S^\tau \cap A^\tau = \emptyset \mid Q \subseteq S^\tau]$ is lower bounded by a constant. This is because A^τ is a small triangle and there is a positive probability that no candidate was selected inside A^τ in the previous τ rounds, even when the convex hull of the members includes the square Q. Finally, for the third inequality, observe that $S^\tau \subseteq S^i$ when $i \geq \tau$. Thus, by Corollary 1, $\Pr[Z(A^t) \mid S^i \cap A^t = \emptyset \wedge (Q \subseteq S^\tau)] \leq \Pr[Z(A^t) \mid Q \,]$.

Next, we claim that if Q is the convex hull then a candidate can be accepted inside A^t only if both candidates are in A^t. In particular, this gives the following useful inequality:

$$\Pr[\, Z(A^t) \mid Q] \;\leq\; \mathrm{vol}(A^t)^2 \;=\; \left(\frac{1}{2}\left(\frac{1}{4t^{\frac{1}{4}}}\right)^2\right)^2 \;=\; \frac{1}{2^{10} \cdot t} \qquad (8)$$

To see this, suppose the claim is false. That is, $w_2 \in A^t$ is selected when Q is the convex hull but $w_1 \notin A^t$. Then, by Lemma 1, it must be the case that

$$w_2 \in B\left(\left(\frac{1}{4}, \frac{1}{4}\right), w_1\right) \cap B\left(\left(\frac{3}{4}, \frac{1}{4}\right), w_1\right) \cap B\left(\left(\frac{1}{4}, \frac{3}{4}\right), w_1\right) \cap B\left(\left(\frac{3}{4}, \frac{3}{4}\right), w_1\right)$$

Observe that $w_1 \in A^t$ if and only if $\|w_1\|_1 = |(w_1)_1| + |(w_1)_2| \le \frac{1}{4t^{\frac{1}{4}}}$, where $(w_1)_i$ denotes the i'th component of w_1. Since $w_1 \notin A^t$, we have $\|w_1\|_1 > \frac{1}{4t^{\frac{1}{4}}}$. As illustrated in Fig. 1, the region $B((\frac{1}{4}, \frac{3}{4}), w_1) \cap B((\frac{3}{4}, \frac{1}{4}), w_1)$ does not intersect A^t if $w_1 \notin A^t$. Thus, the winner w_2 cannot be inside A^t and the claim is verified.

Finally, applying the inequalities (5), (7) and (8), we have

$$\Pr[X^{t+1} = 1] \gtrsim \frac{1}{t} \prod_{i=\tau}^{t-1} (1 - \Pr[Z(A^t)|Q]) \gtrsim \frac{1}{t} \prod_{i=\tau}^{t-1} \left(1 - \frac{1}{2^{10} \cdot t}\right) \gtrsim \frac{1}{t}$$

Ergo, we have $\mathbb{E}[G^T] = \sum_{t=1}^{T} \Pr[X^t] \gtrsim \sum_{t=1}^{T} \frac{1}{t} \cong \ln(T)$. □

5 Upper Bounds on Expected Group Size

We now apply the techniques developed in Sects. 3 and 4 to upper bound the expected cardinality of the group for the unit ball and the unit square. Specifically, we apply Corollary 2 to these metric spaces using the f_K obtained from Lemma 3 and Lemma 4, respectively.

5.1 Upper Bound for the Unit Ball

Observe that exactly one of the two Voronoi regions corresponds to a circular segment. Furthermore, since a circular segment fits inside its complement, $\arg\min_i \Pr[Z(H_i(w_1, w_2))|\overline{H_i(w_1, w_2)}]$ is attained by the H_i corresponding to a circular segment. Let $\delta(w_1, w_2)$ denote the height of the circular segment for this Voronoi region H_i. Then, by Lemma 3, we have $\Pr[Z(H_i(w_1, w_2))|\overline{H_i(w_1, w_2)}] \gtrsim \delta(w_1, w_2)^4$. Thus $f_{\mathbb{B}}(w_1, w_2) = \delta(w_1, w_2)^4$, satisfies the conditions of Corollary 2. However when using Corollary 2 we need to understand $\Phi(\lambda)$; Lemma 5 allows us to do exactly that.

Lemma 5. *Let $\delta(w_1, w_2)$ be the height of the circular segment $H_i(w_1, w_2)$ formed by the Voronoi regions. Then, for all $\lambda \le \frac{1}{10^4}$, we have*

$$\Phi(\lambda) = \int\int \mathbb{I}\left[\delta(w_1, w_2)^4 \le \lambda\right] dw_1\, dw_2 \lesssim \lambda^{\frac{7}{8}}$$

Theorem 7. *For the unit ball $\mathbb{B}$, the expected cardinality of the group after T rounds satisfies $\mathbb{E}[|G^T|] \lesssim T^{\frac{1}{8}}$.*

Proof. When applying Corollary 2, we also need to bound on $\Phi(\lambda)$ for $0 \le \lambda \le \frac{\ln(t)}{t}$. Let t_0 be a constant such that such that $\frac{\ln(t_0)}{t_0} \le \frac{1}{10^4}$, then for round $t \ge t_0$ we have $0 \le \lambda \le \frac{\ln(t)}{t} \le \frac{\ln(t_0)}{t_0} \le \frac{1}{10^4}$. Hence we may apply the bound $\Phi(\lambda) \lesssim \lambda^{\frac{7}{8}}$ for any round $t \ge t_0$, by Lemma 5 using $f_{\mathbb{B}}(w_1, w_2) = \delta(w_1, w_2)^4$. Thus combining Lemma 5 and Corollary 2 for $t \ge t_0$, we see that

$$\Pr[X^{t+1}] \lesssim \frac{1}{t} + \int_0^{\frac{\ln(t)}{t}} te^{-t\lambda} \lambda^{\frac{7}{8}} d\lambda = \frac{1}{t} + \frac{1}{t^{\frac{7}{8}}} \int_0^{\ln(t)} e^{-u} \cdot u^{\frac{7}{8}} du \lesssim \frac{1}{t^{\frac{7}{8}}}$$

Here the equality holds by the substitution $u = t\lambda$. The theorem follows as $\mathbb{E}[|G^T|] = \sum_{t=1}^{T} \Pr[X^t] \lesssim t_0 + \sum_{t=t_0+1}^{T} 1/t^{\frac{7}{8}} \lesssim T^{\frac{1}{8}}$ by applying integral bounds. □

5.2 Upper Bound for the Unit Square

Similar to the unit ball case, we must find an appropriate function $f_{\mathbb{H}}$ satisfying the conditions of Corollary 2. For a cap A with $A \subseteq H_i(w_1, w_2)$ by Corollary 1,

$$\Pr[Z(H_i(w_1,w_2))|\overline{H_i(w_1,w_2)}] \geq \Pr[Z(A)|\overline{H_i(w_1,w_2)}] \geq \Pr[Z(A)|\overline{A}] \quad (9)$$

Let $a(w_1, w_2) \leq b(w_1, w_2)$ be the two side lengths of the triangular cap of greatest area that fits inside both $H_i(w_1, w_2)$. Applying Lemma 4, along with (9) gives

$$\min_i \Pr[Z(H_i(w_1,w_2)) \mid \overline{H_i(w_1,w_2)}] \gtrsim a(w_1,w_2)^4 \cdot \ln\left(e \cdot \frac{b(w_1,w_2)}{a(w_1,w_2)}\right)$$

Thus $f_{\mathbb{H}}(w_1, w_2) = a(w_1, w_2)^4 \ln\left(e\frac{b(w_1,w_2)}{a(w_1,w_2)}\right)$ satisfies the conditions of Corollary 2.

Lemma 6. *Let $a(w_1, w_2) \leq b(w_1, w_2)$ be the two side lengths of the triangular cap of greatest area that fits inside both $H_i(w_1, w_2)$. Then, for any $\lambda \leq \frac{1}{20}$,*

$$\Phi(\lambda) = \int\int \mathbb{I}\left[a(w_1,w_2)^4 \cdot \ln\left(e \cdot \frac{b(w_1,w_2)}{a(w_1,w_2)}\right) \leq \lambda\right] dw_1\, dw_2 \lesssim \lambda \cdot \ln\left(\ln\left(\frac{1}{\lambda}\right)\right)$$

Proof. We need only consider pairs w_1 and w_2 that satisfy the indicator function. As $\lambda \leq \frac{1}{20}$, this implies $a(w_1, w_2) \leq \lambda^{\frac{1}{4}} \leq \left(\frac{1}{20}\right)^{\frac{1}{4}} \leq \frac{1}{2}$ and, without loss of generality, $H_1(w_1, w_2)$ is the smallest Voronoi region and fits (under symmetries) into $H_2(w_1, w_2)$. Furthermore, applying rotational and diagonal symmetries, any pair of points can be transformed into a pair of the form $w_1 = (x, y)$ and $w_2 = (x + \Delta_x, y + \Delta_y)$, with $s = \Delta_y/\Delta_x \leq 1$ and $\Delta_x, \Delta_y \geq 0$. Hence, we lose only a constant factor in making the following assumptions on w_1 and w_2: the triangular cap of greatest area that fits inside both the $H_i(w_1, w_2)$ is contained in $H_1(w_1, w_2)$; the cap contains the origin; the larger side corresponding to $b(w_1, w_2)$ is along the y-axis; the smaller side corresponding to $a(w_1, w_2)$ is along the x-axis.

Recall that $H_1(w_1, w_2)$ is either a right-angled triangle or a right-angled trapezoid. In the former case, the triangular cap of greatest area which fits inside both of the Voronoi regions is H_1 itself. Let $\mathcal{H}$ be the hyperplane separating the two Voronoi regions, it holds that $\mathcal{H}(w_1, w_2) = \{\xi \in \mathbb{R}^2 : (w_2 - w_1) \cdot (\xi - \frac{w_2+w_1}{2}) = 0\}$. The side lengths of H_1 are then the intercepts of $\mathcal{H}$ along the axes. In the latter case, the triangular cap of greatest area satisfies $b(w_1, w_2) = 1$ and $a(w_1, w_2)$ is the intercept of $\mathcal{H}$ on the x-axis. We can then compute explicit expressions for both terms $a(w_1, w_2)$ and $b(w_1, w_2)$. In particular,

$$a(w_1,w_2) = \frac{\|w_2\|^2 - \|w_1\|^2}{2(w_2 - w_1)_1} = x + sy + \frac{\Delta_x}{2}(1 + s^2) = x + sy + \frac{\Delta_y}{2}\left(\frac{1+s^2}{s}\right) \quad (10)$$

Now, because this is the unit square, we have $b(w_1, w_2) \leq 1$. Thus, $b(w_1, w_2) = \min\left(1, \frac{a(w_1,w_2)}{s}\right)$. Hence,

$$a(w_1, w_2)^4 \ln\left(e \cdot \frac{b(w_1, w_2)}{a(w_1, w_2)}\right) = a(w_1, w_2)^4 \ln\left(e \min\left(\frac{1}{a(w_1, w_2)}, \frac{1}{s}\right)\right) \quad (11)$$

For a fixed $w_1 = (x, y)$, let $R(x, y)$ be a rectangle containing all the points $w_2 = (x + \Delta_x, y + \Delta_y)$ satisfying the condition of the indicator function. Thus, it will suffice to show that we can select $R(x, y)$ to have small area. To do this we must show that Δ_x and Δ_y cannot be too large. Again, recall that if the indicator function is true then $a(w_1, w_2) \leq \lambda^{\frac{1}{4}}$. So (10) implies $x \leq \lambda^{\frac{1}{4}}$ and $s \leq \frac{\lambda^{\frac{1}{4}}}{y}$. If $\lambda^{\frac{1}{4}} \leq y \leq 1$ then $\min\left(\frac{1}{a(w_1,w_2)}, \frac{1}{s}\right) \geq \frac{y}{\lambda^{\frac{1}{4}}}$. Plugging into (11) gives $a(w_1, w_2)^4 \cdot \ln\left(e \cdot \frac{b(w_1,w_2)}{a(w_1,w_2)}\right) \geq a(w_1, w_2)^4 \cdot \ln\left(e \cdot \frac{y}{\lambda^{\frac{1}{4}}}\right)$. It follows that $a(w_1, w_2) \leq \frac{\lambda^{1/4}}{\ln^{1/4}\left(\frac{ey}{\lambda^{1/4}}\right)}$. Therefore, by (10):

$$\Delta_x \leq 2 \cdot \left(\frac{\lambda^{\frac{1}{4}}}{\ln^{\frac{1}{4}}\left(\frac{ey}{\lambda^{\frac{1}{4}}}\right)} - ys - x\right) \leq 2 \cdot \left(\frac{\lambda^{\frac{1}{4}}}{\ln^{\frac{1}{4}}\left(\frac{ey}{\lambda^{\frac{1}{4}}}\right)} - x\right) \quad (12)$$

$$\Delta_y \leq 2s \cdot \left(\frac{\lambda^{\frac{1}{4}}}{\ln^{\frac{1}{4}}\left(\frac{ey}{\lambda^{\frac{1}{4}}}\right)} - ys - x\right) \leq \frac{1}{2y} \cdot \left(\frac{\lambda^{\frac{1}{4}}}{\ln^{\frac{1}{4}}\left(\frac{ey}{\lambda^{\frac{1}{4}}}\right)} - x\right)^2 \quad (13)$$

Final inequalities in (12) and (13) were obtained by optimizing over $s \in [0, 1]$. Then noting that $x \leq \lambda^{\frac{1}{4}}$, we have

$$\begin{aligned}
\Phi(\lambda) &\lesssim \int_0^1 \int_0^{\lambda^{\frac{1}{4}}} |R(x, y)|\, dx\, dy \\
&= \int_0^{\lambda^{\frac{1}{4}}} \int_0^{\lambda^{\frac{1}{4}}} |R(x, y)|\, dx\, dy + \int_{\lambda^{\frac{1}{4}}}^1 \int_0^{\lambda^{\frac{1}{4}}} |R(x, y)|\, dx\, dy \\
&\lesssim \lambda + \int_{\lambda^{1/4}}^1 \frac{1}{y} \cdot \int_0^{\frac{\lambda^{1/4}}{\ln^{1/4}(\frac{ey}{\lambda^{1/4}})}} \left(\frac{\lambda^{1/4}}{\ln^{1/4}(\frac{ey}{\lambda^{1/4}})} - x\right)^3 dx\, dy \\
&\lesssim \lambda \cdot \int_{\lambda^{1/4}}^1 \frac{1}{y} \cdot \frac{1}{\ln(\frac{ey}{\lambda^{1/4}})} dy \\
&\lesssim \lambda \cdot \ln\left(\ln\left(\frac{1}{\lambda}\right)\right)
\end{aligned}$$

For the second inequality, since Δ_x must be positive, (12) implies that the limit of the integral becomes $x = \frac{\lambda^{1/4}}{\ln^{1/4}(\frac{ey}{\lambda^{1/4}})}$. To bound the area $|R(x, y)|$ of the rectangles we have two cases. When $0 \leq y \leq \lambda^{\frac{1}{4}}$, observe, by (10), that $a(w_1, w_2) \leq \lambda^{\frac{1}{4}}$ implies $\Delta_x \leq 2\lambda^{\frac{1}{4}}$ and $\Delta_y \leq 2\lambda^{\frac{1}{4}}$. Thus $|R(x,y)| \leq \Delta_x \cdot \Delta_y \lesssim \sqrt{\lambda}$. When $\lambda^{\frac{1}{4}} \leq y \leq 1$, the bound on $|R(x,y)|$ holds by (12) and (13). □

Theorem 8. *For the unit square $\mathbb{H}$, the expected cardinality of the group after T rounds satisfies $\mathbb{E}[|G^T|] \lesssim \ln T \cdot \ln\ln T$.*

Proof. Note that for any round $t \geq 100$ we have $0 \leq \lambda \leq \frac{\ln(t)}{t} \leq \frac{1}{20}$. Thus, combining Corollary 2 and Lemma 6 we get

$$\begin{aligned} \Pr[X^{t+1}] &\lesssim \frac{1}{t} + \int_0^{\frac{\ln(t)}{t}} te^{-t\lambda} \cdot \lambda \ln\left(\ln\left(\frac{1}{\lambda}\right)\right) d\lambda \\ &= \frac{1}{t} + \frac{1}{t}\int_0^1 u \ln\left(\ln\left(\frac{t}{u}\right)\right) du + \frac{1}{t}\int_1^{\ln(t)} ue^{-u} \ln\left(\ln\left(\frac{t}{u}\right)\right) du \end{aligned}$$

Here the equality holds via the substitution $u = t\lambda$. Note that $u \ln\left(\ln\left(\frac{t}{u}\right)\right) \leq \ln\ln t$, when $t \geq 100$ and $0 \leq u \leq 1$, and

$$\int_1^{\ln(t)} ue^{-u} \cdot \ln\left(\ln\left(\frac{t}{u}\right)\right) du \leq \ln\ln t \cdot \int_1^{\infty} ue^{-u}\, du \leq \ln\ln t$$

Hence it follows $\Pr[X^{t+1}] \lesssim \frac{\ln\ln t}{t}$ for all $t \geq 100$. Finally, we see that $\mathbb{E}[|G^T|] = \sum_{t=1}^{T} \Pr[X^t] \lesssim 100 + \sum_{t=101}^{T} \frac{\ln\ln t}{t} \lesssim \ln T \cdot \ln\ln T$, where the last inequality was obtained using integral bounds. □

6 Conclusion

In this paper we presented techniques for studying the evolution of an exclusive social group in a metric space, under the consensus voting mechanism. A natural open problem is to close the gap between the $\Omega(\ln T)$ lower bound and the $O(\ln T \cdot \ln\ln T)$ upper bound on the expected cardinality of the group, after T rounds, in the unit square. Interesting further directions include the study of higher dimensional metric spaces, and allowing for more than two candidates per round. In either direction, our analytic tools may prove useful.

Acknowledgement. We thank the anonymous reviewers for many helpful comments and suggestions.

References

1. Alon, N., Feldman, M., Mansour, Y., Oren, S., Tennenholtz, M.: Dynamics of evolving social groups. ACM Trans. Econ. Comput. **7**(3), 1–27 (2019)
2. Anshelevitch, E., Bhardwaj, O., Postl, J.: Approximating optimal social choice under metric preferences. In: Proceedings of the 29th Conference on Artificial Intelligence (AAAI), pp. 777–783 (2015)

3. Anshelevitch, E., Postl, J.: Randomized social choice functions under metric preferences. J. Artif. Intell. Res. **58**(1), 797–827 (2017)
4. Baddeley, A., Bárány, I., Schneider, R.: Random polytopes, convex bodies, and approximation. In: Weil, W. (ed.) Stochas. Geom., pp. 77–118. Springer, Heidelberg (2007). https://doi.org/10.1007/978-3-540-38175-4_2
5. Black, D.: On the rationale of group decision-making. J. Polit. Econ. **56**, 23–34 (1948)
6. Borodin, A., Lev, O., Shah, N., Strangway, T.: Primarily about primaries. In: Proceedings of the 33rd Conference on Artificial Intelligence (AAAI), pp. 1804–1811 (2019)
7. Claassen, R.: Direction versus proximity: amassing experimental evidence. Am. Polit. Res. **37**(2), 227–253 (2009)
8. Colomer, J.: On the geometry of unanimity rule. J. Theoret. Polit. **11**(4), 543–553 (1999)
9. Davis, O., Hinich, M., Ordeshook, P.: An expository development of a mathematical model of the electoral process. Am. Polit. Sci. Rev. **64**, 426–448 (1970)
10. Downs, A.: An Economic Theory of Democracy. Harper Collins, New York (1957)
11. Enelow, J., Hinich, M.: The Spatial Theory of Voting: An Introduction. Cambridge University Press, Cambridge (1984)
12. Enelow, J., Hinich, M. (eds.): Advances in the Spatial Theory of Voting. Cambridge University Press, Cambridge (1990)
13. Feldman, M., Fiat, A., Golomb, I.: On voting and facility location. In: Proceedings of 17th Conference on Economics and Computation (EC), pp. 269–286 (2016)
14. Grofman, B.: The neglected role of the status quo in models of issue voting. J. Polit. **47**, 230–237 (1985)
15. Hare, P.: Group decision by consensus: reaching unity in the society of friends. Sociol. Inq. **43**(1), 75–84 (1973)
16. Hotelling, H.: Stability in competition. Econ. J. **39**(153), 41–57 (1929)
17. Lacy, D., Paolino, P.: Testing proximity versus directional voting using experiments. Electoral Stud. **29**(3), 460–471 (2010)
18. Lewis, J., King, G.: No evidence on directional vs. proximity voting. Polit. Anal. **8**(1), 21–33 (2000)
19. Matthews, S.: A simple direction model of electoral competition. Public Choice **34**, 141–156 (1979)
20. Merrill III, S., Merrill, S., Grofman, B.: A Unified Theory of Voting: Directional and Proximity Spatial Models. Cambridge University Press, Cambridge (1999)
21. Poole, K.: Spatial Models of Parliamentary Voting. Cambridge University Press, Cambridge (2005)
22. Rabinowitz, G., Stuart, E.: A directional theory of issue voting. Am. Polit. Sci. Rev. **83**, 93–121 (1989)
23. Schofield, N.: The Spatial Models of Politics. Routledge, Abingdon (2007)
24. Skowron, P., Elkind, E.: Social choice under metric preferences: scoring rules and STV. In: Proceedings of the 31st Conference on Artificial Intelligence (AAAI), pp. 706–712 (2017)
25. Tomz, M., Van Houweling, R.: Candidate position and voter choice. Am. Polit. Sci. Rev. **102**(3), 303–318 (2008)
26. Vogel, E. (ed.): Modern Japanese Organization and Decision-Making. University of California Press, Berkeley (1975)

Abstracts

Computing Approximate Equilibria in Weighted Congestion Games via Best-Responses

Yiannis Giannakopoulos[1](✉), Georgy Noarov[2], and Andreas S. Schulz[1]

[1] TU Munich, Munich, Germany
{yiannis.giannakopoulos,andreas.s.schulz}@tum.de
[2] Princeton University, Princeton, NJ, USA
gnoarov@princeton.edu

Abstract. We present a deterministic polynomial-time algorithm for computing $d^{d+o(d)}$-approximate (pure) Nash equilibria in weighted congestion games with polynomial cost functions of degree at most d. This is an exponential improvement of the approximation factor with respect to the previously best deterministic algorithm. An appealing additional feature of our algorithm is that it uses only best-improvement steps in the actual game, as opposed to earlier approaches that first had to transform the game itself. Our algorithm is an adaptation of the seminal algorithm by Caragiannis et al. [FOCS'11, TEAC 2015], but we utilize an approximate potential function directly on the original game instead of an exact one on a modified game.

A critical component of our analysis, which is of independent interest, is the derivation of a novel bound of $[d/\mathcal{W}(d/\rho)]^{d+1}$ for the Price of Anarchy (PoA) of ρ-approximate equilibria in weighted congestion games, where $\mathcal{W}$ is the Lambert-W function. More specifically, we show that this PoA is *exactly* equal to $\Phi_{d,\rho}^{d+1}$, where $\Phi_{d,\rho}$ is the unique positive solution of the equation $\rho(x+1)^d = x^{d+1}$. Our upper bound is derived via a smoothness-like argument, and thus holds even for mixed Nash and correlated equilibria, while our lower bound is simple enough to apply even to singleton congestion games.

Keywords: Atomic congestion games · Computation of equilibria · Price of anarchy · Approximate equilibria · Potential games

Supported by the Alexander von Humboldt Foundation with funds from the German Federal Ministry of Education and Research (BMBF).
Y. Giannakopoulos and A. S. Schulz are associated researchers with the Research Training Group GRK 2201 "Advanced Optimization in a Networked Economy", funded by the German Research Foundation (DFG).
A significant part of this work was done while G. Noarov was a visiting student at the Operations Research group of TU Munich.

The full paper can be found at: https://arxiv.org/abs/1810.12806.

T. Harks and M. Klimm (Eds.): SAGT 2020, LNCS 12283, p. 339, 2020.
https://doi.org/10.1007/978-3-030-57980-7

On the Integration of Shapley–Scarf Housing Markets

Rajnish Kumar[1], Kriti Manocha[2], and Josué Ortega[1](✉)

[1] Queen's Management School, Queen's University Belfast, Belfast, UK
j.ortega@qub.ac.uk
[2] Indian Statistical Institute, Delhi, India

Abstract. We study the consequences of merging Shapley–Scarf markets assuming that the core allocation is implemented before and after the merge occurs. We focus on (i) the number of agents who obtain a better allocation, and (ii) the size of the welfare gains, measured by the rank of the assigned house. We present worst- and average-case results.

In the worst-case scenario, we show that the merge of k markets with n_j agents each and with n agents in total, may harm the vast majority of agents (up to, but no more than, $n - k$ agents). Furthermore, the average rank of an agent's house can decrease asymptotically by, but not more than, 50% of the length of their preference list. These results are substantially worse than those for Gale–Shapley markets [3, 4]. On the other side, our average-case results are more optimistic. We prove that the expected gains from integration in random markets equal $\frac{(n+1)[(n_j+1)H_{n_j}-n_j]}{n_j(n_j+1)n} - \frac{(n+1)H_n-n}{n^2}$, where H_n is the n-th harmonic number. Our computation shows that the expected welfare gains from integration are positive for all agents, and larger for agents that are initially in smaller markets. We also provide an upper bound on the expected number of agents harmed by integration, that allows us to guarantee that a majority of agents benefit from integration when all markets are of equal size and this is below 26. Our work builds on previous probabilistic analysis of Shapley–Scarf markets in the computer science literature [1, 2].

The full article is available at https://arxiv.org/abs/2004.09075.

Keywords: Shapley–Scarf markets · Gains from integration · Random markets

References

1. Frieze, A., Pittel, B.G.: Probabilistic analysis of an algorithm in the theory of markets in indivisible goods. The Annals of Applied Probability, pp. 768–808 (1995)
2. Knuth, D.E.: An exact analysis of stable allocation. J. Algorithms **20**(2), 431–442 (1996)

We acknowledge support by British Council Grant UGC-UKIERI 2016-17-059.

T. Harks and M. Klimm (Eds.): SAGT 2020, LNCS 12283, pp. 340–341, 2020.
https://doi.org/10.1007/978-3-030-57980-7

3. Ortega, J.: Social integration in two-sided matching markets. J. Math. Econ. **78**, 119–126 (2018)
4. Ortega, J.: The losses from integration in matching markets can be large. Econ. Lett. **174**, 48–51 (2019)

The Stackelberg Kidney Exchange Problem is Σ_2^p-complete

Bart Smeulders, Danny Blom(✉), and Frits C. R. Spieksma

Department of Mathematics and Computer Science,
Eindhoven University of Technology, Eindhoven, The Netherlands
d.a.m.p.blom@tue.nl

Kidney Exchange Programmes (KEPs) play a growing role in treatment of end stage renal disease, offering living donor kidneys to recipients with a willing but incompatible donor. Scale is important to KEPs, as combining different pools (owned by separate agents) of patient-donor pairs allows for more transplants compared to the separate pools optimizing independently. The organization of such collaborations is a delicate matter, as the benefits of cooperation may be shared unequally. In some cases, individual agents may even lose transplants when combining patient-donor pools. This observation has motivated research into mechanisms for planning transplants in combined pools. An important result is that, assuming individual agents are rational, no socially optimal mechanism exists. Thus, any mechanism that maximizes transplants in the combined pool runs the risk that agents will not contribute some or all of their pairs.

The risk of agents not cooperating fully depends in part on the agents ability to identify situations where withholding pairs is beneficial to them. We study a relatively simple situation, where the agent has perfect information on pairs of all other agents, knows the fixed strategies of all other agents and knows how the mechanism selects an optimal solution in the common pool. The agent has to decide which pairs to contribute and which to withhold. Her goal is to maximize the total number of her recipients receiving a transplant, either in the common pool or by internal matches in the withheld set. We call this problem the Stackelberg KEP game. Our main result is:

Theorem 1. *Stackelberg KEP game is a Σ_2^p-complete problem, for each fixed maximum cycle length $K \geq 3$.*

For $K = 2$, we rely on results in [1] to prove the following:

Theorem 2. *The Stackelberg KEP game is polynomially solvable if the maximum cycle length $K = 2$.*

A full working paper can be found at https://arxiv.org/abs/2007.03323.

Reference

1. Carvalho, M., Lodi, A.: Game theoretical analysis of kidney exchange programs. arXiv preprint arXiv:1911.09207 (2019)

T. Harks and M. Klimm (Eds.): SAGT 2020, LNCS 12283, p. 342, 2020.
https://doi.org/10.1007/978-3-030-57980-7

Author Index

Printed in the United States
By Bookmasters